Methods in Molecular Biology

Series Editor
John M. Walker
School of Life and Medical Sciences
University of Hertfordshire
Hatfield, Hertfordshire, UK

For further volumes:
http://www.springer.com/series/7651

For over 35 years, biological scientists have come to rely on the research protocols and methodologies in the critically acclaimed *Methods in Molecular Biology* series. The series was the first to introduce the step-by-step protocols approach that has become the standard in all biomedical protocol publishing. Each protocol is provided in readily-reproducible step-by-step fashion, opening with an introductory overview, a list of the materials and reagents needed to complete the experiment, and followed by a detailed procedure that is supported with a helpful notes section offering tips and tricks of the trade as well as troubleshooting advice. These hallmark features were introduced by series editor Dr. John Walker and constitute the key ingredient in each and every volume of the *Methods in Molecular Biology* series. Tested and trusted, comprehensive and reliable, all protocols from the series are indexed in PubMed.

Applications of Genome Modulation and Editing

Edited by

Paul John Verma

Aquatics & Livestock Sciences, South Australian Research and Development Institute, Roseworthy, SA, Australia

Huseyin Sumer

Faculty of Science, Engg & Tech, Swinburne University of Technology, Hawthorn, VIC, Australia

Jun Liu

Dept of Jobs, Precincts and Regions, Agriculture Victoria Research, Bundoora, VIC, Australia

Humana Press

Editors
Paul John Verma
Aquatics & Livestock Sciences
South Australian Research
and Development Institute
Roseworthy, SA, Australia

Huseyin Sumer
Faculty of Science, Engg & Tech
Swinburne University of Technology
Hawthorn, VIC, Australia

Jun Liu
Dept of Jobs, Precincts and Regions
Agriculture Victoria Research
Bundoora, VIC, Australia

ISSN 1064-3745 ISSN 1940-6029 (electronic)
Methods in Molecular Biology
ISBN 978-1-0716-2300-8 ISBN 978-1-0716-2301-5 (eBook)
https://doi.org/10.1007/978-1-0716-2301-5

This Humana imprint is published by the registered company Springer Science+Business Media, LLC, part of Springer Nature.
The registered company address is: 1 New York Plaza, New York, NY 10004, U.S.A.

Preface

Applications of Genome Modulation and Editing is a comprehensive review of protocols for genome editing methods in cell lines and embryos. The applications of genome editing include the generation of cell lines for fundamental research, generation of animal models for human diseases, pest control, and large animal welfare and production outcomes. The emphasis here is on providing readily reproducible techniques for gene regulation, editing and screening of cell lines, as well as real-world outcomes of gene targeting of embryos. Additional chapters provide a historical overview of genome editing and an overview of the tools for efficient genome editing, including ZFNs, TALENs, and CRISPR. This is complemented with a discussion of regulatory and policy considerations around the applications of genome editing in agriculture.

Applications of Genome Modulation and Editing also provides an understanding of genome editing and screening techniques, which is imperative for the successful genome modulation. This volume will prove beneficial to molecular biologists, stem cell biologists, biotechnologists, students, veterinarians, clinicians, and animal care technicians involved with genome modulation, gene editing, and transgenesis.

Roseworthy, SA, Australia
Hawthorn, VIC, Australia
Bundoora, VIC, Australia

Paul John Verma
Huseyin Sumer
Jun Liu

Contents

Contributors

VIRGILIA ALBERIO • *Facultad de Agronomía, Departamento de Producción Animal, Buenos Aires, Laboratorio Biotecnología Animal (LabBA), Universidad de Buenos Aires, Buenos Aires, Argentina; Instituto de Investigaciones en Producción Animal (INPA), CONICET-Universidad de Buenos Aires, Buenos Aires, Argentina*

IGNACIO ANEGÓN • *Inserm, Centre de Recherche en Transplantation et Immunologie, UMR 1064, Nantes, France; Transgenesis Rat ImmunoPhenomic Facility (TRIP), Nantes, France; GenoCellEdit Facility, Nantes, France*

HANNAH AUCH • *Large Animal Models in Cardiovascular Research, Internal Medical Department I, TU Munich, Munich, Germany; Center for Innovative Medical Models, LMU Munich, Munich, Germany*

ALLA BONDAREVA • *Comparative Biology & Experimental Medicine, Faculty of Veterinary Medicine Biochemistry & Molecular Biology, Cumming School of Medicine , University of Calgary, Calgary, AB, Canada*

MARK D. BUNTING • *School of Medicine, University of Adelaide, North Terrace, Adelaide, SA, Australia*

DANIEL F. CARLSON • *Recombinetics, Inc., St. Paul, MN, USA*

VANESSA CHENOUARD • *INSERM Centre de Recherche en Transplantation et Immunologie UMR 1064, Transgenesis Rat ImmunoPhenomic Facility (TRIP), Nantes, France*

YULIN CHEN • *Key Laboratory of Animal Genetics, Breeding and Reproduction of Shaanxi Province, College of Animal Science and Technology, Northwest A&F University, Yangling, China*

AMANDA CHOO • *School of Biological Sciences, University of Adelaide, Adelaide, SA, Australia*

MARTINA CRISPO • *Transgenic and Experimental Animal Unit, Institut Pasteur de Montevideo, Montevideo, Uruguay*

PETER CRISP • *South Australian Research and Development Institute, Adelaide, SA, Australia; School of Agricultural, Food and Wine, University of Adelaide, Adelaide, SA, Australia*

BRETT CROMER • *Department of Chemistry and Biotechnology, School of Science, Computing and Engineering Technologies, Swinburne University of Technology, Hawthorn, VIC, Australia*

FEDERICO CUADRO • *Instituto de Reproducción Animal Uruguay, Fundación IRAUy, Montevideo, Uruguay*

FABIEN DELERUE • *Genome Editing at Macquarie (GEM), Dementia Research Centre, Macquarie University, Sydney, NSW, Australia*

INA DOBRINSKI • *Comparative Biology & Experimental Medicine, Faculty of Veterinary Medicine Biochemistry & Molecular Biology, Cumming School of Medicine, University of Calgary, Calgary, AB, Canada*

PEDRO DOS SANTOS-NETO • *Instituto de Reproducción Animal Uruguay, Fundación IRAUy, Montevideo, Uruguay*

STEFFI FRIEDRICHS • *AcumenIST SPRL, Etterbeek, Belgium*

ELISABETH FUNG • *South Australian Research and Development Institute, Adelaide, SA, Australia*

LUKE GIERUS • *School of Medicine, University of Adelaide, North Terrace, Adelaide, SA, Australia*

TAYLOR GOLDSMITH • *Recombinetics, Inc., St. Paul, MN, USA*

JINCHAO GU • *Department of Chemistry and Biotechnology, School of Science, Computing and Engineering Technologies, Swinburne University of Technology, Hawthorn, VIC, Australia*

JEAN-MARIE HESLAN • *INSERM Centre de Recherche en Transplantation et Immunologie UMR 1064, Transgenesis Rat ImmunoPhenomic facility (TRIP), Nantes, France*

LARS M. ITTNER • *Genome Editing at Macquarie (GEM), Dementia Research Centre, Macquarie University, Sydney, NSW, Australia*

PETER KALDS • *Key Laboratory of Animal Genetics, Breeding and Reproduction of Shaanxi Province, College of Animal Science and Technology, Northwest A&F University, Yangling, China; Department of Animal and Poultry Production, Faculty of Environmental Agricultural Sciences, Arish University, El-Arish, Egypt*

PETER KEARNS • *SwiftEST SARL, Boulogne-Billancourt, France*

NIKOLAI KLYMIUK • *Large Animal Models in Cardiovascular Research, Internal Medical Department I, TU Munich, Munich, Germany; Center for Innovative Medical Models, LMU Munich, Munich, Germany*

STEFANIE KURTZ • *Institute of Farm Animal Genetics, Friedrich-Loeffler-Institut, Neustadt am Ruebenberge, Germany*

CHAO LI • *Key Laboratory of Animal Genetics, Breeding and Reproduction of Shaanxi Province, College of Animal Science and Technology, Northwest A&F University, Yangling, China*

JUN LIU • *Stem Cells and Genome Editing, Genomics and Cellular Sciences, Agriculture Victoria Research, Bundoora, VIC, Australia*

KARINNE ANNE LUDLOW • *Faculty of Law, Monash University, Clayton, VIC, Australia*

LUIS F. MALAVER-ORTEGA • *Monash Functional Genomics Platform, Monash University, Clayton, VIC, Australia; Department of Biochemistry and Molecular Biology, Monash University, Clayton, VIC, Australia*

ALEJO MENCHACA • *Instituto de Reproducción Animal Uruguay, Fundación IRAUy, Montevideo, Uruguay; Instituto Nacional de Investigación Agropecuaria (INIA), Montevideo, Uruguay*

LLUIS MONTOLIU • *National Centre for Biotechnology (CNB-CSIC) and Center for Biomedical Network Research on Rare Diseases (CIBERER-ISCIII), Madrid, Spain*

THU N. M. NGUYEN • *University of Melbourne, Bio21 Institute, School of BioSciences, Melbourne, VIC, Australia*

ANZU OKADA • *School of Biological Sciences, University of Adelaide, Adelaide, SA, Australia*

DANIEL F. PAULO • *Department of Plant and Environmental Protection Sciences (PEPS), The University of Hawaiʻi at Mānoa, Honolulu, HI, USA*

BJÖRN PETERSEN • *Institute of Farm Animal Genetics, Friedrich-Loeffler-Institut, Neustadt am Ruebenberge, Germany*

CHANDRAN PFITZNER • *School of Medicine, University of Adelaide, North Terrace, Adelaide, SA, Australia*

SANDRA PILTZ • *School of Medicine, University of Adelaide, North Terrace, Adelaide, SA, Australia; South Australian Genome Editing Facility, North Terrace, Adelaide, SA, Australia*

BEN ROLLO • *Department of Neuroscience, Central Clinical School, Monash University, Melbourne, VIC, Australia*

JOSEPH ROSENBLUH • *Monash Functional Genomics Platform, Monash University, Clayton, VIC, Australia; Department of Biochemistry and Molecular Biology, Monash University, Clayton, VIC, Australia*

PETRA RUNA-VOCHOZKOVA • *Large Animal Models in Cardiovascular Research, Internal Medical Department I, TU Munich, Munich, Germany; Center for Innovative Medical Models, LMU Munich, Munich, Germany*

DANIEL F. SALAMONE • *Facultad de Agronomía, Departamento de Producción Animal, Buenos Aires, Laboratorio Biotecnología Animal (LabBA), Universidad de Buenos Aires, Buenos Aires, Argentina; Instituto de Investigaciones en Producción Animal (INPA), CONICET-Universidad de Buenos Aires, Buenos Aires, Argentina*

VIRGINIA SAVY • *Facultad de Agronomía, Departamento de Producción Animal, Universidad de Buenos Aires, Buenos Aires, Argentina; Laboratorio Biotecnología Animal (LabBA), Buenos Aires, Argentina; Instituto de Investigaciones en Producción Animal (INPA), CONICET-Universidad de Buenos Aires, Buenos Aires, Argentina*

MAXWELL J. SCOTT • *Department of Entomology and Plant Pathology, North Carolina State University, Raleigh, NC, USA*

YASAMAN SHAMSHIRGARAN • *Laboratory of Clinical Chemistry, Sahlgrenska University Hospital, Gothenburg, Sweden*

MARCELA SOUZA-NEVES • *Instituto de Reproducción Animal Uruguay, Fundación IRAUy, Montevideo, Uruguay*

LIN SU • *Comparative Biology & Experimental Medicine, Faculty of Veterinary Medicine Biochemistry & Molecular Biology, Cumming School of Medicine, University of Calgary, Calgary, AB, Canada*

HUSEYIN SUMER • *Department of Chemistry and Biotechnology, School of Science, Computing and Engineering Technologies, Swinburne University of Technology, Hawthorn, VIC, Australia*

AMIR TAHERI-GHAHFAROKHI • *Quantitative Biology, Discovery Sciences, Biopharmaceuticals R&D, AstraZeneca, Gothenburg, Sweden*

LAURENT TESSON • *INSERM Centre de Recherche en Transplantation et Immunologie UMR 1064, Transgenesis Rat ImmunoPhenomic Facility (TRIP), Nantes, France*

PAUL Q. THOMAS • *School of Medicine, University of Adelaide, North Terrace, Adelaide, SA, Australia; Medical Research Institute, North Terrace, Adelaide, SA, Australia; South Australian Genome Editing Facility, North Terrace, Adelaide, SA, Australia; South Australian Health and Medical Research Institute, North Terrace, Adelaide, SA, Australia*

PAUL J. VERMA • *Livestock Sciences, South Australian Research and Development Institute, Government of South Australia, University of Adelaide, Adelaide, SA, Australia*

ANNA LAURA VOIGT • *Comparative Biology & Experimental Medicine, Faculty of Veterinary Medicine Biochemistry & Molecular Biology, Cumming School of Medicine, University of Calgary, Calgary, AB, Canada*

XIAOLONG WANG • *Key Laboratory of Animal Genetics, Breeding and Reproduction of Shaanxi Province, College of Animal Science and Technology, Northwest A&F University, Yangling, China*

DENNIS WEBSTER • *Recombinetics, Inc., St. Paul, MN, USA*

MELISSA WHITE • *School of Medicine, University of Adelaide, North Terrace, Adelaide, SA, Australia; South Australian Genome Editing Facility, North Terrace, Adelaide, SA, Australia*
MEGAN E. WILLIAMSON • *Department of Entomology and Plant Pathology, North Carolina State University, Raleigh, NC, USA*

Part I

Gene Editing Approaches

Chapter 1

Historical DNA Manipulation Overview

Lluis Montoliu

Abstract

The history of DNA manipulation for the creation of genetically modified animals began in the 1970s, using viruses as the first DNA molecules microinjected into mouse embryos at different preimplantation stages. Subsequently, simple DNA plasmids were used to microinject into the pronuclei of fertilized mouse oocytes and that method became the reference for many years. The isolation of embryonic stem cells together with advances in genetics allowed the generation of gene-specific knockout mice, later on improved with conditional mutations. Cloning procedures expanded the gene inactivation to livestock and other non-model mammalian species. Lentiviruses, artificial chromosomes, and intracytoplasmic sperm injections expanded the toolbox for DNA manipulation. The last chapter of this short but intense history belongs to programmable nucleases, particularly CRISPR-Cas systems, triggering the development of genomic-editing techniques, the current revolution we are living in.

Key words Transgenesis, Cloning, Nuclear transfer, Knockout animals, Genomic editing, CRISPR

1 Introduction

Most textbooks summarizing the history of genetic modification of animals refer to the 1980s when talking about the origins of these methods. And certainly, I will credit those researchers that launched the mouse embryo microinjection techniques. But it has to be stated clearly that DNA manipulation and the generation of genetically modified animals began a decade earlier, thanks to the studies and methods developed by the German embryologist Rudolf Jaenisch, initially educated as a virologist, which is the real, often unrecognized, pioneer of these techniques.

In fact, Jaenisch began his experiments with mouse embryos in the laboratory of Beatrice Mintz, another giant in mouse developmental biology. In 1974, Jaenisch and Mintz reported the successful finding of SV40 viral DNA in adult mice after injecting this virus in the blastocoel cavity of mouse blastocysts [1]. They found the viral DNA in some tissues of the resulting adult animals, although, at the time, they could not confirm whether this was a consequence

Paul John Verma, Huseyin Sumer and Jun Liu (eds.), *Applications of Genome Modulation and Editing*,
Methods in Molecular Biology, vol. 2495, https://doi.org/10.1007/978-1-0716-2301-5_1,

of a DNA integration or the manifestation of an episomal event. This is the first study that deserves to be listed as the origin of DNA manipulation efforts towards the generation of genetically modified animals.

Jaenisch continued on his own, further developing this nascent technique for transferring SV40 DNA to mice through infecting blastocysts [2]. He continued his work exploring another virus, the Moloney murine leukemia virus (M-MuLV), used to infect mouse preimplantation embryos at earlier stages (4- to 8-cell embryos) and in the birth of some mice developing leukemia, thereby proving the efficient integration of this exogenous DNA molecules into the host mouse genome [3]. Eventually, germ line transmission of M-MuLV DNA was confirmed in 1976 by Jaenisch and his coworkers [4]. Subsequent experiments consolidate the approach of using DNA integrative viruses to infect mouse developing embryos to obtain genetically modified animals [5]. These experiments represented the foundation of a new discipline: animal transgenesis, even though the words transgenesis or transgene would not be used until 6 years later, in 1982, by Jon Gordon and Frank Ruddle [6].

Gordon and Ruddle are usually referred to as the fathers of animal (mouse) transgenesis, often ignoring previous Jaenisch contributions. Actually, they contributed extensively to the popularization of the initial mouse microinjection techniques, since their first study, published in 1980, demonstrating how a popular plasmid, pBR322, could be microinjected into the pronuclei of fertilized mouse oocytes and later found in the resulting newborn mice [7]. One year later, they reported the effective germ line integration, and hence, transmission of the microinjected DNA molecules to the progeny of founder transgenic mice [8]. Gordon and Ruddle are to be recognized for writing the first manual with detailed procedures for producing transgenic mice by pronuclear microinjection of 1-cell stage embryos in 1983 [9].

Whereas Gordon and Ruddle were the first to report exogenous recombinant DNA molecules microinjected into mouse embryos and successfully transmitted through the germ line, another embryologist must be underlined for his outstanding contributions to these initial years of mouse transgenesis. Ralph Brinster is probably the key researcher that contributed the most, alone and in collaboration with the molecular biologist Richard Palmiter, his long-time collaborator, by investigating, optimizing and fine tuning all the required techniques and adequate procedures for the successful generation of transgenic mice.

Brinster was, and still is, an embryologist with a genuine interest in exploring culture conditions for investigating mouse embryo development in vitro, as it was nicely reviewed by one of his students, Richard Hammer [10]. Brinster learnt the basic methods from John Biggers and defined the classical culture media conditions, in a series of seminal studies published in the 1960s and the 1970s [11]. Therefore, Brinster was in an excellent position to develop and optimize further the pronuclear microinjection method described by Gordon and Ruddle. His first opportunity to join the nascent field of mouse transgenesis came through a first of many collaborations with Richard Palmiter. In 1981, they showed how a plasmid containing the structural gene for thymidine kinase from herpes simplex virus (HSV), fused to the promoter and regulatory region of the mouse metallothionein-I gene, could be microinjected into the pronuclear of mouse embryos, resulting in the birth of several transgenic mice carrying the plasmid transgene [12]. In this initial study, they realized and first reported the integration of multiple copies of transgene DNA organized as tandem arrays. That seminal article was followed by other complementary studies with similar plasmid transgenes where Brinster and Palmiter used mouse transgenesis to investigate the role of regulatory elements found in promoter genomic fragments [13], or the instability of expression of some transgenes integrated in different host genomic locations [14].

However, the public exposure and popularity of mouse transgenesis techniques had yet to come. A milestone in the generation of transgenic mice occurred by the end of 1982. Brinster and Palmiter generated transgenic mice with a transgene carrying the promoter of the metallothionein-I (MT-I) gene fused to the rat growth hormone (rGH) gene. The resulting transgenic mice were dramatically (~2.5 times) bigger than their non-transgenic littermates, and, not surprisingly that picture was chosen for the cover of the *Nature* issue where this experiment was published [15]. Numerous other examples of transgenesis followed and both researchers published the second comprehensive review on the subject in 1984 [16]. Hammer, Palmiter, and Brinster should be commended also for envisaging the first biomedical application of transgenesis, upon partially correcting the dwarfism of mutant mice with their MT-I-rGH transgene [17]. The same three researchers, along with other collaborators, must be credited as well for generating the first transgenic rabbits, sheep, and pigs in 1985 by pronuclear microinjection of these very different mammalian embryos, although with lower efficiencies, as compared with those observed in mice [18]. Similar experiments were reproduced some year later in Europe, by Gottfried Brem and his collaborators [19]. These early days of mammalian transgenesis have been nicely reviewed by Richard Palmiter in 1998 [20].

Brinster and Palmiter very soon anticipated the extraordinary potential of mouse (and, in general, animal) transgenesis for basic research and biotechnological applications, but also they realized about the need to optimize every single step of the pronuclear microinjection technique, in an attempt to increase the otherwise disappointing limited efficiencies observed (<5-10%) [21]. With this aim, they began reporting how numerous endogenous and external factors (e.g., site of microinjection, cytoplasmic or nuclear; the mouse strain; the microinjection of linear versus circular DNA molecules; the presence or absence of vector prokaryotic DNA in transgenes; the microinjection buffer, etc.) influenced the success rate of generating transgenic mice [22, 23]. And these optimized parameters have remained, without significant changes, essentially valid to our days, more than 35 years later. Brinster and Palmiter were also the first to report the benefits of adding non-coding sequences (e.g., introns) to transgenes for increasing their expression levels [24].

2 Targeted Inactivation of Genes

Classical mouse transgenesis by pronuclear injection was an instrumental technique to assess the expression of numerous transgenes and their effect in the phenotype of the resulting transgenic mice. The applications were nonetheless limited to overexpress the transgenes in all or in a subset of cells or tissues of the host. Those gain-of-function experiments included cell-specific ablation studies, for example, upon fusing tissue-specific promoter sequences with the diphtheria toxin gene [25]. However, the integration of transgenes occurred randomly, associated with undesired insertional mutagenesis events and chromosomal position effects. Thus, the gene-specific inactivation or modification had not been achieved. In a unique manifestation of self-confidence "you can do it" (in his words), Ralph Brinster decided to demonstrate that gene targeting could also occur by pronuclear transgenesis. With this aim, he microinjected more than 10,000 mouse embryos and generated more than 500 transgenic mice with a genomic construct before coming across one single mouse founder compatible with an integration that could have occurred through homologous recombination [26]. That mega-experiment would not be possible nowadays, due to the animal welfare considerations, the observance of the 3R rules (replacement, reduction, and refinement), and the mandatory ethical reviews that all animal experimentation projects must undergo before being authorized. However, that experiment nicely illustrated the need for developing alternative technologies that would enable gene-specific genomic alterations.

The solution came through the isolation of the first mouse embryonic stem (ES) cells, a group of pluripotent cells found in the inner cell mass of blastocysts contributing to all somatic and germ line tissues of the developing embryo. The mouse ES cells were first isolated and described by Martin Evans and Matthew Kaufmann in 1981 [27], and independently confirmed, a couple of months later, by Gail Martin, using a different and simpler method that became the reference in the field [28].

Isolating and cultivating ES cells was only part of the protocols required to target-specific genes. Injected ES cells into blastocysts had to demonstrate the potential to colonize the germ line. This was first explored by Allan Bradley, in Martin Evans laboratory, with embryo-derived teratocarcinoma (EC) cells [29] and, eventually, in 1986 they confirmed the generation of germline transmitting chimeras using retroviral vectors inserted into mouse ES cells injected into blastocysts [30]. This seminal experiment confirmed the uniqueness and utility of mouse ES cells as a tool for elucidating developmental genetics of the mouse [31].

Independently, the geneticist Oliver Smithies was investigating the process of homologous recombination in mammalian cells, aiming to substitute endogenous DNA sequences by exogenously provided ones, according to their homology [32]. Eventually, Smithies achieved his goal in 1985 and managed to target an endogenous gene with a plasmid construction previously digested with a suitable restriction enzyme cutting within the stretch of DNA homologous to the murine chromosome, and with a significant efficiency, in about one per thousand transfected clones [33].

With the ES cells available and a procedure to trigger homologous recombination, there was only a need to group all these independently devised methods into a robust procedure that would allow universal gene targeting in mice. This was the role of Mario Capecchi, a smart molecular and developmental biologist who had also been investigating homologous recombination events upon microinjecting DNA into mammalian cells [34, 35]. This tedious and cumbersome microinjection methods led Capecchi and his coworkers to demonstrate in 1986 that these targeted modifications could be used to introduce mutations in selected genes in mammalian cells in culture [36], and with a high frequency [37], similar to the ratio observed by Smithies upon transforming his cells [33].

Eventually, in 1987, Capecchi integrated all these previous concepts and developed a method for targeting specific genes in ES cells through homologous recombination, using a positive-negative selection method [38]. The first version of this method was meant to target selectable genes, such as the *HPRT* gene, whose mutations could be selected positively in the presence of the antibiotic analogous G418 with a neomycin-resistance (*neoR*) gene, and, negatively, upon growing the transformed cells in a culture medium containing the base analog 6-thioguanine, where only HPRT mutant clones would survive [38]. Smithies replicated Capecchi's observations correcting HPRT mutations [39] or introducing HPRT mutations [40] by homologous recombination in mouse ES cells.

Capecchi published in 1988 an improved version of his method, for non-selectable genes, where enrichment for homologous recombination events in mouse ES cells was driven by *neoR* surrounded by homologous sequences with the endogenous locus, and where the random integration was counteracted by the inclusion of the *HSV-tk* gene in the construct, outside the region of homology, whose integration rendered the transformed clones sensitive to nucleoside analogs such as FIAU or ganciclovir, resulting in the interference with DNA replication [41]. This was a robust universal method for targeting any murine gene in mouse ES cells by homologous recombination. The first (*Hprt*) mutant mice obtained from gene-targeted ES cells were obtained by Smithies in 1989 [42], followed by *Int-1* mouse mutant mice by Capecchi in 1990 [43], the same locus by Bradley also in 1990 [44] and other laboratories.

Ten years after establishing the DNA pronuclear microinjection technique for generating transgenic mice, an innovative, and far more sophisticated method, involving the culture and transformation of ES cells, and the generation of germline transmitting chimeras with those cells, allowed the generation of thousands of gene-specific mutant mice, also known as knockout (KO) mice [45]. The extraordinary relevance of this achievement was recognized with the Nobel Prize in Physiology or Medicine awarded jointly to Martin Evans, Mario Capecchi, and Oliver Smithies in 2007, 20 years after the outstanding contributions of these three researchers were integrated by Capecchi [46].

3 Overcoming the Limitations of Standard Transgenic and Mutant Mice

The toolbox for modifying the mouse genome at will, by adding or inactivating genes, was not as perfect as initially envisaged. Transgenic mice heavily suffered from the random integration events, often leading to non-expressing or ectopically expressing transgenes, beyond the expected pattern [47]. These chromosomal position effects were usually addressed by generating multiple transgenic mouse lines and by confirming similar phenotypes associated with transgenes integrated at different genome locations. In addition, some large genes, beyond the cloning capacity of standard plasmid DNA molecules, could not be easily analyzed.

On the other hand, many knockout mouse lines could not be established because the inactivation of some genes resulted in embryonic lethality [48]. Therefore, beyond confirming the relevance of those genes for proper embryonic development there were no animals available to further investigate the function of these genes in adult animals.

To overcome the observed suboptimal, variable, and often ectopic transgene expression associated with standard transgenic mice, an innovative approach was envisaged. The use of very large genomic fragments, in the range of hundreds to 1–2 thousand kilobases, appeared as a putative solution for inserting a locus into the mouse genome, along with all known and unknown surrounding regulatory elements likely to be required for the faithful expression of the transgene, similar to that of the endogenous locus. In the early 1990s, there was only one vector type compatible with such requirements: the yeast artificial chromosomes (YAC), whose size could be anywhere from a few dozens of kilobases up to 1–2 megabases. These YACs were capable of encompassing very large and complex loci, such as the Hox genes [49].

In 1992, Günther Schütz laboratory was the first to report the generation of transgenic mice with a relatively modest YAC (35 kb), carrying a tyrosinase (*Tyr*) transgene, a minigene, with fragments derived from the mouse genome and from cDNA [50]. In early 1993, three groups succeeded independently and simultaneously in generating transgenic mice with larger (>250 kb) YAC transgenes for the first time [51, 52, 53], using different technological approaches, and in all three cases optimal transgene expression patterns was reported. YAC transgenesis became rapidly popular and very large genomic fragments, including some from the human genome, were transferred and inserted into the mouse genome, including the human immunoglobin loci [54] and the human *APP* gene [55]. The take home message from all these pioneer experiments with YAC transgenes in mice was that "bigger was probably better" [56]. The use of large genomic fragments encompassing the locus of interest cloned within a YAC was compatible with a reproducible and robust transgene expression pattern, irrespective of the integration site, and often copy-number dependent.

However, YACs were often unstable and chimeric and its manipulation and modification required specific skills to grow yeast cells and alter the yeast genome, where these YACs had to be constructed. Therefore, other types of artificial chromosomes were explored, including P1 bacteriophage-derived artificial chromosomes (PACs) and bacterial artificial chromosomes (BACs), both with a smaller cloning capacity (<300 kb) as compared to YACs, but easier to handle with the standard bacterial procedures available in almost every molecular biology laboratory [57]. The first transgenic animal generated with a PAC was reported in 1993 [58], whereas the first BAC transgenic mice were published in 1997 [59].

BAC transgenesis soon became the method of choice, due to the ease by which these vectors could be manipulated and the development of operative homologous recombination methods in bacteria, such as recombineering [60].

An overview of BAC transgenic mice was reported by Thomas Saunders in 2009 [61]. In this review, through the analysis of 707 transgenic founder mice generated from 86 BAC transgenes purified by different methods, they concluded that most (80%), but not all, constructs produced transgenes with the expected expression pattern. In addition, a careful analysis was required to discriminate between intact BAC transgenes and fragments of them integrated, prone to produce unexpected expression results.

An alternative method was also required to overcome the embryonic lethality observed in some knockout mouse experiments. Such desired method should prevent gene inactivation during development and should allow triggering this inactivation of the targeted gene in the adulthood, once the mouse would have been born. Such method was known as "conditional mutagenesis" and was first developed by Klaus Rajewsky in 1994 [62]. In this seminal publication, they reported the use of a prokaryotic site-specific recombination system, the Cre-loxP system, composed of a Cre recombinase and a target loxP sequence, a short (34 bp) directional DNA fragment recognized by Cre recombinase and used to remove any DNA sequence found in between two consecutive loxP sequences in cis and oriented in the same direction.

The approach was smart and innovative. Two loxP sites had to be cloned in a DNA construct of the target locus to be inactivated, usually inside consecutive introns, delimiting an exon ready to be excised. Thereafter, using ES cells and homologous recombination, such construct was integrated in the mouse genome, thereby generating the so-called floxed (flanked by loxP sites) allele. This resulting allele was ready to be inactivated as soon it would be exposed to Cre recombinase. This nuclease was brought in through a standard transgene, in a different mouse, driving the expression of the Cre recombinase in a given cell type or tissue of choice. Then, upon mating the floxed mouse with the transgenic Cre mouse, the planned gene would only be inactivated where loxP sites and Cre recombinase coincided, thereby generating a tissue- or cell-specific knockout. Rajewsky demonstrated this method with a T-cell-specific knockout of a DNA polymerase gene beta, whose ubiquitous deletion would have been lethal [62].

Conditional mutagenesis boosted mouse knockout projects and multiplied the number of experiments and phenotypes one could obtain from a single floxed mouse allele, upon exposing them to a variety of transgenic mice expressing Cre recombinase gene in different tissues or cell types [63]. The function of mouse genes could be investigated, for example, in the brain [64] or the immune system [65], thanks to these conditional mutagenesis approach.

The ease by which these mice carrying floxed alleles could be generated triggered some international initiatives aiming to generate and phenotype mouse knockout lines representative for all mouse (~20,000) genes. Initially, there were different consortia organized around the world, including the International Knockout Mouse Consortium (IKMC) [66] and the European Conditional Mouse Mutagenesis Program (EUCOMM) [67], but eventually all projects merged into the International Mouse Phenotyping Consortium (IMPC) [68], currently active through its web portal (https://www.mousephenotype.org/), and where up to 7590 mouse knockout lines have been already generated and phenotyped (May 2021 release), with their corresponding data fully available to anyone interested.

Conditional mutagenesis was also further boosted with the production of inducible chimeric Cre recombinases, fused to hormone-binding domain from nuclear receptors, thereby enabling the activation of Cre recombinase activity at will, on a timely manner, through the administration of hormone analogs. These are used to release the nuclease, bound to chaperones in the cytoplasm, and allow its nuclear translocation for the expected recombinase activity to occur. The first inducible Cre recombinases were produced by Pierre Chambon Laboratory in 1995 [69]. Other than the Cre/loxP system, additional site-specific recombinases (e.g., Flp, Dre, PhiC31) have been described and used in genetically modified animals for controlling the activation/inactivation of selected genes [70].

4 From Cloning to Genome Engineering Beyond Mice. Somatic Cell Nuclear Transfer Techniques

For the first 20 years of genetic manipulation of animals, the techniques used were applied essentially in mice. Not even in rats, another rodent model but not as suitable for genetic manipulation as mice. Standard pronuclear microinjection of DNA was working in other mammalian species, but at lower efficiencies, often <1% [71]. Furthermore, with the uncertainties associated with standard transgenesis, due to the randomness of transgene integration, very few among the produced transgenic animals in livestock delivered the expected expression levels, with some unique and celebrated exceptions, such as one extraordinary transgenic sheep born in Scotland, which was able to produce high levels of a recombinant protein in the milk [72].

On the other hand, the isolation of *bona fide* ES cells had only been achieved in mice in 1981 [27, 28], and they would not be described in humans by J.A. Thomson until 1998 [73], and in rats by Austin Smith until 2008 [74]. This was a major hurdle to overcome in order to extend the possibility of gene targeting to other species, beyond mice. Therefore, it was clear that a disruptive, totally independent, method was urgently needed in animal transgenesis.

The solution came from the Roslin Institute, in Edinburg (UK), where the team of Ian Wilmut, with Angelika Schnieke and Keith Campbell, managed to clone the first mammal from and adult cell. Nuclear transfer techniques had been successfully established in frogs by John Gurdon in the late 1950s and thereafter [75, 76], but exporting similar procedures to mammals have repeatedly failed. The Scottish team pioneered in 1996 the use of some sheep embryonic pluripotent cells, called TNT (for Totipotent for Nuclear Transfer), and using their nuclei they were able to produce live lambs, upon reconstituting with them 1-cell embryos by first removing nuclear genetic material from unfertilized oocytes [77]. Dolly the sheep, the first mammal cloned from an adult cell, was also born in 1996. The publication reporting the birth of Dolly, derived from nuclei of a sheep cell line established from mammary gland tissue, appeared in 1997 and immediately generated enormous expectations and controversies of all kinds [78].

The cloning procedure by somatic cell nuclear transfer (SCNT), originally described in sheep, was subsequently replicated in several species, with some species-specific methodological adaptations. The series of cloned animals included cows [79, 80], mice [81], goats [82], pigs [83], rats [84], rabbits [85], horses [86], dogs [87], and several other species.

The relevance of SCNT methods were promptly recognized. The nuclei used to reconstruct embryos could also be derived from genetically modified cells in culture. Hence, the resulting animal upon reconstructing embryos with these nuclei would be a clone and a transgenic animal. Furthermore, the genetic manipulations could include not only adding new genes but also inactivating genes. In other words, thanks to SCNT methods it was possible to produce knockout animals in different species, beyond mice, overcoming the lack of ES cells, through an alternative route.

The first transgenic and cloned animals were also sheep, produced at the Roslin Institute in 1997, illustrating a more robust and efficient way of generating transgenic animals in livestock, as compared with DNA pronuclear microinjection [88]. A similar success was reported in bovine in 1998, with transgenic and cloned calves born from genetically modified fetal fibroblasts [89], followed by comparable successful reports in other livestock species, such as pigs [90].

Gene targeting was the next challenge in livestock, using SCNT techniques. This was achieved in 2000, with the description of the first knockout sheep, inactivating the *COL1A1* locus [91]. The application of these methods was extended to pigs, where the first animals carrying the inactivated gene-encoding alpha-1,3-galactosyltransferase, to be used for xenotransplantation projects, were born in 2002 [92]. SCNT methods have been, until the revolution triggered by the programmable nucleases in the genome-editing era, the reference methods for generating both transgenic and knockout animals in mammalian species beyond mice, in species where DNA pronuclear microinjection was not efficient enough, or where *bona fide* ES cells could not be obtained [93].

For human beings, reproductive cloning (routinely applied in other animals) triggered enormous ethical dilemmas, mainly because the technique required the potential destruction and manipulation of human embryos. On the other hand, the proposed therapeutic cloning route generated many expectations for its potential application in regenerative medicine. However, therapeutic cloning was never achieved in humans, rather only described in a seminal paper from Rudolf Jaenisch, demonstrating the complexity of the process in mice [94].

Most of the ethical dilemmas were resolved in 2006 by Shinya Yamanaka, when he reported the identification of the four transcription factors that were required to transform and reprogram a somatic cell into a pluripotent cell [95]. These findings enabled the establishment of inducible pluripotent stem cells (iPS) which could be used in combination with transgenic techniques [96]. Since then, the description and establishment of iPS cells in livestock has been attempted with mixed success, indicating a great potential of these inducible pluripotent stems cells in farm animals [97].

Shinya Yamanaka and John Gurdon were jointly awarded the Nobel Prize in Physiology or Medicine in 2012, for the discovery that mature cells can be reprogrammed to become pluripotent. Unfortunately, none of the members of the Roslin team, responsible for the generation of Dolly the sheep, the first mammal to be cloned from adult cells, were included in this public recognition.

5 Additional Transgenic Technologies: Transposons, Lentivirus, ICSI, SMGT, and RNA Interference

In the history of genetic modification of animal genomes, there are other technologies that have been explored and presented. They have been used in several experiments and served to overcome some of the known limitations of classical transgenesis with a mixed success. These are complementary techniques the researchers have available in their toolbox before considering a given genetic alteration experiment. In some cases, these additional techniques might be the only method available for obtaining the desired genetically modified animal.

Transposons are mobile DNA elements with the extraordinary capacity to jump between different genomic locations propelled by an associated enzyme, a transposase, which usually recognizes specific DNA sequences with inverted repeats delimiting these elements. Historically, transposons have also been used to trigger animal transgenesis in a variety of species. For example, the P-elements in *Drosophila* have been instrumental for functional genetics analyses in flies [98]. And transposition is also the method of choice for transgenesis in zebrafish [99].

Progressively, transposons were adapted to be used in mouse functional genomics. At first, as an ancillary technique to prepare, in vitro, the complex DNA constructs to be used later for homologous recombination in mouse ES cells [100].

Probably, the first transposition system to succeed in vertebrates for transgenic experiments was *Sleeping Beauty* (SB), whose mutated DNA sequence was rescued in vitro thereby resurrecting this mobile element from the genome of salmons and proven to work in mammalian cells [101]. Indeed, the SB transposition system was implemented as a rapid and highly efficient system to trigger genetic modifications in several mammalian species [102]. In mice, the SB system was shown to sustain a capacity for insertional mutagenesis and confirmed to be transmitted through the germline [103]. The development of hyperactive SB transposition systems enhanced their functionality for insertional mutagenesis [104], and its use was confirmed in mice, rats, and rabbits [105]. However, chromosomal rearrangements are not rare at the transposition genomic locations, which can compromise the expected expression patterns of the transgenes. This is one of the known limitations of transposons in transgenic animals [106].

The use of viral vectors in animal transgenesis was proven already in the 1970s by Jaenisch, the pioneer in this field. However, the use of viral vectors for shuttling genes of interest into animal genomes was implemented many years later. In 2002, David Baltimore and collaborators showed how lentiviral vectors carrying reporter genes would generate transgenic mice and rats at much higher efficiencies than those obtained with DNA pronuclear microinjection techniques [107]. These lentiviral vectors were immediately applied for the generation of transgenic livestock and demonstrated higher level of transgenesis efficiencies [108]. The advantages of lentivirus included its use in a wider variety of species and mouse strains, where DNA pronuclear injection as very inefficient or not possible, and the high number of founder animals this technique could generate. In contrast, lentiviral transgenesis had a limited cloning capacity (~7 kb) and was often associated with multiple integration sites, thus making the establishment and characterization of stable transgenic animal strains a real challenge. The use of lentivirus for transient experiments in transgenesis is probably the method of choice, the one that ensures a higher number of transgenic newborns. Furthermore, the microinjection is easier because viral particles can be directly microinjected under the zona pellucida. These advantages and disadvantages of lentiviral transgenesis have probably prevented a wider success for this method. Nonetheless, lentiviral transgenesis remains available in the toolbox for animal genetic modification, and it can rescue a transgenic experiment where other techniques might fail [109].

The intracytoplasmic sperm injection (ICSI) technique is a well-established method in human-assisted reproduction, available in most fertility clinics. Primate (including human) embryos are robust and direct microinjection of sperm heads into unfertilized oocytes can be achieved routinely and has resulted in the birth of many babies around the world [110]. In contrast, ICSI techniques in other mammalian species, particularly in mice, have been most challenging for many years. Eventually, success was achieved thanks to a complementary technique, piezo-injection, which softens the introduction of the ICSI needle into the cytoplasm of mouse unfertilized oocytes, clearly more fragile than the human oocytes. Rizo Yanagimachi showed how ICSI could be applied in mice [111] and rats [112] and popularized the use of piezo-injectors for this purpose.

ICSI can also be combined with exposure of partially damaged sperm heads to DNA molecules, thus resulting in an alternative method for efficient delivery of DNA into the host genome of choice. The method was reported by Teruhiko Wakayama in 1999 and demonstrated high efficiencies of transgenesis in mice [113]. This technique of ICSI transgenesis was later on exported and validated in other species, for example, in non-human primates [114]. ICSI was also challenged for delivering large transgenes, such as BACs [115] or YACs [116], into the mouse genome, although with a lower efficiency for the integration of intact and entire DNA constructs [116]. ICSI was also instrumental for cloning mice since the introduction of somatic cell nuclei was achieved with this method [81].

The combination of ICSI with DNA delivery techniques could also be considered a sperm-mediated gene transfer (SMGT) method although the origins of SMGT were rather controversial. A report by Claudio Spadafora in 1989 announced the use of sperm as a vector for introducing DNA into mouse eggs, resulting in the production of transgenic mice [117]. This experiment could not be reproduced initially in many laboratories, likely due to some undefined genetic factors playing an important role in the success of this technique [118, 119]. The SMGT method was also applied for the generation of transgenic pigs and reported by the same Italian group to be highly efficient method [120]. However, contradictory results were obtained by other laboratories [121], thus suggesting that the SMGT method was still not adequately defined, nor a robust technique to apply, too exposed to variable results.

RNA interference is also a method that was developed, initially in nematodes, to alter the expression of some genes by interfering with its translation. The production of antisense RNA sequences would result in double-strand RNA molecules that will trigger an endogenous mechanism of clearance, resulting in the knockdown or silencing of the selected gene. Andrew Fire and Craig Mello described this technique in the worm *Caenorhabditis elegans* in 1998, as a powerful alternative to inactivate a given gene [122]. Their efforts were recognized in 2006 with a Nobel Prize in Physiology or Medicine which they were awarded jointly.

Since the initial description of the RNA interference concept, several attempts were undertaken to demonstrate the application of this method for silencing or inactivating the expression of genes without interrupting the genome. As opposed to knockout animals (where the genome had been altered), these animals were known as knockdown because often the expression of the targeted gene was significantly reduced, but it did not disappear. Different RNA molecules had been used for his purpose, including antisense RNA and distinct subtypes of RNA interference (RNAi, siRNA, shRNA, miRNA, etc.) molecules in transgenic animals. For example, in zebrafish the use of modified oligonucleotides (morpholinos) have been widely used to interfere with RNA translation and produce gene-specific knockdowns [123]. However, the use of RNA interference approach in zebrafish must be undertaken with great care, taken into account the required controls and experimental conditions that were nicely detailed by a consortium of researchers in 2017 [124].

The summary of events and technological innovations in DNA modification techniques in animals can be viewed in Fig. 1, representing the transgenesis timeline, where all different methods referred in this chapter and the year in which they were implemented are illustrated.

6 Genome-Editing Techniques: The Use of Programmable Nucleases for the Generation of Transgenic and Knockout Animals

Although genome-editing techniques are currently associated almost exclusively to CRISPR-Cas tools, there are different types of programmable nucleases that have been used in the past to trigger gene-specific alterations in the genome of animals.

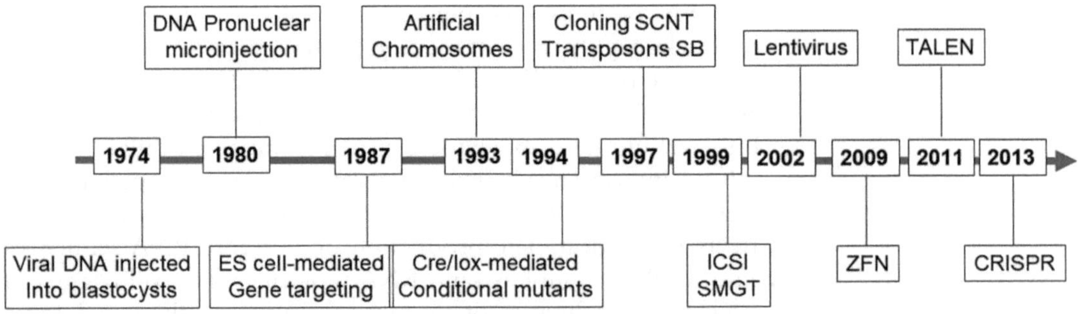

Fig. 1 Transgenesis timeline. A summary of milestones and key technological advances that occurred during the development of different methods for the genetic modification of animals. Abbreviations: *SCNT* somatic cell nuclear transfer, *SB* sleeping beauty, *ES* embryonic stem, *ICSI* intracytoplasmic sperm injection, *SMGT* sperm-mediated gene transfer, *ZFN* zinc-finger nuclease, *TALEN* transcription activator-like effector nucleases (TALEN), *CRISPR* clustered regularly interspaced short palindromic repeats. Scheme prepared by Lluis Montoliu

At least, we know of four different types of genome-editing tools that have been applied for this purpose, namely: yeast meganucleases, zinc-finger nucleases (ZFN), transcription activator-like effector nucleases (TALEN), and clustered regularly interspaced short palindromic repeats (CRISPR) and Cas (CRISPR associated) nucleases [125]. In all these cases, the underlying mechanism responsible for the specific inactivation of genes was the same. In brief, the generation of a double-strand break (DSB) in the gene of interest triggers the endogenous repair mechanisms. In the absence of donor DNA to be used as template for homology-driven repair, the mechanism acting by default in all cells is the non-homologous end joining (NHEJ) pathway, resulting in the random insertion and deletion (INDEL) of nucleotides, hoping to generate a local micro-homology between the two DNA strands that could be used for restoring the physical continuity of the chromosome. The likely consequence associated with this DNA repair pathway is the inactivation of the gene, hence promoting a specific cut in a gene of interest usually results in the corresponding gene knockout [125].

Yeast-derived meganucleases were the first genome-editing tools explored to trigger gene-specific modifications through the generation of DSBs in the DNA [126]. However, these first tools had no versatility. The restriction enzyme activity recognized a given DNA sequence, 20–40 bp long, which had to be present in the genome in order to use the corresponding nuclease. This lack of versatility essentially resulted in a limited use of these proteins for genome-editing purposes. Years later, recombinant meganucleases were eventually produced, detecting DNA sequences of choice and used for the generation of gene-targeted animals [127], but this late improvement coincided with the explosion of newer and more versatile genome-editing tools, and hence the use of engineered meganucleases was mostly abandoned.

ZFNs were first envisaged in 2001, as chimeric proteins, containing DNA-binding domains, associated with several zing-finger modules, and the DNA endonuclease activity from a bacterial enzyme, *Fok* I, producing DSB at the genome locations dictated by the zinc-finger domains [128]. The recognition of DNA sequences was again on the protein side, similar to meganucleases. Therefore, a new ZFN had to be generated for each gene to be targeted. Furthermore, the technological platform for creating ZFNs was proprietary and any potential user had to purchase these engineered ZFNs. Nonetheless, the capacity of ZFNs for specifically inactivating genes in animals was first demonstrated in rats in 2009 by Aron Geurts and collaborators [129], with impressively high efficiencies, never seen before with homologous recombination approaches in mouse ES cells. This experiment triggered the use of ZFNs for targeting genes in a variety of species including rats [130], mice [131], cattle [132], and pigs [133], with great success in all cases.

ZFNs were particularly used in livestock for generating animals with additional properties such as increased resilience to diseases. Examples of these applications were ZFN genome-edited cattle with increased resistance to mastitis [134], or ZFN genome-edited pigs carrying a warthog-derived genetic variant meant to confer resistance to African Swine Fever [135].

The limited availability of suitable ZFNs, essentially to be obtained from commercial sources, triggered the search for alternative genome-editing tools that could be engineered in every single molecular biology laboratory. The TALENs were the next programmable nucleases to be explored. These were partially derived from prokaryotic transcription factors of bacteria-infecting plant cells fused with the same *Fok* I restriction enzyme domain. The first gene inactivation reported with TALEN was described by Ignacio Anegon in rats in 2011 [136]. This initial success and the fact that TALENs can be fabricated by the researchers themselves quickly prompted many laboratories to jump from using ZFNs to TALENs. These new tools were used for generating genome-edited mice [137], and their reliability was extended to livestock [138], including cattle [139], pigs [140], and goats [141].

TALENs could be engineered in the laboratory in contrast to ZFNs, but the process was not simple. The construction of individual TALENs was rather laborious and prone to errors due to the high number of plasmids (~100) required for generating specific proteins [142]. Furthermore, the use of plasmids carrying the TALEN gene constructs could lead to the unnoticed integration of these DNA molecules in the host genome [143], as it was unfortunately demonstrated in cattle generated with TALEN to reproduce the animal welfare hornless trait, associated with the *Polled* mutation [144].

Both ZFNs and TALENs were capable of producing genome-edited animals with high efficiency, robustly, and reliably. However, the generation of the ZFNs and TALENs themselves was not an easy, rather a sophisticated task, thereby preventing a universal and massive use of these DNA genetic modification techniques. There was a need for an alternative genome-editing tool that could be more amenable, cheaper, versatile, and easier to implement.

It did not take long until the CRISPR-Cas tools, and ancient prokaryotic defense system used by bacteria to fight bacteriophages [145], were proposed to be used for genome-editing purposes. The proposal came from a scientific collaboration between Emmanuelle Charpentier and Jennifer Doudna, who published a seminal article in Science in 2012 describing the compounds of the CRISPR-Cas9 system from the pathogenic bacteria, *Streptococcus pyogenes* [146]. Eight years later, in 2020, they were jointly awarded the Nobel Prize in Chemistry for the development of a genome-editing method.

In contrast to meganucleases, ZFNs or TALENs, the CRISPR-Cas genome-editing tools were highly versatile and fully programmable. In their simplest constitution, they were composed of a DNA endonuclease (usually Cas9, but many other had been derived and described since 2012) and a small RNA molecule (synthetic guide RNA, sgRNA) that would drive the Cas9 to generate a DSB precisely at the gene whose complementary sequence is represented in one of the ends of the sgRNA. The system has full capacity for adapting to match any desired target genomic sequence, with only the limitations imposed by the corresponding Cas nuclease (e.g., the protospacer adjacent motif (PAM) a short DNA sequence that should appear next to the complementary DNA sequence represented in the sgRNA) [145, 146].

Even though the CRISPR-Cas technological revolution was launched in 2012, it was not until 2013 when the first genome-edited animals were produced. At first, zebrafish was the first animal species to be genetically modified by CRISPR-Cas9 tools [147], closely followed by genome-edited mice [148]. In mice, the generation of multiple simultaneous gene inactivation, mediated by CRISPR-Cas9 tools, as well and the precise introduction of mutations at desired loci was expanded to the production of conditional mutated alleles, floxed, with significant efficiency [149].

The possibility of inactivating multiple genes simultaneously with CRISPR tools was one of the first features to be underlined by researchers. This capacity was brought to the extreme by George Church upon inactivating all inserted porcine endogenous retroviruses, PERVs, in the pig genome (>60 integrated copies of these retroviruses) by CRISPR-Cas9 strategy, first in porcine cells [150], and later on, using the genome-edited cells, in genome-edited pigs derived through SCNT [151]. A similar multiple mutation approach was undertaken with pigs generated for xenotransplantation purposes [152].

Another field often unexplored, due to its complexity, which benefited from the advent of CRISPR-Cas9 techniques was the functional analysis of the non-coding genome. The intergenic DNA, full of repetitive sequences and mobile transposon and retrotransposon elements, was often a nightmare for any scientific project aiming to establish the relevance of some DNA regulatory elements. With CRISPR-Cas9 strategies, it was much easier to target those DNA regulatory elements and excise them with two CRISPR-Cas reagents targeting contiguous genomic sequences in cis separated by the desired DNA element whose relevance had to be investigated [153, 154]. Furthermore, CRISPR technology allowed to test the association of some non-coding genome alterations involving DNA regulatory elements with diseases [155].

In livestock, CRISPR-Ca9 methods boosted the production of genome-edited animals for biomedical and biotechnological applications in nearly all species, with an extraordinary efficiency that had previously not been observed. Alone, by delivering CRISPR-Cas9 reagents inside 1-cell embryos, or in combination with SCNT methods, by first using CRISPR-Cas9 on cultured cells and thereafter using their nuclei to derive the corresponding genome-edited animals, the CRISPR tools have been instrumental for the generation and analysis of many new genetically modified animals in multiple species.

Some of the first examples of CRISPR-Cas9 use in mammalian species beyond mice included genome-edited pigs [156], sheep [157], cattle [158], rabbits and goats [159], and dogs [160], among many other genome-edited animal species.

Genome editing is not always clean. The known limitations include the possibility of targeting similar genome sequences elsewhere, other than the target locus, a problem referred to as "off-targets," and the genetic diversity of alleles commonly generated by the NHEJ repair pathway, among which one hopes to find the planned allele. Both limitations have been seriously preventing the universal application of these genome-editing techniques on human beings, for obvious reasons, and they are of course also found in genome-edited animals. However, these limitations can be reasonably and efficiently handled in animals, simply by selecting the most appropriate founder and segregating the multiple alleles of these often mosaic animals among the progeny. One or two rounds of breeding to wild-type animals will usually be sufficient to get rid of any undesired off-target, while selecting for the desired genetic allele at the target locus [153].

The future of CRISPR-Cas variants goes beyond genome editing with the development of base editors and prime editors, associated with deaminase and reverse transcriptase activities, respectively, for the more precise genetic alteration without DSB. Most of these recent and innovative CRISPR variants have been produced by David R Liu [161]. Base editors have been already successfully used in mouse models of progeria [162] and in a non-human primate model for studying cardiovascular diseases [163]. Prime editors have just begun to be used in mice, with an impressive outcome, maintaining the efficiency of gene-targeting and reducing to non-detectable levels the mosaicism and the off-target effects [164].

This is the current end of the transgenesis timeline (Fig. 1), but I am sure it is not the end of the development of further technologies for the efficient and precise genetic modification of animals. From 1974 to 2021 almost 50 years have passed, where numerous innovations have been subsequently proposed by researchers, always aiming to improve the techniques for altering the genome of animals at will. The initial random procedures have been replaced by precise tools that trigger the desired alteration only at the chosen locus. Still, some technical limitations must be overcome, and protocols need to be further improved. We do not know what the next 50 years of DNA modification will bring us, in terms of new methods, but what is clear is that the evolution of transgenesis techniques will continue generating surprises and interesting results.

Acknowledgments

This work was supported by the Spanish Ministry of Science and Innovation (MICINN) through the Grant [RTI2018-101223-B-I00] to L.M.

References

1. Jaenisch R, Mintz B (1974) Simian virus 40 DNA sequences in DNA of healthy adult mice derived from preimplantation blastocysts injected with viral DNA. Proc Natl Acad Sci U S A 71(4):1250–1254
2. Jaenisch R (1975) Infection of mouse blastocysts with SV40 DNA: normal development of the infected embryos and persistence of SV40-specific DNA sequences in the adult animals. Cold Spring Harb Symp Quant Biol 39(Pt 1):375–380
3. Jaenisch R, Fan H, Croker B (1975) Infection of preimplantation mouse embryos and of newborn mice with leukemia virus: tissue distribution of viral DNA and RNA and leukemogenesis in the adult animal. Proc Natl Acad Sci U S A 72(10):4008–4012
4. Jaenisch R (1976) Germ line integration and Mendelian transmission of the exogenous Moloney leukemia virus. Proc Natl Acad Sci U S A 73(4):1260–1264
5. Jaenisch R (1980) Retroviruses and embryogenesis: microinjection of Moloney leukemia virus into midgestation mouse embryos. Cell 19(1):181–188
6. Gordon JW, Ruddle FH (1982) Germ line transmission in transgenic mice. Prog Clin Biol Res 85(Pt B):111–124
7. Gordon JW, Scangos GA, Plotkin DJ, Barbosa JA, Ruddle FH (1980) Genetic transformation of mouse embryos by microinjection of purified DNA. Proc Natl Acad Sci U S A 77(12):7380–7384
8. Gordon JW, Ruddle FH (1981) Integration and stable germ line transmission of genes injected into mouse pronuclei. Science 214(4526):1244–1246
9. Gordon JW, Ruddle FH (1983) Gene transfer into mouse embryos: production of transgenic mice by pronuclear injection. Methods Enzymol 101:411–433
10. Hammer RE (1998) Egg culture: the foundation. Int J Dev Biol 42(7):833–839
11. Biggers JD (1998) Reflections on the culture of the preimplantation embryo. Int J Dev Biol 42(7):879–884
12. Brinster RL, Chen HY, Trumbauer M, Senear AW, Warren R, Palmiter RD (1981) Somatic expression of herpes thymidine kinase in mice following injection of a fusion gene into eggs. Cell 27(1 Pt 2):223–231
13. Brinster RL, Chen HY, Warren R, Sarthy A, Palmiter RD (1982) Regulation of metallothionein—thymidine kinase fusion plasmids injected into mouse eggs. Nature 296(5852):39–42
14. Palmiter RD, Chen HY, Brinster RL (1982) Differential regulation of metallothionein-thymidine kinase fusion genes in transgenic mice and their offspring. Cell 29(2):701–710
15. Palmiter RD, Brinster RL, Hammer RE et al (1982) Dramatic growth of mice that develop from eggs microinjected with metallothionein-growth hormone fusion genes. Nature 300(5893):611–615
16. Brinster RL, Palmiter RD (1984) Introduction of genes into the germ line of animals. Harvey Lect 80:1–38
17. Hammer RE, Palmiter RD, Brinster RL (1984) Partial correction of murine hereditary growth disorder by germ-line incorporation of a new gene. Nature 311(5981):65–67
18. Hammer RE, Pursel VG, Rexroad CE Jr, Wall RJ, Bolt DJ, Ebert KM, Palmiter RD, Brinster RL (1985) Production of transgenic rabbits, sheep and pigs by microinjection. Nature 315(6021):680–683
19. Weidle UH, Lenz H, Brem G (1991) Genes encoding a mouse monoclonal antibody are

expressed in transgenic mice, rabbits and pigs. Gene 98(2):185–191

20. Palmiter RD (1998) Transgenic mice--the early days. Int J Dev Biol 42(7):847–854
21. Palmiter RD, Brinster RL (1985) Transgenic mice. Cell 41(2):343–345
22. Brinster RL, Chen HY, Trumbauer ME, Yagle MK, Palmiter RD (1985) Factors affecting the efficiency of introducing foreign DNA into mice by microinjecting eggs. Proc Natl Acad Sci U S A 82(13):4438–4442
23. Palmiter RD, Brinster RL (1986) Germ-line transformation of mice. Annu Rev Genet 20: 465–499
24. Brinster RL, Allen JM, Behringer RR, Gelinas RE, Palmiter RD (1988) Introns increase transcriptional efficiency in transgenic mice. Proc Natl Acad Sci U S A 85(3):836–840
25. Behringer RR, Mathews LS, Palmiter RD, Brinster RL (1988) Dwarf mice produced by genetic ablation of growth hormone-expressing cells. Genes Dev 2(4):453–461
26. Brinster RL, Braun RE, Lo D, Avarbock MR, Oram F, Palmiter RD (1989) Targeted correction of a major histocompatibility class II E alpha gene by DNA microinjected into mouse eggs. Proc Natl Acad Sci U S A 86(18): 7087–7091
27. Evans MJ, Kaufman MH (1981) Establishment in culture of pluripotential cells from mouse embryos. Nature 292(5819):154–156
28. Martin GR (1981) Isolation of a pluripotent cell line from early mouse embryos cultured in medium conditioned by teratocarcinoma stem cells. Proc Natl Acad Sci U S A 78(12): 7634–7638
29. Bradley A, Evans M, Kaufman MH, Robertson E (1984) Formation of germ-line chimaeras from embryo-derived teratocarcinoma cell lines. Nature 309(5965):255–256
30. Robertson E, Bradley A, Kuehn M, Evans M (1986) Germ-line transmission of genes introduced into cultured pluripotential cells by retroviral vector. Nature 323(6087): 445–448
31. Bradley A, Robertson E (1986) Embryo-derived stem cells: a tool for elucidating the developmental genetics of the mouse. Curr Top Dev Biol 20:357–371
32. Smithies O, Koralewski MA, Song KY, Kucherlapati RS (1984) Homologous recombination with DNA introduced into mammalian cells. Cold Spring Harb Symp Quant Biol 49:161–170
33. Smithies O, Gregg RG, Boggs SS, Koralewski MA, Kucherlapati RS (1985) Insertion of DNA sequences into the human chromosomal beta-globin locus by homologous recombination. Nature 317(6034):230–234
34. Capecchi MR (1980) High efficiency transformation by direct microinjection of DNA into cultured mammalian cells. Cell 22(2 Pt 2):479–488
35. Folger KR, Wong EA, Wahl G, Capecchi MR (1982) Patterns of integration of DNA microinjected into cultured mammalian cells: evidence for homologous recombination between injected plasmid DNA molecules. Mol Cell Biol 2(11):1372–1387
36. Thomas KR, Capecchi MR (1986) Introduction of homologous DNA sequences into mammalian cells induces mutations in the cognate gene. Nature 324(6092):34–38
37. Thomas KR, Folger KR, Capecchi MR (1986) High frequency targeting of genes to specific sites in the mammalian genome. Cell 44(3): 419–428
38. Thomas KR, Capecchi MR (1987) Site-directed mutagenesis by gene targeting in mouse embryo-derived stem cells. Cell 51(3):503–512
39. Doetschman T, Gregg RG, Maeda N, Hooper ML, Melton DW, Thompson S, Smithies O (1987) Targetted correction of a mutant HPRT gene in mouse embryonic stem cells. Nature 330(6148):576–578
40. Doetschman T, Maeda N, Smithies O (1988) Targeted mutation of the Hprt gene in mouse embryonic stem cells. Proc Natl Acad Sci U S A 85(22):8583–8587
41. Mansour SL, Thomas KR, Capecchi MR (1988) Disruption of the proto-oncogene int-2 in mouse embryo-derived stem cells: a general strategy for targeting mutations to non-selectable genes. Nature 336(6197): 348–352
42. Koller BH, Hagemann LJ, Doetschman T, Hagaman JR, Huang S, Williams PJ, First NL, Maeda N, Smithies O (1989) Germ-line transmission of a planned alteration made in a hypoxanthine phosphoribosyltransferase gene by homologous recombination in embryonic stem cells. Proc Natl Acad Sci U S A 86(22): 8927–8931
43. Thomas KR, Capecchi MR (1990) Targeted disruption of the murine int-1 proto-oncogene resulting in severe abnormalities in midbrain and cerebellar development. Nature 346(6287):847–850
44. McMahon AP, Bradley A (1990) The Wnt-1 (int-1) proto-oncogene is required for development of a large region of the mouse brain. Cell 62(6):1073–1085

45. Capecchi MR (1989) Altering the genome by homologous recombination. Science 244(4910):1288–1292
46. Vogel G (2007) Nobel Prizes. A knockout award in medicine. Science 318(5848): 178–179
47. Farini E, Whitelaw CB (1995) Ectopic expression of beta-lactoglobulin transgenes. Mol Gen Genet 246(6):734–738
48. Stephens LE, Sutherland AE, Klimanskaya IV, Andrieux A, Meneses J, Pedersen RA, Damsky CH (1995) Deletion of beta 1 integrins in mice results in inner cell mass failure and peri-implantation lethality. Genes Dev 9(15): 1883–1895
49. Rubock MJ, Larin Z, Cook M, Papalopulu N, Krumlauf R, Lehrach H (1990) A yeast artificial chromosome containing the mouse homeobox cluster Hox-2. Proc Natl Acad Sci U S A 87(12):4751–4755
50. Schedl A, Beermann F, Thies E, Montoliu L, Kelsey G, Schütz G (1992) Transgenic mice generated by pronuclear injection of a yeast artificial chromosome. Nucleic Acids Res 20(12):3073–3077
51. Schedl A, Montoliu L, Kelsey G, Schütz G (1993) A yeast artificial chromosome covering the tyrosinase gene confers copy number-dependent expression in transgenic mice. Nature 362(6417):258–261
52. Strauss WM, Dausman J, Beard C, Johnson C, Lawrence JB, Jaenisch R (1993) Germ line transmission of a yeast artificial chromosome spanning the murine alpha 1 (I) collagen locus. Science 259(5103): 1904–1907
53. Jakobovits A, Moore AL, Green LL, Vergara GJ, Maynard-Currie CE, Austin HA, Klapholz S (1993) Germ-line transmission and expression of a human-derived yeast artificial chromosome. Nature 362(6417):255–258
54. Choi TK, Hollenbach PW, Pearson BE, Ueda RM, Weddell GN, Kurahara CG, Woodhouse CS, Kay RM, Loring JF (1993) Transgenic mice containing a human heavy chain immunoglobulin gene fragment cloned in a yeast artificial chromosome. Nat Genet 4(2): 117–123
55. Lamb BT, Sisodia SS, Lawler AM, Slunt HH, Kitt CA, Kearns WG, Pearson PL, Price DL, Gearhart JD (1993) Introduction and expression of the 400 kilobase amyloid precursor protein gene in transgenic mice [corrected]. Nat Genet 5(1):22–30
56. Forget BG (1993) YAC transgenes: bigger is probably better. Proc Natl Acad Sci U S A 90(17):7909–7911
57. Giraldo P, Montoliu L (2001) Size matters: use of YACs, BACs and PACs in transgenic animals. Transgenic Res 10(2):83–103
58. Linton MF, Farese RV Jr, Chiesa G, Grass DS, Chin P, Hammer RE, Hobbs HH, Young SG (1993) Transgenic mice expressing high plasma concentrations of human apolipoprotein B100 and lipoprotein(a). J Clin Invest 92(6):3029–3037
59. Yang XW, Model P, Heintz N (1997) Homologous recombination based modification in Escherichia coli and germline transmission in transgenic mice of a bacterial artificial chromosome. Nat Biotechnol 15(9):859–865
60. Testa G, Zhang Y, Vintersten K, Benes V, Pijnappel WW, Chambers I, Smith AJ, Smith AG, Stewart AF (2003) Engineering the mouse genome with bacterial artificial chromosomes to create multipurpose alleles. Nat Biotechnol 21(4):443–447
61. Van Keuren ML, Gavrilina GB, Filipiak WE, Zeidler MG, Saunders TL (2009) Generating transgenic mice from bacterial artificial chromosomes: transgenesis efficiency, integration and expression outcomes. Transgenic Res 18(5):769–785
62. Gu H, Marth JD, Orban PC, Mossmann H, Rajewsky K (1994) Deletion of a DNA polymerase beta gene segment in T cells using cell type-specific gene targeting. Science 265(5168):103–106
63. Plück A (1996) Conditional mutagenesis in mice: the Cre/loxP recombination system. Int J Exp Pathol 77(6):269–278
64. Morozov A, Kellendonk C, Simpson E, Tronche F (2003) Using conditional mutagenesis to study the brain. Biol Psychiatry 54(11):1125–1133
65. Sharma S, Zhu J (2014) Immunologic applications of conditional gene modification technology in the mouse. Curr Protoc Immunol 105:10.34.1–10.34.13
66. Bradley A, Anastassiadis K, Ayadi A, Battey JF, Bell C, Birling MC, Bottomley J, Brown SD, Bürger A, Bult CJ, Bushell W, Collins FS, Desaintes C, Doe B, Economides A, Eppig JT, Finnell RH, Fletcher C, Fray M, Frendewey D, Friedel RH, Grosveld FG, Hansen J, Hérault Y, Hicks G, Hörlein A, Houghton R, Hrabé de Angelis M, Huylebroeck D, Iyer V, de Jong PJ, Kadin JA, Kaloff C, Kennedy K, Koutsourakis M, Lloyd KC, Marschall S, Mason J, McKerlie C, McLeod MP, von Melchner H, Moore M, Mujica AO, Nagy A, Nefedov M, Nutter LM, Pavlovic G, Peterson JL, Pollock J, Ramirez-Solis R, Rancourt DE, Raspa M, Remacle JE, Ringwald M,

Rosen B, Rosenthal N, Rossant J, Ruiz Noppinger P, Ryder E, Schick JZ, Schnütgen F, Schofield P, Seisenberger C, Selloum M, Simpson EM, Skarnes WC, Smedley D, Stanford WL, Stewart AF, Stone K, Swan K, Tadepally H, Teboul L, Tocchini-Valentini GP, Valenzuela D, West AP, Yamamura K, Yoshinaga Y, Wurst W (2012) The mammalian gene function resource: the International Knockout Mouse Consortium. Mamm Genome 23(9-10): 580–586
67. Friedel RH, Seisenberger C, Kaloff C, Wurst W (2007) EUCOMM—The European conditional mouse mutagenesis program. Brief Funct Genomic Proteomic 6(3):180–185
68. Brown SD, Moore MW (2012) The International Mouse Phenotyping Consortium: past and future perspectives on mouse phenotyping. Mamm Genome 23(9-10):632–640
69. Metzger D, Clifford J, Chiba H, Chambon P (1995) Conditional site-specific recombination in mammalian cells using a ligand-dependent chimeric Cre recombinase. Proc Natl Acad Sci U S A 92(15):6991–6995
70. Birling MC, Gofflot F, Warot X (2009) Site-specific recombinases for manipulation of the mouse genome. Methods Mol Biol 561: 245–263
71. Chan AW (1999) Transgenic animals: current and alternative strategies. Cloning 1(1): 25–46
72. Wright G, Carver A, Cottom D, Reeves D, Scott A, Simons P, Wilmut I, Garner I, Colman A (1991) High level expression of active human alpha-1-antitrypsin in the milk of transgenic sheep. Biotechnology (N Y) 9(9): 830–834
73. Thomson JA, Itskovitz-Eldor J, Shapiro SS, Waknitz MA, Swiergiel JJ, Marshall VS, Jones JM (1998) Embryonic stem cell lines derived from human blastocysts. Science 282(5391): 1145–1147
74. Buehr M, Meek S, Blair K, Yang J, Ure J, Silva J, McLay R, Hall J, Ying QL, Smith A (2008) Capture of authentic embryonic stem cells from rat blastocysts. Cell 135(7): 1287–1298
75. Gurdon JB, Elsdale TR, Fischberg M (1958) Sexually mature individuals of Xenopus laevis from the transplantation of single somatic nuclei. Nature 182(4627):64–65
76. Gurdon JB (1962) Adult frogs derived from the nuclei of single somatic cells. Dev Biol 4: 256–273
77. Campbell KH, McWhir J, Ritchie WA, Wilmut I (1996) Sheep cloned by nuclear transfer from a cultured cell line. Nature 380(6569): 64–66
78. Wilmut I, Schnieke AE, McWhir J, Kind AJ, Campbell KH (1997) Viable offspring derived from fetal and adult mammalian cells. Nature 385(6619):810–813
79. Cibelli JB, Stice SL, Golueke PJ, Kane JJ, Jerry J, Blackwell C, Ponce de León FA, Robl JM (1998) Cloned transgenic calves produced from nonquiescent fetal fibroblasts. Science 280(5367):1256–1258
80. Kato Y, Tani T, Sotomaru Y, Kurokawa K, Kato J, Doguchi H, Yasue H, Tsunoda Y (1998) Eight calves cloned from somatic cells of a single adult. Science 282(5396): 2095–2098
81. Wakayama T, Perry AC, Zuccotti M, Johnson KR, Yanagimachi R (1998) Full-term development of mice from enucleated oocytes injected with cumulus cell nuclei. Nature 394(6691):369–374
82. Baguisi A, Behboodi E, Melican DT, Pollock JS, Destrempes MM, Cammuso C, Williams JL, Nims SD, Porter CA, Midura P, Palacios MJ, Ayres SL, Denniston RS, Hayes ML, Ziomek CA, Meade HM, Godke RA, Gavin WG, Overström EW, Echelard Y (1999) Production of goats by somatic cell nuclear transfer. Nat Biotechnol 17(5):456–461
83. Polejaeva IA, Chen SH, Vaught TD, Page RL, Mullins J, Ball S, Dai Y, Boone J, Walker S, Ayares DL, Colman A, Campbell KH (2000) Cloned pigs produced by nuclear transfer from adult somatic cells. Nature 407(6800): 86–90
84. Zhou Q, Renard JP, Le Friec G, Brochard V, Beaujean N, Cherifi Y, Fraichard A, Cozzi J (2003) Generation of fertile cloned rats by regulating oocyte activation. Science 302(5648):1179
85. Challah-Jacques M, Chesne P, Renard JP (2003) Production of cloned rabbits by somatic nuclear transfer. Cloning Stem Cells 5(4):295–299
86. Galli C, Lagutina I, Crotti G, Colleoni S, Turini P, Ponderato N, Duchi R, Lazzari G (2003) Pregnancy: a cloned horse born to its dam twin. Nature 424(6949):635
87. Lee BC, Kim MK, Jang G, Oh HJ, Yuda F, Kim HJ, Hossein MS, Kim JJ, Kang SK, Schatten G, Hwang WS (2005) Dogs cloned from adult somatic cells. Nature 436(7051): 641
88. Schnieke AE, Kind AJ, Ritchie WA, Mycock K, Scott AR, Ritchie M, Wilmut I, Colman A, Campbell KH (1997) Human factor IX transgenic sheep produced by transfer

of nuclei from transfected fetal fibroblasts. Science 278(5346):2130–2133

89. Cibelli JB, Stice SL, Golueke PJ, Kane JJ, Jerry J, Blackwell C, Ponce de León FA, Robl JM (1998) Transgenic bovine chimeric offspring produced from somatic cell-derived stem-like cells. Nat Biotechnol 16(7): 642–646
90. Polejaeva IA (2001) Cloning pigs: advances and applications. Reprod Suppl 58:293–300
91. McCreath KJ, Howcroft J, Campbell KH, Colman A, Schnieke AE, Kind AJ (2000) Production of gene-targeted sheep by nuclear transfer from cultured somatic cells. Nature 405(6790):1066–1069
92. Lai L, Kolber-Simonds D, Park KW, Cheong HT, Greenstein JL, Im GS, Samuel M, Bonk A, Rieke A, Day BN, Murphy CN, Carter DB, Hawley RJ, Prather RS (2002) Production of alpha-1,3-galactosyltransferase knockout pigs by nuclear transfer cloning. Science 295(5557):1089–1092
93. Polejaeva IA, Campbell KH (2000) New advances in somatic cell nuclear transfer: application in transgenesis. Theriogenology 53(1):117–126
94. Rideout WM 3rd, Hochedlinger K, Kyba M, Daley GQ, Jaenisch R (2002) Correction of a genetic defect by nuclear transplantation and combined cell and gene therapy. Cell 109(1): 17–27
95. Takahashi K, Yamanaka S (2006) Induction of pluripotent stem cells from mouse embryonic and adult fibroblast cultures by defined factors. Cell 126(4):663–676
96. Hanna J, Markoulaki S, Schorderet P, Carey BW, Beard C, Wernig M, Creyghton MP, Steine EJ, Cassady JP, Foreman R, Lengner CJ, Dausman JA, Jaenisch R (2008) Direct reprogramming of terminally differentiated mature B lymphocytes to pluripotency. Cell 133(2):250–264
97. Su Y, Zhu J, Salman S, Tang Y (2020) Induced pluripotent stem cells from farm animals. J Anim Sci 98(11):skaa343
98. Rubin GM, Spradling AC (1983) Vectors for P element-mediated gene transfer in Drosophila. Nucleic Acids Res 11(18): 6341–6351
99. Kawakami K (2005) Transposon tools and methods in zebrafish. Dev Dyn 234(2): 244–254
100. Westphal CH, Leder P (1997) Transposon-generated 'knock-out' and 'knock-in' gene-targeting constructs for use in mice. Curr Biol 7(7):530–533
101. Ivics Z, Hackett PB, Plasterk RH, Izsvák Z (1997) Molecular reconstruction of Sleeping Beauty, a Tc1-like transposon from fish, and its transposition in human cells. Cell 91(4): 501–510
102. Izsvák Z, Ivics Z, Plasterk RH (2000) Sleeping Beauty, a wide host-range transposon vector for genetic transformation in vertebrates. J Mol Biol 302(1):93–102
103. Dupuy AJ, Clark K, Carlson CM, Fritz S, Davidson AE, Markley KM, Finley K, Fletcher CF, Ekker SC, Hackett PB, Horn S, Largaespada DA (2002) Mammalian germ-line transgenesis by transposition. Proc Natl Acad Sci U S A 99(7):4495–4499
104. Zayed H, Izsvák Z, Walisko O, Ivics Z (2004) Development of hyperactive sleeping beauty transposon vectors by mutational analysis. Mol Ther 9(2):292–304
105. Katter K, Geurts AM, Hoffmann O, Mátés L, Landa V, Hiripi L, Moreno C, Lazar J, Bashir S, Zidek V, Popova E, Jerchow B, Becker K, Devaraj A, Walter I, Grzybowksi M, Corbett M, Filho AR, Hodges MR, Bader M, Ivics Z, Jacob HJ, Pravenec M, Bosze Z, Rülicke T, Izsvák Z (2013) Transposon-mediated transgenesis, transgenic rescue, and tissue-specific gene expression in rodents and rabbits. FASEB J 27(3):930–941
106. Geurts AM, Collier LS, Geurts JL, Oseth LL, Bell ML, Mu D, Lucito R, Godbout SA, Green LE, Lowe SW, Hirsch BA, Leinwand LA, Largaespada DA (2006) Gene mutations and genomic rearrangements in the mouse as a result of transposon mobilization from chromosomal concatemers. PLoS Genet 2(9):e156
107. Lois C, Hong EJ, Pease S, Brown EJ, Baltimore D (2002) Germline transmission and tissue-specific expression of transgenes delivered by lentiviral vectors. Science 295(5556): 868–872
108. Fässler R (2004) Lentiviral transgene vectors. EMBO Rep 5(1):28–29
109. Park F (2007) Lentiviral vectors: are they the future of animal transgenesis? Physiol Genomics 31(2):159–173
110. Palermo GD, Cohen J, Rosenwaks Z (1996) Intracytoplasmic sperm injection: a powerful tool to overcome fertilization failure. Fertil Steril 65(5):899–908
111. Kimura Y, Yanagimachi R (1995) Intracytoplasmic sperm injection in the mouse. Biol Reprod 52(4):709–720

112. Dozortsev D, Wakaiama T, Ermilov A, Yanagimachi R (1998) Intracytoplasmic sperm injection in the rat. Zygote 6(2):143–147
113. Perry AC, Wakayama T, Kishikawa H, Kasai T, Okabe M, Toyoda Y, Yanagimachi R (1999) Mammalian transgenesis by intracytoplasmic sperm injection. Science 284(5417): 1180–1183
114. Chan AW, Luetjens CM, Dominko T, Ramalho-Santos J, Simerly CR, Hewitson L, Schatten G (2000) TransgenICSI reviewed: foreign DNA transmission by intracytoplasmic sperm injection in rhesus monkey. Mol Reprod Dev 56(2 Suppl):325–328
115. Perry AC, Rothman A, de las Heras JI, Feinstein P, Mombaerts P, Cooke HJ, Wakayama T. (2001) Efficient metaphase II transgenesis with different transgene archetypes. Nat Biotechnol 19(11):1071–1073
116. Moreira PN, Giraldo P, Cozar P, Pozueta J, Jiménez A, Montoliu L, Gutiérrez-Adán A (2004) Efficient generation of transgenic mice with intact yeast artificial chromosomes by intracytoplasmic sperm injection. Biol Reprod 71(6):1943–1947
117. Lavitrano M, Camaioni A, Fazio VM, Dolci S, Farace MG, Spadafora C (1989) Sperm cells as vectors for introducing foreign DNA into eggs: genetic transformation of mice. Cell 57(5):717–723
118. Maione B, Lavitrano M, Spadafora C, Kiessling AA (1998) Sperm-mediated gene transfer in mice. Mol Reprod Dev 50(4):406–409
119. Smith K, Spadafora C (2005) Sperm-mediated gene transfer: applications and implications. BioEssays 27(5):551–562
120. Lavitrano M, Busnelli M, Cerrito MG, Giovannoni R, Manzini S, Vargiolu A (2006) Sperm-mediated gene transfer. Reprod Fertil Dev 18(1-2):19–23
121. García-Vázquez FA, Ruiz S, Grullón LA, de Ondiz A, Gutiérrez-Adán A, Gadea J (2011) Factors affecting porcine sperm mediated gene transfer. Res Vet Sci 91(3):446–453
122. Fire A, Xu S, Montgomery MK, Kostas SA, Driver SE, Mello CC (1998) Potent and specific genetic interference by double-stranded RNA in Caenorhabditis elegans. Nature 391(6669):806–811
123. Nasevicius A, Ekker SC (2000) Effective targeted gene 'knockdown' in zebrafish. Nat Genet 26(2):216–220
124. Stainier DYR, Raz E, Lawson ND, Ekker SC, Burdine RD, Eisen JS, Ingham PW, Schulte-Merker S, Yelon D, Weinstein BM, Mullins MC, Wilson SW, Ramakrishnan L, Amacher SL, Neuhauss SCF, Meng A, Mochizuki N, Panula P, Moens CB (2017) Guidelines for morpholino use in zebrafish. PLoS Genet 13(10):e1007000
125. Fernández A, Josa S, Montoliu L (2017) A history of genome editing in mammals. Mamm Genome 28(7-8):237–246
126. Choulika A, Perrin A, Dujon B, Nicolas JF (1995) Induction of homologous recombination in mammalian chromosomes by using the I-SceI system of Saccharomyces cerevisiae. Mol Cell Biol 15(4):1968–1973
127. Ménoret S, Fontanière S, Jantz D, Tesson L, Thinard R, Rémy S, Usal C, Ouisse LH, Fraichard A, Anegon I (2013) Generation of Rag1-knockout immunodeficient rats and mice using engineered meganucleases. FASEB J 27(2):703–711
128. Bibikova M, Carroll D, Segal DJ, Trautman JK, Smith J, Kim YG, Chandrasegaran S (2001) Stimulation of homologous recombination through targeted cleavage by chimeric nucleases. Mol Cell Biol 21(1):289–297
129. Geurts AM, Cost GJ, Freyvert Y, Zeitler B, Miller JC, Choi VM, Jenkins SS, Wood A, Cui X, Meng X, Vincent A, Lam S, Michalkiewicz M, Schilling R, Foeckler J, Kalloway S, Weiler H, Ménoret S, Anegon I, Davis GD, Zhang L, Rebar EJ, Gregory PD, Urnov FD, Jacob HJ, Buelow R (2009) Knockout rats via embryo microinjection of zinc-finger nucleases. Science 325(5939):433
130. Rémy S, Tesson L, Ménoret S, Usal C, Scharenberg AM, Anegon I (2010) Zinc-finger nucleases: a powerful tool for genetic engineering of animals. Transgenic Res 19(3): 363–371
131. Meyer M, de Angelis MH, Wurst W, Kühn R (2010) Gene targeting by homologous recombination in mouse zygotes mediated by zinc-finger nucleases. Proc Natl Acad Sci U S A 107(34):15022–15026
132. Yu S, Luo J, Song Z, Ding F, Dai Y, Li N (2011) Highly efficient modification of beta-lactoglobulin (BLG) gene via zinc-finger nucleases in cattle. Cell Res 21(11): 1638–1640
133. Hauschild J, Petersen B, Santiago Y, Queisser AL, Carnwath JW, Lucas-Hahn A, Zhang L, Meng X, Gregory PD, Schwinzer R, Cost GJ, Niemann H (2011) Efficient generation of a biallelic knockout in pigs using zinc-finger nucleases. Proc Natl Acad Sci U S A 108(29):12013–12017
134. Liu X, Wang Y, Tian Y, Yu Y, Gao M, Hu G, Su F, Pan S, Luo Y, Guo Z, Quan F, Zhang Y (2014) Generation of mastitis resistance in cows by targeting human lysozyme gene to

β-casein locus using zinc-finger nucleases. Proc Biol Sci 281(1780):20133368

135. Lillico SG, Proudfoot C, King TJ, Tan W, Zhang L, Mardjuki R, Paschon DE, Rebar EJ, Urnov FD, Mileham AJ, McLaren DG, Whitelaw CB (2016) Mammalian interspecies substitution of immune modulatory alleles by genome editing. Sci Rep 6:21645
136. Tesson L, Usal C, Ménoret S, Leung E, Niles BJ, Remy S, Santiago Y, Vincent AI, Meng X, Zhang L, Gregory PD, Anegon I, Cost GJ (2011) Knockout rats generated by embryo microinjection of TALENs. Nat Biotechnol 29(8):695–696
137. Panda SK, Wefers B, Ortiz O, Floss T, Schmid B, Haass C, Wurst W, Kühn R (2013) Highly efficient targeted mutagenesis in mice using TALENs. Genetics 195(3): 703–713
138. Carlson DF, Tan W, Lillico SG, Stverakova D, Proudfoot C, Christian M, Voytas DF, Long CR, Whitelaw CB, Fahrenkrug SC (2012) Efficient TALEN-mediated gene knockout in livestock. Proc Natl Acad Sci U S A 109(43):17382–17387
139. Wu H, Wang Y, Zhang Y, Yang M, Lv J, Liu J, Zhang Y (2015) TALE nickase-mediated SP110 knockin endows cattle with increased resistance to tuberculosis. Proc Natl Acad Sci U S A 112(13):E1530–E1539
140. Xin J, Yang H, Fan N, Zhao B, Ouyang Z, Liu Z, Zhao Y, Li X, Song J, Yang Y, Zou Q, Yan Q, Zeng Y, Lai L (2013) Highly efficient generation of GGTA1 biallelic knockout inbred mini-pigs with TALENs. PLoS One 8(12):e84250
141. Cui C, Song Y, Liu J, Ge H, Li Q, Huang H, Hu L, Zhu H, Jin Y, Zhang Y (2015) Gene targeting by TALEN-induced homologous recombination in goats directs production of β-lactoglobulin-free, high-human lactoferrin milk. Sci Rep 5:10482
142. Seruggia D, Montoliu L (2014) The new CRISPR-Cas system: RNA-guided genome engineering to efficiently produce any desired genetic alteration in animals. Transgenic Res 23(5):707–716
143. Norris AL, Lee SS, Greenlees KJ, Tadesse DA, Miller MF, Lombardi HA (2020) Template plasmid integration in germline genome-edited cattle. Nat Biotechnol 38(2):163–164
144. Carlson DF, Lancto CA, Zang B, Kim ES, Walton M, Oldeschulte D, Seabury C, Sonstegard TS, Fahrenkrug SC (2016) Production of hornless dairy cattle from genome-edited cell lines. Nat Biotechnol 34(5): 479–481
145. Mojica FJM, Montoliu L (2016) On the origin of CRISPR-Cas technology: from prokaryotes to mammals. Trends Microbiol 24(10): 811–820
146. Jinek M, Chylinski K, Fonfara I, Hauer M, Doudna JA, Charpentier E (2012) A programmable dual-RNA-guided DNA endonuclease in adaptive bacterial immunity. Science 337(6096):816–821
147. Hwang WY, Fu Y, Reyon D, Maeder ML, Tsai SQ, Sander JD, Peterson RT, Yeh JR, Joung JK (2013) Efficient genome editing in zebrafish using a CRISPR-Cas system. Nat Biotechnol 31(3):227–229
148. Wang H, Yang H, Shivalila CS, Dawlaty MM, Cheng AW, Zhang F, Jaenisch R (2013) One-step generation of mice carrying mutations in multiple genes by CRISPR/Cas-mediated genome engineering. Cell 153(4): 910–918
149. Yang H, Wang H, Shivalila CS, Cheng AW, Shi L, Jaenisch R (2013) One-step generation of mice carrying reporter and conditional alleles by CRISPR/Cas-mediated genome engineering. Cell 154(6):1370–1379
150. Yang L, Güell M, Niu D, George H, Lesha E, Grishin D, Aach J, Shrock E, Xu W, Poci J, Cortazio R, Wilkinson RA, Fishman JA, Church G (2015) Genome-wide inactivation of porcine endogenous retroviruses (PERVs). Science 350(6264):1101–1104
151. Niu D, Wei HJ, Lin L, George H, Wang T, Lee IH, Zhao HY, Wang Y, Kan Y, Shrock E, Lesha E, Wang G, Luo Y, Qing Y, Jiao D, Zhao H, Zhou X, Wang S, Wei H, Güell M, Church GM, Yang L (2017) Inactivation of porcine endogenous retrovirus in pigs using CRISPR-Cas9. Science 357(6357): 1303–1307
152. Fischer K, Kraner-Scheiber S, Petersen B, Rieblinger B, Buermann A, Flisikowska T, Flisikowski K, Christan S, Edlinger M, Baars W, Kurome M, Zakhartchenko V, Kessler B, Plotzki E, Szczerbal I, Switonski M, Denner J, Wolf E, Schwinzer R, Niemann H, Kind A, Schnieke A (2016) Efficient production of multi-modified pigs for xenotransplantation by 'combineering', gene stacking and gene editing. Sci Rep 6:29081
153. Seruggia D, Fernández A, Cantero M, Pelczar P, Montoliu L (2015) Functional validation of mouse tyrosinase non-coding regulatory DNA elements by CRISPR-Cas9-mediated mutagenesis. Nucleic Acids Res 43(10):4855–4867
154. Seruggia D, Fernández A, Cantero M, Fernández-Miñán A, Gomez-Skarmeta JL,

Pelczar P, Montoliu L (2020) Boundary sequences flanking the mouse tyrosinase locus ensure faithful pattern of gene expression. Sci Rep 10(1):15494

155. Lupiáñez DG, Kraft K, Heinrich V, Krawitz P, Brancati F, Klopocki E, Horn D, Kayserili H, Opitz JM, Laxova R, Santos-Simarro F, Gilbert-Dussardier B, Wittler L, Borschiwer M, Haas SA, Osterwalder M, Franke M, Timmermann B, Hecht J, Spielmann M, Visel A, Mundlos S (2015) Disruptions of topological chromatin domains cause pathogenic rewiring of gene-enhancer interactions. Cell 161(5): 1012–1025

156. Hai T, Teng F, Guo R, Li W, Zhou Q (2014) One-step generation of knockout pigs by zygote injection of CRISPR/Cas system. Cell Res 24(3):372–375

157. Crispo M, Mulet AP, Tesson L, Barrera N, Cuadro F, dos Santos-Neto PC, Nguyen TH, Crénéguy A, Brusselle L, Anegón I, Menchaca A (2015) Efficient generation of myostatin knock-out sheep using CRISPR/Cas9 technology and microinjection into zygotes. PLoS One 10(8):e0136690

158. Wang Z (2015) Genome engineering in cattle: recent technological advancements. Chromosom Res 23(1):17–29

159. Guo R, Wan Y, Xu D, Cui L, Deng M, Zhang G, Jia R, Zhou W, Wang Z, Deng K, Huang M, Wang F, Zhang Y (2016) Generation and evaluation of Myostatin knock-out rabbits and goats using CRISPR/Cas9 system. Sci Rep 6:29855

160. Amoasii L, Hildyard JCW, Li H, Sanchez-Ortiz E, Mireault A, Caballero D, Harron R, Stathopoulou TR, Massey C, Shelton JM, Bassel-Duby R, Piercy RJ, Olson EN (2018) Gene editing restores dystrophin expression in a canine model of Duchenne muscular dystrophy. Science 362(6410):86–91

161. Anzalone AV, Koblan LW, Liu DR (2020) Genome editing with CRISPR-Cas nucleases, base editors, transposases and prime editors. Nat Biotechnol 38(7):824–844

162. Koblan LW, Erdos MR, Wilson C, Cabral WA, Levy JM, Xiong ZM, Tavarez UL, Davison LM, Gete YG, Mao X, Newby GA, Doherty SP, Narisu N, Sheng Q, Krilow C, Lin CY, Gordon LB, Cao K, Collins FS, Brown JD, Liu DR (2021) In vivo base editing rescues Hutchinson-Gilford progeria syndrome in mice. Nature 589(7843):608–614

163. Musunuru K, Chadwick AC, Mizoguchi T, Garcia SP, DeNizio JE, Reiss CW, Wang K, Iyer S, Dutta C, Clendaniel V, Amaonye M, Beach A, Berth K, Biswas S, Braun MC, Chen HM, Colace TV, Ganey JD, Gangopadhyay SA, Garrity R, Kasiewicz LN, Lavoie J, Madsen JA, Matsumoto Y, Mazzola AM, Nasrullah YS, Nneji J, Ren H, Sanjeev A, Shay M, Stahley MR, Fan SHY, Tam YK, Gaudelli NM, Ciaramella G, Stolz LE, Malyala P, Cheng CJ, Rajeev KG, Rohde E, Bellinger AM, Kathiresan S (2021) In vivo CRISPR base editing of PCSK9 durably lowers cholesterol in primates. Nature 593(7859): 429–434

164. Gao P, Lyu Q, Ghanam AR, Lazzarotto CR, Newby GA, Zhang W, Choi M, Slivano OJ, Holden K, Walker JA 2nd, Kadina AP, Munroe RJ, Abratte CM, Schimenti JC, Liu DR, Tsai SQ, Long X, Miano JM (2021) Prime editing in mice reveals the essentiality of a single base in driving tissue-specific gene expression. Genome Biol 22(1):83

Chapter 2

Tools for Efficient Genome Editing; ZFN, TALEN, and CRISPR

Yasaman Shamshirgaran, Jun Liu, Huseyin Sumer, Paul J. Verma, and Amir Taheri-Ghahfarokhi

Abstract

The last two decades have marked significant advancement in the genome editing field. Three generations of programmable nucleases (ZFNs, TALENs, and CRISPR-Cas system) have been adopted to introduce targeted DNA double-strand breaks (DSBs) in eukaryotic cells. DNA repair machinery of the cells has been exploited to introduce insertion and deletions (indels) at the targeted DSBs to study function of any gene-of-interest. The resulting indels were generally assumed to be "random" events produced by "error-prone" DNA repair pathways. However, recent advances in computational tools developed to study the Cas9-induced mutations have changed the consensus and implied the "non-randomness" nature of these mutations. Furthermore, CRISPR-centric tools are evolving at an unprecedented pace, for example, base- and prime-editors are the newest developments that have been added to the genome editing toolbox. Altogether, genome editing tools have revolutionized our way of conducting research in life sciences. Here, we present a concise overview of genome editing tools and describe the DNA repair pathways underlying the generation of genome editing outcome.

Key words ZFN, TALEN, Cas9, Genome editing, DNA repair

1 Introduction

Genome editing consists of two sequential events, i.e., cut and repair. In the first step, an engineering nuclease is employed to introduce a precise DNA lesion at the targeted site, which is, in most of the cases, a double-strand break (DSB). In the second step, the DNA repair pathways of the cell are exploited to repair the DNA lesions, enabling scientists to introduce "desired" changes at the targeted sites. Each of the DNA repair pathways requires the contribution of different sets of proteins; consequently, they may leave different footprints at the repaired site and often the result is a heterogenous collection of genetic alterations. Cut and repair are not completely independent events, the identity of the DNA lesion—dictated by inherent characteristics of the used nuclease and the sequence context of the targeted site—may rather promote

Paul John Verma, Huseyin Sumer and Jun Liu (eds.), *Applications of Genome Modulation and Editing*,
Methods in Molecular Biology, vol. 2495, https://doi.org/10.1007/978-1-0716-2301-5_2,

the repair by one of the repair pathways. Therefore, the outcome of genome editing depends on several factors including the context of the target site, used genome editing tool (engineering nuclease), activated DNA repair pathways, and the interplay between all of these factors.

Higher eukaryotic genomes are large, complex, and organized in chromosomal structures. The fundamental characteristic shared among all genome editing tools is being able to find their target site in such a sea of A, T, C, and G combinations in a "programmable" fashion. Each of the genome editing classes offers a different approach to tackle this "seek and destroy" challenge. The fundamentals of DNA binding via three major classes of genome editing tools, i.e., Zinc-Finger Proteins (ZFPs), Transcription Activator-Like Effectors (TALEs), and **C**lustered **R**egularly **I**nterspaced **S**hort **P**alindromic **R**epeats (CRISPR)–**as**sociated (CRISPR-Cas) proteins are briefly described in this chapter. Being able to program DNA-binding proteins to seek a target sequence in a genome opened up an avenue to expose the targeted gene to a variety of effector proteins that could be fused to the DNA-binding proteins. Effector proteins include, but not limited to, nucleases, transcription activators, and inhibitors, epigenetic modulators (e.g., histone acetyl transferases and DNA methyl transferases), base-editors (e.g., cytosine and adenosine deaminases), reverse transcriptase, and fluorescent proteins. Even though some of these effectors have only been used with CRISPRs, nevertheless, the toolbox for gene targeting is fairly expanded nowadays.

In this chapter, we focus on describing the nucleases, namely, zinc-finger nucleases (ZFNs), TALE nucleases (TALENs), and CRISPR-Cas9 nuclease. Wherever appropriate throughout the text, we point to other review articles that would provide additional resources for readers. A concise overview of the DNA repair pathways and how they have been exploited for genome editing purposes is also provided. Finally, we present recent progress and challenges that remain to be addressed in the CRISPR field.

2 Programmable Nucleases

Advances in genome editing tools (ZFNs, TALENs, and CRISPR-Cas nucleases) have enabled manipulation of genes at their natural chromosomal context in a cell and have transformed many aspects of life sciences including agriculture, biotechnology, and medicine. The key characteristic common among the gene editing tools is the programmability of their DNA-binding mechanisms, which provides a capacity for these proteins to specifically identify and bind to a genetic location of interest and exert the intended changes (Fig. 1a–c).

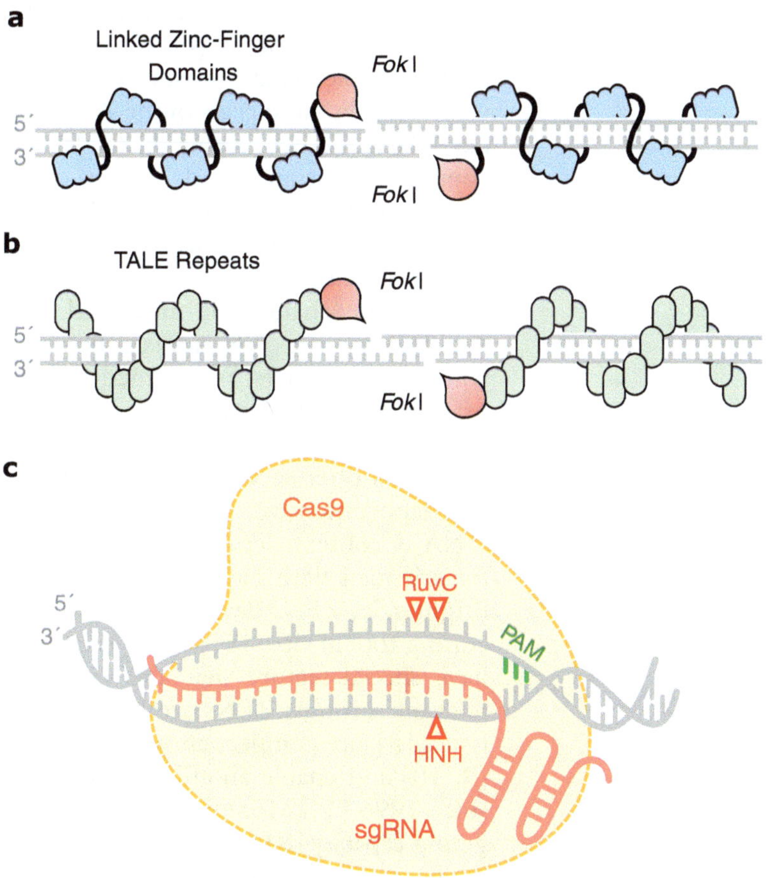

Fig. 1 Major classes of programmable nucleases. Schematic representation of DNA-binding domains in ZFNs (**a**) and TALENs (**b**) are shown. Dimerized *FokI* nucleases introduce a DNA cut producing a 4-bp 5′-overhang. (**c**) sketch of the Cas9 nuclease protein and its single guideRNA (sgRNA) bound to its target sequence. Cas9 introduces blunt and/or 1-bp staggered cuts in a context-dependent manner. Cas9 genomic target sites require a short motif called Protospacer Adjacent Motif (PAM), which for instance is NGG for the mostly used Cas9 from *Streptococcus pyogenes* (spCas9)

2.1 ZFNs

ZFNs are chimeric proteins artificially built by fusing two protein domains, i.e., zinc-finger (ZF) DNA-binding modules and *Fok*I nuclease domain [1]. The DNA-binding domain in ZFNs consists of a tandem array of ZF modules, each recognizing a triple nucleotide on the DNA (Fig. 1a). An array of 3 to 6 modules of the ZFs enables ZFNs to specifically bind a target sequence. ZFs were first identified as relatively small DNA-binding domains (about 30 amino acids) in transcription factors, with a finger-like structure stabilized by one or more zinc ions [2]. However, now ZF proteins are recognized as a diverse family of proteins interacting with a broader range of macromolecules including DNA, RNA, and protein [3]. A range of user-specified catalytic or functional domains could be fused to the ZF arrays; nevertheless, the most commonly used domain in ZFNs is the nuclease domain of *Fok*I restriction enzyme. A key characteristic of the *Fok*I nuclease domain, which was crucial in the success of the ZFNs, is that *Fok*I must dimerize to cut DNA. As such, a pair of ZFNs must bind to the DNA in an appropriate distance and correct orientation to make the *FokI* dimerization possible (Fig. 1a). Hence, a longer sequence of the DNA can be targeted which increases the chance of the target sequence being unique within a large genome, and concurrently reduces the chance for unwanted "off-target" cleavage events. Additionally, the *Fok*I domain has been engineered to cut only when heterodimers are formed [4], which further reduces the off-target events that could potentially arise from using *Fok*I homodimers.

ZFNs offer several pros and cons. Engineering ZFNs is time and labor-intensive. The interaction between ZF domains and DNA is context dependent which rather complicates the design for scientists with basic knowledge in molecular biology. On the bright side, ZFNs have been proven to be an efficacious genome editing tool in vitro [5] as well as in vivo [6, 7]. Remarkably, ZFNs have special feature that are particularly important in clinically relevant settings; ZFNs are less likely to trigger immunogenicity issues [8], are compact enough to fit into adeno-associated viruses [9, 10] and enable an efficacious delivery, and have been studied since 1996 [1]. Moreover, recent efforts have expanded both targeting capacity [8] and specificity [11] of ZFNs. Protein engineering has been used to diversify ZFNs' architectures, which allows the fusion of the *FokI* cleavage domain to both amino and carboxyl terminus of zinc-finger proteins, as well as providing options to target non-adjacent triplet nucleotides by skipping bases between them [8]. The specificity of ZFNs have been improved via attenuating the cleavage activity of *Fok*I cleavage domain [11]. Despite these recent advances, ZFNs are not the method of choice for many scientists in biology-related fields.

2.2 TALENs

Similar to ZFNs, TALENs are also chimeric proteins produced by the attachment of a tandem array of DNA-binding modules and *Fok*I nuclease domain (Fig. 1b). However, the DNA-binding modules in TALENs were originally adopted from a bacterium called *Xanthamonas* which causes diseases in plants [12, 13]. Each module is about 33–35 amino acids long corresponding to one base on the targeted DNA. TALE repeats differ from each other only at two amino acid residues (known as Repeat Variable Di-residues, RVDs); hence, the whole array of the repeats in TALENs can be constructed from four modules [14–16]. As such, designing TALENs is much simpler than designing ZFNs, and this enables scientists with basic knowledge of molecular biology to design and generate TALEN-encoding plasmids in less than a week. Similar to ZFNs, a pair of TALENs requires to be properly positioned at close proximity and correct orientation on the DNA strands to enable the dimerization of the *Fok*I cleavage domain. TALENs and ZFNs would benefit equally from the efforts spent on engineering the *Fok*I domain, for instance, attenuating its catalytic activity [11] or obliging its heterodimerization [4, 9] to reduce the off-target activities.

Deciphering the code of TALEs in 2009 simplified the generation of new engineering nucleases [12, 13] and dramatically reduced the required time and cost to perform gene manipulations. Several cloning strategies have been developed to facilitate cloning of TALENs [15, 17, 18]. Moreover, learnings from over a decade of research on ZFNs speeded up TALENs' applications in a variety of cells and organisms. TALENs, similar to ZFNs, offer a limited multiplexability which hinders conducting high-throughput functional genomics studies that require targeting a large number of genes. An advantage of ZFNs and TALENs over Cas9 is their all-protein structure, which might enable non-viral delivery methods. Both ZFNs and TALENs are currently used in clinical trials [19].

TALENs and ZFNs introduce a staggered DNA cleavage into their targeted sites (Fig. 1a, b). Even though both TALENs and ZFNs use *Fok*I cleavage domain in their structure, they generate distinct mutational signatures [20]. Noteworthy, the majority of insertions in both cases are duplications of four nucleotides, which are likely to be the results of gap-filling polymerization during the repair of 4-nucleotides 5′-overhangs generated by *Fok*I [21].

2.3 Cas9

DNA recognition in both ZFNs and TALENs is via protein engineering, for example, a new rearrangement of the TALE or ZF modules in TALENs or ZFNs proteins is needed for every new target sequence. In contrast, Cas9 offers a simpler paradigm for being programmed [22–25]. Cas9 is a multi-functional protein which uses a single guideRNA (sgRNA) molecule to recognize its target sequence within a genome (Fig. 1c).

A Cas9 target sequence is typically about 20 nucleotides complementary to the 20 nucleotides at the 5′-end of the sgRNA molecule. Another requirement for a site to be targetable by Cas9 is the presence of a short sequence called protospacer adjacent motif (PAM) on the non-targeting strand of the DNA. PAM is Cas9 dependent, for instance, PAM for Cas9 from *Streptococcus pyogenes* (SpCas9) is "NGG" [22]. Initially, Cas9 recognizes the sgRNA scaffold and together with sgRNA forms a ribonucleoprotein (RNP) complex. The length of sgRNA molecule is about 100 nucleotides single-stranded RNA, of which 80 nucleotides is constant and contributes to the formation of a particular stem-and-loop structure that could be recognized by Cas9. The other 20 nucleotides are gene specific and is customized according to a target site of interest. Next, Cas9 searches for PAM sequences on the DNA strands for which the upstream sequence matches the sgRNA sequence. Invasion of the sgRNA into the DNA double strand is facilitated by Cas9 protein structure. Finally, Cas9 cuts the DNA at both non-targeting (PAM strand) and targeting strands, using its two nuclease domains, RuvC and HNH, respectively.

Characteristics of the Cas9 cuts are important to understand as they might dictate the repair events at the broken site and aid in our prediction of the outcome. RuvC is flexible and context dependent, frequently cuts at 3–4 nucleotides before PAM, while HNH is more rigid and dominantly cuts three nucleotides before PAM [21, 26]. For example, presence of "T" or "G" at the fourth position before PAM promotes Cas9 to generate staggered or blunt cuts, respectively. Therefore, Cas9 introduces a mixture of blunt and single-nucleotide staggered cleavages at the targeted site (Fig. 1c). The result is that the majority of the insertions at Cas9 target sites are indeed duplications of the fourth nucleotide before PAM [21, 26, 27].

It is possible to mutate either or both nuclease domains, which renders Cas9 into a nickase or dead versions [28, 29]. Cas9 and its sgRNA scaffold are strikingly flexible. It is easy to manipulate Cas9 protein and/or fuse it to other functional domains. Similarly, sgRNA scaffold also has been modified with the addition of protein-binding scaffolds [30, 31].

3 Origin of CRISPR-Cas9 System

Nowadays, CRISPR-Cas9 technology is a well-established genome editing platform. Despite its relatively short history, CRISPR-Cas9 has revolutionized our ability to precisely edit or modulate any genomic sequence in a broad range of cultured cells, plants, animals, and industrial microbes. Many of CRISPR technologies have high potential to be translated into agricultural and therapeutical applications. Knowing the prokaryotic origin of the CRISPR-Cas technology, and its huge diversity in bacteria and archaea, would offer us a better perspective on how diverse this technology will potentially be in near future.

Prokaryotic organisms (bacteria and archaea) have evolved numerous defense strategies to protect themselves from viral infections. Among these strategies, **C**lustered **R**egularly **I**nterspaced **S**hort **P**alindromic **R**epeats (CRISPR) and **as**sociated proteins (Cas) provide adaptive immunity against viral predators [32–34]. Upon viral infection, a short sequence of the viral or plasmid genome, termed spacer, is integrated into the CRISPR locus. The spacers are transcribed and processed into small CRISPR RNAs (crRNAs) that bind to trans-activating CRISPR-RNA (tracrRNA) and direct Cas proteins to the invading DNA in a sequence-specific

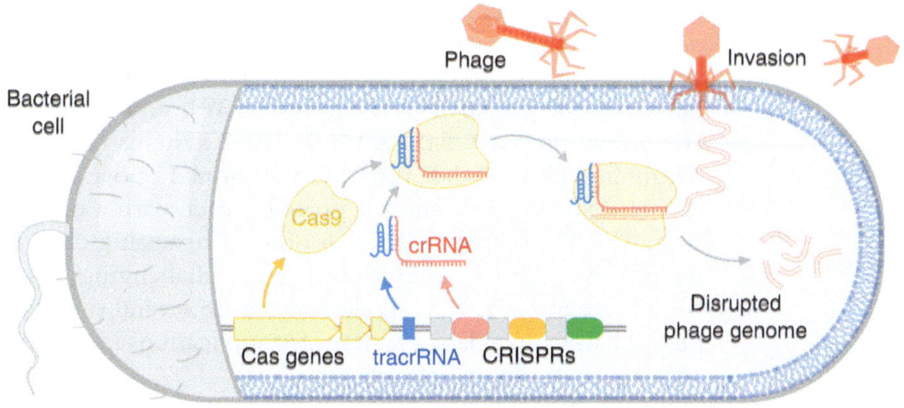

Fig. 2 Sketch of bacterial adaptive immune system, CRISPR-Cas9. Bacteria are constantly exposed to the phages in their environment. Upon invasion, phage's genome hijacks transcription and translation machinery of the cell to propagate phage particles and ultimately destroy the host. In response, bacteria have evolved the CRISPR-Cas systems that use the Cas9 protein complexed with CRISPR-RNA (crRNA) and trans-acting RNA (tracrRNA) to seek and destroy the invaded genetic materials. The components of the CRISPR-Cas systems vary among different species of bacteria

manner (Fig. 2). The nuclease activity of Cas proteins (e.g., Cas9) cuts and destroy the invader genome. The ability of immunization through CRISPR-Cas system not only enable cells to memorize, recognize, and clear reoccurring infection but also passes this resistance mechanism to their offspring.

CRISPR-Cas systems are remarkably diverse and have been found in the genome of nearly all archaea and half of the bacteria. As of December 2020, there were a total number of 288 thousand prokaryotic genomes available at the National Center for Biotechnology Information (NCBI) online database. World-wide efforts to study CRISPR-Cas systems have substantially increased in recent years and have resulted in the identification and characterization of several distinct CRISPR proteins. The most recent classification of CRISPR-Cas systems includes 2 classes, 6 types, and 33 subtypes [35]. CRISPR-Cas systems are divided into two main classes, i.e., class-1 and class-2, and the main difference between them is the number of Cas molecules. Class-1 systems include multiple small Cas genes each have a different function, while in class-2 systems a single large protein possesses several key functions.

Class-2 systems are more suitable for being harnessed towards genome editing applications. One key consideration in regard to harnessing new systems is the existence of single proteins with multiple functions. For example, Cas9 is a single protein with all required functions including interacting with PAM, binding and unwinding the DNA, pairing crRNA and DNA target strand, and finally nuclease activity. Accumulation of all these functions into a single protein as amazing as it is and has also been key in the success of CRISPR-Cas9 technology. To date, the mostly used Cas9 in the genome editing context is SpCas9 derived from *Streptococcus pyogenes*; however, several other class-2 effectors from a wide range of bacterial species also have been reported, for example, SaCas9 [36], NmCas9 [37], GeoCas9 [38], Cpf1 [39], and many more biochemically diverse CRISPR-Cas9 class-2 systems characterized by Virginijus Siksnys's lab [40]. As mentioned above, CRISPR-Cas class-1 systems have multiple proteins, each with a different function, which cooperate with each other. Harnessing class-1 systems for genome editing applications is more challenging; hence, it has been less explored. Nevertheless, there are several successful examples of harnessing class-1 systems for applications in eukaryotic cells [41–45].

4 DNA Double-Strand Break Repair Pathways

DNA double-strand breaks (DSBs) are the most deleterious types of DNA damage to cells. Hence, eukaryotic cells have conserved several mechanisms to constantly keep the integrity of their genome under surveillance and urgently repair the damaged sites. Failure in this process would either result in cell death or potentially dangerous genetic rearrangement. Gene editing technologies take advantage of the DSB repair mechanisms to either introduce insertion and deletions (indels) or precisely incorporate new genetic information at the targeted site by providing a repair template (Fig. 3). There are four distinct DSB repair pathways in human cells, namely, Non-Homologous End Joining (NHEJ), Microhomology-Mediated End Joining (MMEJ, also knowns as alternative EJ, alt-EJ, a-EJ), Single-Strand Annealing (SSA) and Homologous Recombination (HR) [46–48]. Controlling the contribution of DNA repair pathways in repairing the targeted sites would further enhance our ability to obtain a more homogenous and desirable genome editing outcome. Please see review by Yeh C. D. et al., (2019) for different methods in controlling the DNA repair pathways [49]. Describing the details of DNA repair pathways and their implications for genome editing requires efforts beyond this chapter; however, we concisely explain the essentials for newcomers to understand the principal concepts.

4.1 Non-Homologous End Joining (NHEJ)

NHEJ directly ligates the two broken ends of DNA. The term "non-homologous" was historically used to contrast the mechanistic differences between NHEJ and homologous recombination, which requires hundreds of homology base pairs as template to repair seamlessly. However, it shouldn't be deduced as NHEJ doesn't use homologies during the repair process [46]. Repair by NHEJ could be precise as well as error-prone, which the latter essentially results in the formation of insertions and/or deletions (indels) at the site under repair. A precise repair by NHEJ restores

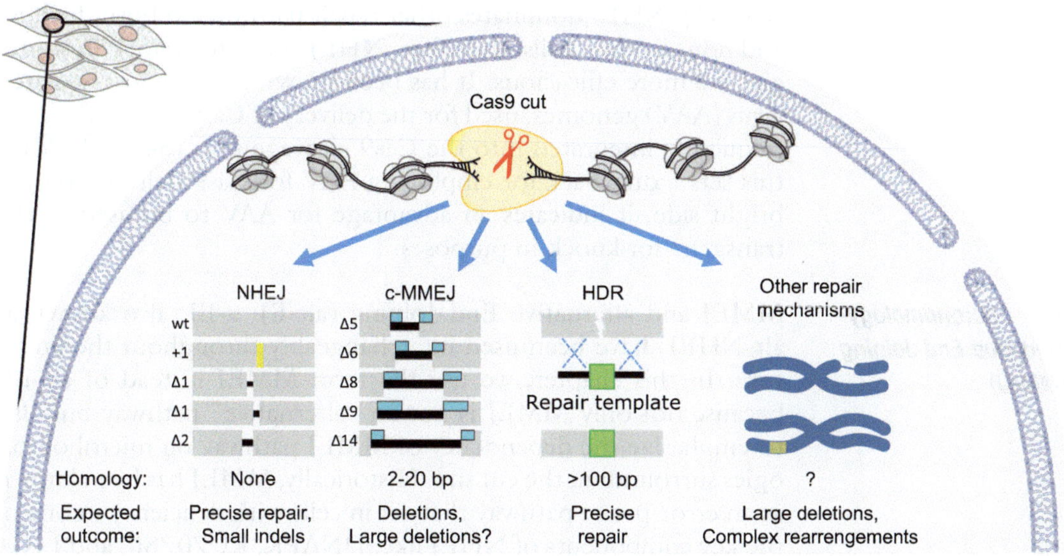

Fig. 3 Schematic representation of the DNA repair pathways and their footprints in mammalian cells. Non-Homologous End Joining (NHEJ) results in either precise repair or small insertion and/or deletions (indels). NHEJ-mediated insertions are mostly duplications of the nucleotides at the Cas9 cut site. Classical Microhomology-Mediated End Joining (c-MMEJ) uses microhomologies at the break sites to re-ligate the broken ends. Therefore, c-MMEJ results in deletions with apparent microhomologies at the junction breakpoints. NHEJ and c-MMEJ are exploited in genome editing field to generate isogenic knockout cells. Homology-directed repair (HDR) precisely repairs the break site based on the genetic materials provided using a donor template (e.g., plasmid DNA). HDR is often used to generate knock-in cells. Besides, translocations, large deletions, and complex rearrangements are among the unintended outcome of genome editing which their prevalence and underlying mechanisms are less explored

the original sequence; therefore, Cas9 could target it again and again until the new mutations render the site unrecognizable by Cas9. Analyzing the mutations at the Cas9 cut sites has revealed that NHEJ acts quickly, with indels detectable within a few hours after transfection of Cas9 and sgRNA [21, 50]. Key genes involved in NHEJ pathway include LIG4, PRKDC, TP53BP1, and XRCC6 (KU70/80). Please see Zhao B. et al. (2020) for a recent review on the molecular basis of NHEJ [48].

As mentioned above, NHEJ can be precise which results in the formation of the original allele and potentially reduces the editing efficiency. It has been shown that coupling Cas9 with over expression of TREX2, an exonuclease, resulted in the higher editing efficiency [51]. A reanalysis of the resulting indel revealed that expression of TREX2 also had changed the mutational patterns [21]. Therefore, forcing repair pathways to be more error-prone could offer a strategy to achieve higher editing efficiency, but this would translate into higher genetic alterations at the unwanted break sites [51], known as off-targets. Another implication of NHEJ precision is that NHEJ can be used for knock-in purposes

[52, 53]. NHEJ dominates other repair pathways in both dividing and non-dividing cells; therefore, NHEJ-mediated knock-in strategies are more efficacious. It has been shown that adeno-associated virus (AAV) genomes, used for the delivery of Cas9 and sgRNA, are frequently integrated into the Cas9 cleavage sites [54, 55]. While this sets a drawback for employing AAV for Cas9 delivery, on the bright side, it indicates an advantage for AAV to be used as the transgene for knock-in purposes.

4.2 Microhomology-Mediated End Joining (MMEJ)

MMEJ and alternative End Joining (alt-EJ, a-EJ, also known as alt-NHEJ) have been used interchangeably throughout the literature. In this chapter, we use the term MMEJ instead of alt-EJ, because not only MMEJ is not an "alternative" pathway but also to emphasize the dependency of MMEJ pathway on microhomologies surrounding the cut site. Historically, MMEJ has been known as an error-prone pathway that act in cells with deficiency in one of the key components of NHEJ like, DNAPK, KU70/80, and LIG4 [56–59]. However, a growing body of evidences support MMEJ as a major repair pathway active in human cells which results in a distinctive mutational signature [21, 60]. Classical MMEJ (c-MMEJ) depends on short microhomology sequences (>1 bp) surrounding the broken site of DNA (Fig. 3). Therefore, c-MMEJ is always error-prone and results in deletions that have apparent microhomologies. Other MMEJ sub-pathways also have been reported, for example, synthesis-dependent MMEJ (SD-MMEJ) results in insertions at the break site [61]. MMEJ is slower than NHEJ and its mutational signature become more detectable at later time points after Cas9-sgRNA transfection [21]. While NHEJ-associated indels were detectable as early as 4 after transfection, MMEJ-associated deletions arise to 48 h post transfection. Slower kinetic of MMEJ might presumably result in translocations in situations where cells encounter multiple DNA breaks at the same time. Key genes involved in the MMEJ repair pathways include PARP1 and POLQ.

MMEJ is context dependent; hence, some target sites are more suitable to be repaired by MMEJ. The DNA repair capacity of cell lines also might differ, for example, it has been shown that MMEJ-associated deletions have a lower occurrence in HCT116 cells compared to HEK293 and K562 cells [21]. Furthermore, mutational signatures produced by MMEJ, like NHEJ products, could potentially change the reading frame of coding sequences in protein-encoding genes. Altogether, our current understanding of Cas9 nuclease activity and NHEJ/MMEJ-associated mutational patterns would help to predict the outcome of gene editing experiments with a significant degree of certainty [27]. It is still not clear how large deletions MMEJ can produce and at what frequency. A recent analysis of Cas9-induced large deletions revealed high prevalence of microhomologies at breakpoint junctions

[62]. Noteworthy, unintended large deletions have been reported at the Cas9-targeted sites in different cell types [63].

4.3 Single-Strand Annealing (SSA)

SSA, like MMEJ, is an error-prone process, and results in large deletions of several kilobases of the DNA. SSA requires homologies above 20 nucleotides. A key protein in SSA is RAD52. It is not clear to what extent SSA contributes to the formation of genetic alteration at the Cas9-targeted sites in human cells. However, a recent analysis revealed highly frequent SSA-mediated deletions in *Leishmania* [64].

4.4 Homologous Recombination (HR)

HR is a high-fidelity repair pathway active in S and G2 phases of the cell cycle. HR and homology-directed repair (HDR) have been interchangeably used in the context of genome editing, referring to error-free repair and precise modification of the targeted site based on the sequences provided either artificially by a donor molecule or by a homologous allele on the sister chromosome. So, unlike NHEJ and MMEJ that introduce a range of insertions and deletions at the targeted site, HDR allows a more precise and controlled edit to occur at the cut site. However, the efficiency of HDR is less than NHEJ and MMEJ, hence generating edited cells using HDR requires extra efforts. The common term for referring to the HDR events is knock-in though the usage of knock-in is not limited to the HDR-mediated modifications. One reason that HDR is less efficient relates to the availability of the competitor pathways, NHEJ and MMEJ, in cells. Many approaches have been proposed to either enhance the efficacy of HDR using compound-mediated inhibition of NHEJ, modified Cas9s, timed expression of Cas9 (*see* Liu M. et al., 2019 for a review [65]), or alternatively utilizing NHEJ/MMEJ pathways for knock-in purposes [52, 53, 66]. For example, compound-mediated inhibition of NHEJ has been reported to achieve higher HDR-mediated knock-ins [49, 67]. Other methodologies have used fusion of proteins to the Cas9 which could recruit other HDR enzymes at the proximity of the cut site. Key genes involve in HDR-mediated repair include RAD51, BRCA1, and BRCA2.

5 Genome Editing Waves

The number of published studies on ZFNs and TALENs was dramatically reduced after the invention of revolutionary CRISPR-Cas9 system (Fig. 4a), this indicates how fast scientists replaced ZFNs and TALENs with the new system. Also, as CRISPR-Cas9 technology democratized the genome editing field as the number of published studies that have used CRISPR is rapidly growing. Moreover, Cas9 and its sgRNA provided a remarkably flexible platform for fusion of other functional proteins

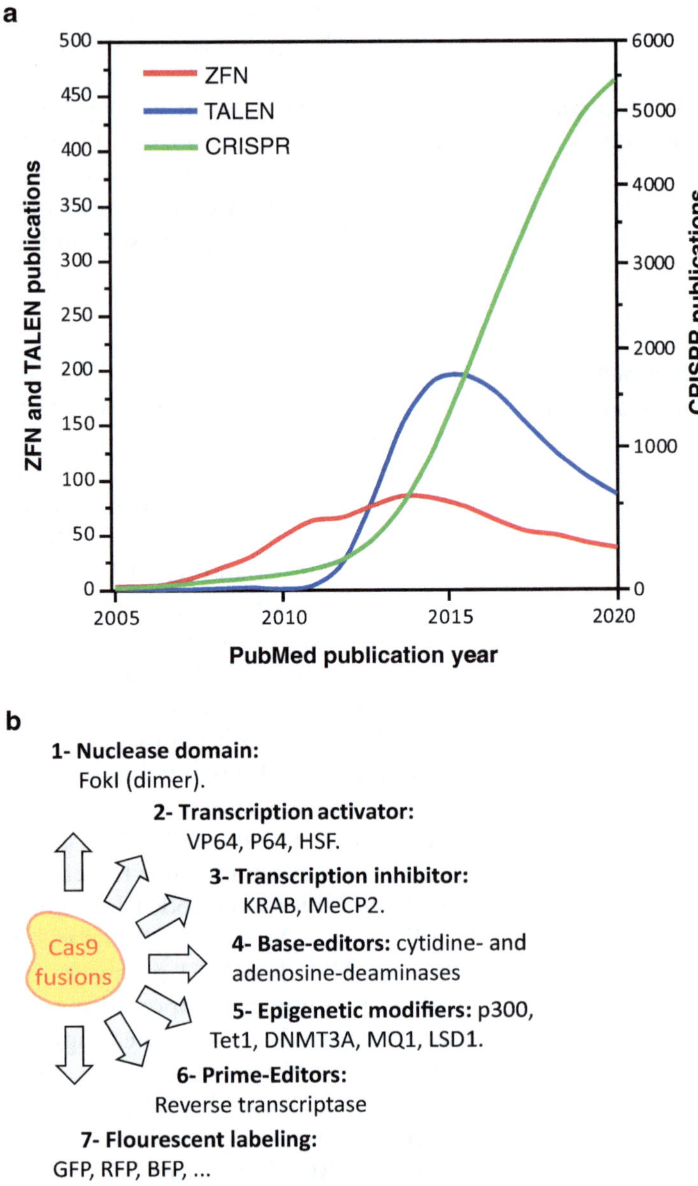

Fig. 4 Trends in genome editing field. (**a**) number of published articles indexed in PubMed database with having ZFN, TALEN, or CRISPR in their title or abstract plotted against the publication year. (**b**) different flavors of Cas9-centric applications enabled using fusion of different functional domains to partially inactivated (Cas9-nickase) or fully inactivated (dead-Cas9) protein

to the Cas9-sgRNA complex, by which new features have been added to the genome editing toolbox (Fig. 4b). For example, mutated version of Cas9 have been used together with functional domains like *Fok*I nuclease [68, 69], transcription modulators [30], base-editors [70–73], epigenetic modifiers (histone acetylation by p300 [74], histone demethylation by LSD1 [75], cytosine methylation by DNMT3A [76, 77] or MQ1 [78], cytosine demethylation by Tet1 [77, 79, 80]), and reverse transcriptase [81]. Each of these additions has enabled more precise and controlled changes on the genome, and even beyond genome on transcriptome, and epigenome. The efficacy and fidelity of Cas9-centric applications have been constantly improved since the invention of CRISPR-Cas9 technology.

Until recently, mutations that were introduced at the targeted sites after a Cas9 cut were generally assumed to be "random" footprints of "error-prone" DNA repair pathways. However, a growing body of evidences indicates that Cas9-induced mutations are non-random [21, 26, 50] and could be predicted by computational tools [27]. Understanding the underlying mechanisms behind the formation of the mutations at the Cas9 break sites is fundamental in the genome editing field. Such an understanding would aid designing better strategies for different applications ranging from simple gene knockout studies to carefully designed therapeutic genome editing programs. Specifically, it is vital in the context of therapeutic genome editing to study the resulting mutations and the possible consequences with an utmost care. If the cut is placed within the coding regions of the targeted genes, then each mutation regardless of causing a frameshift or not, will result in production of a new protein in patients, which might not be possible to predict their effects such as immunogenicity. As such, the heterogeneity of the resulting mutation patterns is important.

6 Concluding Remarks

An unprecedented technology: gene editing technologies have empowered geneticist to study the function of genes. We described three generations of programmable nucleases mostly used in genome editing field over the last two decades; however, the regularly used systems nowadays are the Cas9-centric technologies. The reason for such popularity is the simplicity and multiplexability that Cas9 offers. Groundbreaking studies published back in 2012 [22] and 2013 [23, 25] on harnessing the Cas9 as a robust technology for applications in human cells stirred a genome editing storm during the last decade, which still continues with several new innovative methodologies becoming available every year.

Shifting from genetic vandalism to precise edits: pioneering studies that aid our understanding of mutation patterns at Cas9-targeted sites [21, 26, 50] shed a light on the underlying repair mechanisms behind mutations and enhanced the prediction of genome editing outcome with a significant level of confidence [27]. However, with most efforts focused on the addition of new enzymatic functionalities to the Cas9 variants, several other aspects of genome editing remained less explored. These areas include roles of the DNA repair pathways in occurrence of unwanted and complex genetic rearrangements, context-dependency of DNA repairs and activity of nucleases, plasticity of Cas9 domains in making a range of blunt and staggered-end cuts, and the interplay between the DNA lesions and DNA repair pathways. We expect the genome editing toolbox to be greatly expanded with not only discovery of new CRISPR systems but also addition of new features to the existing Cas9 proteins that would assist genome editing precision.

Challenges with identifying the wide spectrum of genetic alterations: thanks to efficacious DNA repair pathways it seems safe to use nucleases, which are literally like scissors, to cut the genome. But detection of all intended and unintended gene editing outcome is rather complicated, which is often simplified to detection of small indels. Most of the indel detection methodologies [82] are only capable of screening for small genetic changes at a narrow window, while frequent large deletions and complex rearrangements are among the unwanted outcome [62, 63]. Moreover, all nucleases might cut promiscuously resulting in potentially harmful indels and intra-chromosomal translocations. Therefore, a careful analysis for ruling out all unintended genetic changes should be taken as an essential part of conducting gene editing experiments.

Biological plasticity of the cells: if the cut and repair occurs within the coding sequences of a gene, the consequential genetic alterations could potentially render a gene to produce non-functional protein(s). Therefore, gene targeting is used as a tool to generate isogenic knockout models to study functions of a gene-of-interest. However, it is important to keep in mind that genetic mutations, including frameshifting indels, are not essentially equal to gene knockout. For several reasons, a mutated gene could still produce a functional protein. For example, the mutated region might not be in all transcript variants, or simply the genetic alteration causes the targeted exon to be skipped in the transcripts. Another example is the presence of cryptic start codons, i.e., when the targeted region is immediately at the proximal region of the gene and translation could initiate from a downstream in-frame start codon. Recently, a systematic investigation of 193 genetically verified deletions in 136 genes targeted using CRISPR-Cas9 system, revealed a variable level of expression, from low to the original level, in a third of the investigated genes [83]. Therefore, a careful investigation of gene edited cell lines generated using CRISPR-Cas9 is crucial for having robust interpretation of resulting phenotypes.

References

1. Kim YG, Cha J, Chandrasegaran S (1996) Hybrid restriction enzymes: zinc finger fusions to Fok I cleavage domain. Proc Natl Acad Sci U S A 93:1156–1160. https://doi.org/10.1073/pnas.93.3.1156
2. Klug A, Rhodes D (1987) Zinc fingers: a novel protein fold for nucleic acid recognition. Cold Spring Harb Symp Quant Biol 52:473–482. https://doi.org/10.1101/SQB.1987.052.01.054
3. Cassandri M, Smirnov A, Novelli F et al (2017) Zinc-finger proteins in health and disease. Cell Death Discov 3:1–2
4. Doyon Y, Vo TD, Mendel MC et al (2011) Enhancing zinc-finger-nuclease activity with improved obligate heterodimeric architectures. Nat Methods 8:74–79. https://doi.org/10.1038/nmeth.1539
5. Urnov FD, Miller JC, Lee YL et al (2005) Highly efficient endogenous human gene correction using designed zinc-finger nucleases. Nature 435:646–651. https://doi.org/10.1038/nature03556
6. Ou L, DeKelver RC, Rohde M et al (2019) ZFN-mediated in vivo genome editing corrects murine hurler syndrome. Mol Ther 27:178–187. https://doi.org/10.1016/j.ymthe.2018.10.018
7. Li H, Haurigot V, Doyon Y et al (2011) In vivo genome editing restores haemostasis in a mouse model of haemophilia. Nature 475:217–221. https://doi.org/10.1038/nature10177
8. Paschon DE, Lussier S, Wangzor T et al (2019) Diversifying the structure of zinc finger nucleases for high-precision genome editing. Nat Commun 10:1–2. https://doi.org/10.1038/s41467-019-08867-x
9. Miller JC, Holmes MC, Wang J et al (2007) An improved zinc-finger nuclease architecture for highly specific genome editing. Nat Biotechnol 25:778–785. https://doi.org/10.1038/nbt1319
10. Dong JY, Fan PD, Frizzell RA (1996) Quantitative analysis of the packaging capacity of recombinant adeno-associated virus. Hum Gene Ther 7:2101–2112. https://doi.org/10.1089/hum.1996.7.17-2101
11. Miller JC, Patil DP, Xia DF et al (2019) Enhancing gene editing specificity by attenuating DNA cleavage kinetics. Nat Biotechnol 37:945–952. https://doi.org/10.1038/s41587-019-0186-z
12. Moscou MJ, Bogdanove AJ (2009) A simple cipher governs DNA recognition by TAL effectors. Science 326:1501. https://doi.org/10.1126/science.1178817
13. Boch J, Scholze H, Schornack S et al (2009) Breaking the code of DNA binding specificity of TAL-type III effectors. Science 326:1509–1512. https://doi.org/10.1126/science.1178811
14. Li T, Huang S, Jiang WZ et al (2011) TAL nucleases (TALNs): hybrid proteins composed of TAL effectors and FokI DNA-cleavage domain. Nucleic Acids Res 39:359–372. https://doi.org/10.1093/nar/gkq704
15. Cermak T, Doyle EL, Christian M et al (2011) Efficient design and assembly of custom TALEN and other TAL effector-based constructs for DNA targeting. Nucleic Acids Res 39:e82. https://doi.org/10.1093/nar/gkr218
16. Christian M, Cermak T, Doyle EL et al (2010) Targeting DNA double-strand breaks with TAL effector nucleases. Genetics 186:756–761. https://doi.org/10.1534/genetics.110.120717
17. Schmid-Burgk JL, Schmidt T, Kaiser V et al (2013) A ligation-independent cloning technique for high-throughput assembly of transcription activator-like effector genes. Nat Biotechnol 31:76–81. https://doi.org/10.1038/nbt.2460
18. Reyon D, Tsai SQ, Khgayter C et al (2012) FLASH assembly of TALENs for high-throughput genome editing. Nat Biotechnol 30:460–465. https://doi.org/10.1038/nbt.2170
19. Mullard A (2020) Gene-editing pipeline takes off. Nat Rev Drug Discov 19:367–372
20. Kim Y, Kweon J, Kim JS (2013) TALENs and ZFNs are associated with different mutation signatures. Nat Methods 10:185
21. Taheri-Ghahfarokhi A, Taylor BJM, Nitsch R et al (2018) Decoding non-random mutational signatures at Cas9 targeted sites. Nucleic Acids Res 46:8417–8434. https://doi.org/10.1093/nar/gky653
22. Jinek M, Chylinski K, Fonfara I et al (2012) A programmable dual-RNA-guided DNA endonuclease in adaptive bacterial immunity. Science 337:816–821. https://doi.org/10.1126/science.1225829
23. Mali P, Yang L, Esvelt KM et al (2013) RNA-guided human genome engineering via

Cas9. Science 339:823–826. https://doi.org/10.1126/science.1232033

24. Doudna JA, Charpentier E (2014) The new frontier of genome engineering with CRISPR-Cas9. Science 346
25. Cong L, Ran FA, Cox D et al (2013) Multiplex genome engineering using CRISPR/Cas systems. Science 339:819–823. https://doi.org/10.1126/science.1231143
26. Li Y, Park AI, Mou H et al (2015) A versatile reporter system for CRISPR-mediated chromosomal rearrangements. Genome Biol 16:1. https://doi.org/10.1186/s13059-015-0680-7
27. Shen MW, Arbab M, Hsu JY et al (2018) Predictable and precise template-free CRISPR editing of pathogenic variants. Nature 563:646–651. https://doi.org/10.1038/s41586-018-0686-x
28. Mali P, Aach J, Stranges PB et al (2013) CAS9 transcriptional activators for target specificity screening and paired nickases for cooperative genome engineering. Nat Biotechnol 31:833–838. https://doi.org/10.1038/nbt.2675
29. Kim E, Kim S, Kim DH et al (2012) Precision genome engineering with programmable DNA-nicking enzymes. Genome Res 22:1327–1333. https://doi.org/10.1101/gr.138792.112
30. Dahlman JE, Abudayyeh OO, Joung J et al (2015) Orthogonal gene knockout and activation with a catalytically active Cas9 nuclease. Nat Biotechnol 33:1159–1161. https://doi.org/10.1038/nbt.3390
31. Ma H, Tu LC, Naseri A et al (2016) Multiplexed labeling of genomic loci with dCas9 and engineered sgRNAs using CRISPRainbow. Nat Biotechnol 34:528–530. https://doi.org/10.1038/nbt.3526
32. Lillestøl RK, Redder P, Garrett RA, Brügger K (2006) A putative viral defence mechanism in archaeal cells. Archaea 2:59–72. https://doi.org/10.1155/2006/542818
33. Makarova KS, Grishin NV, Shabalina SA et al (2006) A putative RNA-interference-based immune system in prokaryotes: computational analysis of the predicted enzymatic machinery, functional analogies with eukaryotic RNAi, and hypothetical mechanisms of action. Biol Direct 1:1–26
34. Marraffini LA, Sontheimer EJ (2010) CRISPR interference: RNA-directed adaptive immunity in bacteria and archaea. Nat Rev Genet 11:181–190
35. Makarova KS, Wolf YI, Iranzo J et al (2020) Evolutionary classification of CRISPR–Cas systems: a burst of class 2 and derived variants. Nat Rev Microbiol 18:67–83
36. Ran FA, Cong L, Yan WX et al (2015) In vivo genome editing using Staphylococcus aureus Cas9. Nature 520:186–191. https://doi.org/10.1038/nature14299
37. Hou Z, Zhang Y, Propson NE et al (2013) Efficient genome engineering in human pluripotent stem cells using Cas9 from Neisseria meningitidis. Proc Natl Acad Sci U S A 110:15644–15649. https://doi.org/10.1073/pnas.1313587110
38. Harrington LB, Paez-Espino D, Staahl BT et al (2017) A thermostable Cas9 with increased lifetime in human plasma. Nat Commun 8:1–8. https://doi.org/10.1038/s41467-017-01408-4
39. Zetsche B, Gootenberg JS, Abudayyeh OO et al (2015) Cpf1 is a single RNA-guided endonuclease of a class 2 CRISPR-Cas system. Cell 163:759–771. https://doi.org/10.1016/j.cell.2015.09.038
40. Gasiunas G, Young JK, Karvelis T et al (2020) A catalogue of biochemically diverse CRISPR-Cas9 orthologs. Nat Commun 11:1. https://doi.org/10.1038/s41467-020-19344-1
41. Osakabe K, Wada N, Murakami E, Osakabe Y (2020) Genome editing in mammals using CRISPR type I-D nuclease. bioRxiv. https://doi.org/10.1101/2020.03.14.991976
42. Pickar-Oliver A, Black JB, Lewis MM et al (2019) Targeted transcriptional modulation with type I CRISPR–Cas systems in human cells. Nat Biotechnol 37:1493–1501. https://doi.org/10.1038/s41587-019-0235-7
43. Morisaka H, Yoshimi K, Okuzaki Y et al (2019) CRISPR-Cas3 induces broad and unidirectional genome editing in human cells. Nat Commun 10:1–3. https://doi.org/10.1038/s41467-019-13226-x
44. Dolan AE, Hou Z, Xiao Y et al (2019) Introducing a Spectrum of long-range genomic deletions in human embryonic stem cells using type I CRISPR-Cas. Mol Cell 74:936–950.e5. https://doi.org/10.1016/j.molcel.2019.03.014
45. Chen Y, Liu J, Zhi S et al (2020) Repurposing type I–F CRISPR–Cas system as a transcriptional activation tool in human cells. Nat Commun 11:1–4. https://doi.org/10.1038/s41467-020-16880-8
46. Pannunzio NR, Watanabe G, Lieber MR (2018) Nonhomologous DNA end-joining for repair of DNA double-strand breaks. J Biol Chem 293:10512–10523
47. Scully R, Panday A, Elango R, Willis NA (2019) DNA double-strand break repair-

pathway choice in somatic mammalian cells. Nat Rev Mol Cell Biol 20:698–714

48. Zhao B, Rothenberg E, Ramsden DA, Lieber MR (2020) The molecular basis and disease relevance of non-homologous DNA end joining. Nat Rev Mol Cell Biol 21:765–781. https://doi.org/10.1038/s41580-020-00297-8
49. Yeh CD, Richardson CD, Corn JE (2019) Advances in genome editing through control of DNA repair pathways. Nat Cell Biol 21: 1468–1478
50. van Overbeek M, Capurso D, Carter MM et al (2016) DNA repair profiling reveals nonrandom outcomes at Cas9-mediated breaks. Mol Cell 63:633–646. https://doi.org/10.1016/j.molcel.2016.06.037
51. Chari R, Mali P, Moosburner M, Church GM (2015) Unraveling CRISPR-Cas9 genome engineering parameters via a library-on-library approach. Nat Methods 12:823–826. https://doi.org/10.1038/nmeth.3473
52. Maresca M, Lin VG, Guo N, Yang Y (2013) Obligate ligation-gated recombination (ObLiGaRe): custom-designed nuclease-mediated targeted integration through nonhomologous end joining. Genome Res 23:539–546. https://doi.org/10.1101/gr.145441.112
53. Suzuki K, Tsunekawa Y, Hernandez-Benitez R et al (2016) In vivo genome editing via CRISPR/Cas9 mediated homology-independent targeted integration. Nature 540:144–149. https://doi.org/10.1038/nature20565
54. Hanlon KS, Kleinstiver BP, Garcia SP et al (2019) High levels of AAV vector integration into CRISPR-induced DNA breaks. Nat Commun 10:1. https://doi.org/10.1038/s41467-019-12449-2
55. Nelson CE, Wu Y, Gemberling MP et al (2019) Long-term evaluation of AAV-CRISPR genome editing for Duchenne muscular dystrophy. Nat Med 25:427–432. https://doi.org/10.1038/s41591-019-0344-3
56. Perrault R, Wang H, Wang M et al (2004) Backup pathways of NHEJ are suppressed by DNA-PK. J Cell Biochem 92:781–794. https://doi.org/10.1002/jcb.20104
57. Kabotyanski EB, Gomelsky L, Han JO et al (1998) Double-strand break repair in Ku86- and XRCC4-deficient cells. Nucleic Acids Res 26:5333–5342. https://doi.org/10.1093/nar/26.23.5333
58. Boulton SJ, Jackson SP (1996) Saccharomyces cerevisiae Ku70 potentiates illegitimate DNA double-strand break repair and serves as a barrier to error-prone DNA repair pathways. EMBO J 15:5093–5103. https://doi.org/10.1002/j.1460-2075.1996.tb00890.x
59. Boboila C, Jankovic M, Yan CT et al (2010) Alternative end-joining catalyzes robust IgH locus deletions and translocations in the combined absence of ligase 4 and Ku70. Proc Natl Acad Sci U S A 107:3034–3039. https://doi.org/10.1073/pnas.0915067107
60. Ceccaldi R, Rondinelli B, D'Andrea AD (2016) Repair pathway choices and consequences at the double-Strand break. Trends Cell Biol 26:52–64
61. Khodaverdian VY, Hanscom T, Yu AM et al (2017) Secondary structure forming sequences drive SD-MMEJ repair of DNA double-strand breaks. Nucleic Acids Res 45:12848–12861. https://doi.org/10.1093/nar/gkx1056
62. Owens DDG, Caulder A, Frontera V et al (2019) Microhomologies are prevalent at Cas9-induced larger deletions. Nucleic Acids Res 47:7402–7417. https://doi.org/10.1093/nar/gkz459
63. Kosicki M, Tomberg K, Bradley A (2018) Repair of double-strand breaks induced by CRISPR–Cas9 leads to large deletions and complex rearrangements. Nat Biotechnol 36: 765–771. https://doi.org/10.1038/nbt.4192
64. Zhang W-W, Matlashewski G (2019) Single-strand annealing plays a major role in double-strand DNA break repair following CRISPR-Cas9 cleavage in leishmania. mSphere 4: e00408–e00419. https://doi.org/10.1128/msphere.00408-19
65. Liu M, Rehman S, Tang X et al (2019) Methodologies for improving HDR efficiency. Front Genet 9:691
66. Sakuma T, Nakade S, Sakane Y et al (2016) MMEJ-assisted gene knock-in using TALENs and CRISPR-Cas9 with the PITCh systems. Nat Protoc 11:118–133. https://doi.org/10.1038/nprot.2015.140
67. Bischoff N, Wimberger S, Maresca M, Brakebusch C (2020) Improving precise CRISPR genome editing by small molecules: is there a magic potion? Cell 9
68. Tsai SQ, Wyvekens N, Khayter C et al (2014) Dimeric CRISPR RNA-guided FokI nucleases for highly specific genome editing. Nat Biotechnol 32:569–576. https://doi.org/10.1038/nbt.2908
69. Guilinger JP, Thompson DB, Liu DR (2014) Fusion of catalytically inactive Cas9 to FokI nuclease improves the specificity of genome modification. Nat Biotechnol 32:577–582. https://doi.org/10.1038/nbt.2909

70. Komor AC, Kim YB, Packer MS et al (2016) Programmable editing of a target base in genomic DNA without double-stranded DNA cleavage. Nature 533:420–424. https://doi.org/10.1038/nature17946
71. Gaudelli NM, Komor AC, Rees HA et al (2017) Programmable base editing of T to G C in genomic DNA without DNA cleavage. Nature 551:464–471. https://doi.org/10.1038/nature24644
72. Richter MF, Zhao KT, Eton E et al (2020) Phage-assisted evolution of an adenine base editor with improved Cas domain compatibility and activity. Nat Biotechnol 38:883–891. https://doi.org/10.1038/s41587-020-0453-z
73. Sakata RC, Ishiguro S, Mori H et al (2020) Base editors for simultaneous introduction of C-to-T and A-to-G mutations. Nat Biotechnol 38:865–869. https://doi.org/10.1038/s41587-020-0509-0
74. Hilton IB, D'Ippolito AM, Vockley CM et al (2015) Epigenome editing by a CRISPR-Cas9-based acetyltransferase activates genes from promoters and enhancers. Nat Biotechnol 33:510–517. https://doi.org/10.1038/nbt.3199
75. Williams RM, Senanayake U, Artibani M et al (2018) Genome and epigenome engineering CRISPR toolkit for in vivo modulation of cis-regulatory interactions and gene expression in the chicken embryo. Development 145: dev160333. https://doi.org/10.1242/dev.160333
76. Vojta A, Dobrinic P, Tadic V et al (2016) Repurposing the CRISPR-Cas9 system for targeted DNA methylation. Nucleic Acids Res 44: 5615–5628. https://doi.org/10.1093/nar/gkw159
77. Liu XS, Wu H, Ji X et al (2016) Editing DNA methylation in the mammalian genome. Cell 167:233–247.e17. https://doi.org/10.1016/j.cell.2016.08.056
78. Lei Y, Zhang X, Su J et al (2017) Targeted DNA methylation in vivo using an engineered dCas9-MQ1 fusion protein. Nat Commun 8:1. https://doi.org/10.1038/ncomms16026
79. Morita S, Noguchi H, Horii T et al (2016) Targeted DNA demethylation in vivo using dCas9-peptide repeat and scFv-TET1 catalytic domain fusions. Nat Biotechnol 34: 1060–1065. https://doi.org/10.1038/nbt.3658
80. Xu X, Tao Y, Gao X et al (2016) A CRISPR-based approach for targeted DNA demethylation. Cell Discov 2:1–2. https://doi.org/10.1038/celldisc.2016.9
81. Anzalone AV, Randolph PB, Davis JR et al (2019) Search-and-replace genome editing without double-strand breaks or donor DNA. Nature 576:149–157. https://doi.org/10.1038/s41586-019-1711-4
82. Bennett EP, Petersen BL, Johansen IE et al (2020) INDEL detection, the 'Achilles heel' of precise genome editing: a survey of methods for accurate profiling of gene editing induced indels. Nucleic Acids Res 48:11958–11981. https://doi.org/10.1093/nar/gkaa975
83. Smits AH, Ziebell F, Joberty G et al (2019) Biological plasticity rescues target activity in CRISPR knock outs. Nat Methods 16: 1087–1093. https://doi.org/10.1038/s41592-019-0614-5

Part II

Genetic Manipulation Methods

Chapter 3

Efficient Generation of Stable Cell Lines with Inducible Neuronal Transgene Expression Using the piggyBac Transposon System

Jinchao Gu, Huseyin Sumer, and Brett Cromer

Abstract

The piggyBac transposon system has been adapted to be a highly efficient genome engineering tool for transgenesis of eukaryotic cells and organisms. As with other methods of transgenesis, incorporation of an inducible promoter, such as a tetracycline-responsive element, enables inducible transgene expression. Here, we describe an efficient method of using the piggyBac system to create stably transfected mammalian cell lines, including inducible transgene expression. Gibson assembly is used to construct the required vectors as it enables multiple DNA fragments to be seamlessly assembled in a single isothermal reaction. We demonstrate an application of this approach to generate a stably transfected pluripotent stem cell line that can be induced to express a transcription factor transgene and rapidly differentiate into neurons in a single step.

Key words PiggyBac, Gibson assembly, Inducible transgene expression, Stem cells, Neuronal differentiation

1 Introduction

The piggyBac (PB) transposon is a mobile genetic element that efficiently integrates genes of interest from vectors to host cell genomes via a "cut and paste" mechanism [1]. In the PB vectors, there are two transposon-specific inverted terminal repeats (ITRs) that the PB transposase recognizes and the segment in between can be incorporated into TTAA chromosomal sites anywhere within the genome of a cell [1]. Compared with other transposons, such as Tol2 and Sleeping Beauty, PB has shown higher transposition efficiency in mammalian cells [2]. It has also been demonstrated that PB can efficiently mediate multiplex gene transfer in mouse ES cells [2]. Furthermore, the high cargo capacity and footprint-free gene excision of the PB transposon make it a feasible and advantageous system for clinical gene therapy.

Paul John Verma, Huseyin Sumer and Jun Liu (eds.), *Applications of Genome Modulation and Editing*,
Methods in Molecular Biology, vol. 2495, https://doi.org/10.1007/978-1-0716-2301-5_3,

Gibson assembly is a molecular cloning method that allows multiple overlapping DNA fragments to be covalently joined in an isothermal single reaction [3]. The assembly reaction involves a cocktail of a 5'exonuclease, a DNA polymerase, and a DNA ligase along with other buffer components. The T5 exonuclease chews back DNA fragments from 5'end, leaving single-stranded DNA overhangs that enables adjacent fragments to be specifically annealed. once annealed, Phusion DNA polymerase primes from available 3' ends of annealed regions and fills any adjacent single-stranded gaps. Finally, *Taq* DNA ligase repairs the nicks, yielding a circular vector and seamless cloning product [4]. Hence, given the need to assemble a variety of DNA fragments, we have used Gibson assembly to construct appropriate plasmids for generating stably transfected inducible cell lines using the PB system.

Embryonic stem cells (ESCs) and induced pluripotent stem cells (iPSCs) have the ability to differentiate into any cell type of the three primary germ layers [5], which makes them suitable as in vitro cell models for studying human diseases. Recent research has shown that ectopic expression of specific transcription factors (TFs) is able to differentiate ECSs and PSCs into particular neuronal lineages [6]. Neurogenin-2 (Ngn2) is a proneural TF that can specify excitatory neurons from ECSs and PSCs [7, 8]. Instead of the widely used approach of transducing cells with a TF-expressing lentivirus for each differentiation experiment, the creation of stably transfected inducible cell lines can reduce cell variability within and between experiments and facilitate higher throughput. Here, we describe an efficient transgenesis system, using the PB transposon system, for inducible TF-driven neurogenesis of stem cells.

2 Materials

2.1 Plasmids

1. PB-CMV-MCS-EF1α-GreenPuro (System Biosciences #PB513B-1).
2. Super PiggyBac transposase expression vector (System Biosciences #PB210PA-1).
3. CreP2ANgn2 (kind gift from Prof. Dr. Manfred Schartl).
4. FUW-M2rtTA (Addgene #20342).
5. pTet-O-Ngn2-puro (Addgene #52047).
6. DLX2-hygro (Addgene #97330).

2.2 Oligos

1. Oligos (Table 1).

Table 1
Primers for construction of PB plasmids with inducible TF expression

Code	Primer name	Sequence (5′-3′)
S18.16	AmpR FWD	AGCGGTTAGCTCCTTCGGTCCTCCGATCGTTGTC
S18.17	AmpR REV	ACGATCGGAGGACCGAAGGAGCTAACCGCTTTTTTG
S18.34	CMV-rtTA FWD	ATCGGATCCGCGGCCACCATGTCTAGACTGGACAAGA
S18.35	CMV-rtTA REV	CAGTCTAGACATGGTGGCCGCGGATCCGATTTAAATTC
S18.38	Hygro-EF1α FWD	GCAATCACAAGTAGCAAGGATCTGCGATCGCTC
S18.39	Hygro-EF1α REV	CGATCGCAGATCCTTGCTACTTGTGATTGCTCCATGT
S18.40	EF1a-CreERT FWD	GTGACCGGCGCCTACCCATTTCAGGTGGCTAGCAT
S18.41	EF1a-CreERT REV	AGCCACCTGAAATGGGTAGGCGCCGGTCACAGCT
S18.42	CreERT-GFP FWD	GCTGAAGCAGGCCGGCGACGTGGAGGAGAA CCCCGGCCCCATGGAGAGCGACGAGAGCGG
S18.43	CreERT-GFP REV	CTCCACGTCGCCGGCCTGCTTCAGCAGGGAGAAG TTGGTGGCAGCTGTGGCAGGGAAACC
S18.44	Puro-loxP FWD	ACTTCGTATAGCATACATTATACGAAGTTATGGATCC TAACTATGTTGCTCCTTTTACGC
S18.45	Puro-loxP REV	GATCCATAACTTCGTATAATGTATGCTATACGAAG TTATCCAGAGGTTGATTTCAGGCAC
S18.46	Ngn2-Hygro FWD	GACTGTATCTCTAGATAGGGACGATTTTCTAAGGA TCCCTC
S18.47	Ngn2-Hygro REV	CTTAGAAAATCGTCCCTATCTAGAGATACAGTCC CTGGC
S18.51	rtTA-Ngn2 FWD	ATACCGTCGACCTCGCAGGGACAGCAGAGA TCCAG
S18.52	rtTA-Ngn2 REV	TCTCTGCTGTCCCTGCGAGGTCGACGGTA TCGATG

2.3 Molecular Cloning

1. Luria-Bertani (LB) broth: Weigh and transfer 4 g of tryptone, 2 g of yeast extract, and 4 g NaCl to a 500 mL bottle. Add 400 mL distilled water. Sterilize by autoclaving for 20 min on liquid cycle. Store the media at 4 °C.
2. LB agar plates: Weigh and transfer 4 g of tryptone, 2 g of yeast extract, 4 g NaCl, and 6 g of agar to a 500 mL bottle. Add 400 mL distilled water. Sterilize by autoclaving for 20 min on liquid cycle. Cool the melted solution in 55 °C water bath. Add proper concentration of desired antibiotic to the agar solution for selection (e.g., 400 μL of 100 mg/mL ampicillin stock) and swirl to mix. Pour plates near a Bunsen burner or in a fume hood and allow them to solidify. Wrap the plates and store at 4 °C.

3. 0.5 M EDTA stock solution: Weigh out 93.05 g of EDTA disodium salt. Dissolve in 400 mL deionized water. Adjust the pH with NaOH to about 8.0 until EDTA is completely dissolved (*see* **Note 1**). Adjust the solution to a final volume of 500 mL.
4. Tris acetate EDTA (TAE) buffer: Weigh out 242 g of Tris-base. Dissolve in 700 mL deionized water. Add 57.1 mL of 100% glacial acid and 100 mL of 0.5 M EDTA. Top up the solution to a final volume of 1 L to make 50× TAE buffer. Store the buffer at room temperature. Dilute 20 mL of the stock solution with 980 mL deionized water to get 1× TAE buffer for making and running gels. The final concentration in 1× TAE buffer includes 40 mM Tris, 20 mM acetic acid, and 1 mM EDTA.
5. PrimeSTAR Max DNA Polymerase (2×, Takara Bio).
6. Thermocycler.
7. SyBR Safe DNA Gel Stain (10,000× in DMSO, Thermo).
8. Gel tray, comb, tank, and electrophoresis equipment.
9. Gel loading dye and Quick-Load Purple 1 kb Plus DNA Ladder 100 bp–10 kb (NEB).
10. UV transilluminator.
11. NEB 5-alpha Competent E. coli (C2987).
12. Super optimal broth with catabolite repression (SOC medium, NEB).
13. Rapid cracking buffer: prepare stock solution of 10% glycerol and 5 mM EDTA. Take 4.7 mL of the stock and add 50 μL of 5 M NaOH, 250 μL of 10% SDS, and 62 μL of 2% bromocresol green (BCG) (*see* **Note 2**).
14. Shaking incubator.
15. Macherey-Nagel NucleoSpin Plasmid Mini kit.
16. Centrifuge.
17. NanoDrop.
18. Restriction enzymes and CutSmart Buffer (NEB).
19. NEBuilder HiFi DNA Assembly Cloning Kit (NEB).

2.4 ES Cell Culture

1. Mouse ES culture medium: Mix 75 mL of fetal bovine serum (FBS), 5 mL of 100× GlutaMAX, 5 mL of 100× non-essential amino acids (NEAA), 2.5 mL of 100× penicillin/streptomycin, 500 μL of 1000× β-mercaptoethanol, and 500 μL of 1000 U/mL leukemia inhibitory factor (LIF) in a bottle of 500 mL Dulbecco's Modified Eagle Medium (DMEM). Store the supplemented medium at 4 °C.
2. Freezing medium (2×): Dilute 2 mL of DMSO with 8 mL of FBS.

3. Ethanol 70%.
4. Cell culture plastics.
5. Class II biosafety cabinet.
6. Centrifuge.
7. CO_2 incubator (37 °C, 5% CO_2).
8. Dulbecco's PBS.
9. TrypLE Express Enzyme, no phenol red (1×, Gibco).
10. Hemocytometer.
11. Trypan blue solution.
12. Microscope.
13. Mr. Frosty box.

2.5 Transfection

1. Opti-MEM Reduced Serum Medium (Gibco).
2. Lipofectamine 3000 Transfection Reagent (Thermo).
3. Fluorescent microscope.
4. Puromycin (10 mg/mL, Thermo).

2.6 Neuronal Differentiation

1. N2B27 neural medium: To make 50 mL medium, mix 23.75 mL of DMEM/F12 medium, 23.75 mL of Neurobasal medium, 500 μL of 50× B27 supplement, 250 μL of 100× N2 supplement, 250 μL of 100× GlutaMAX, 500 μL of 100× NEAA, 45 μL of β-mercaptoethanol, and 500 μL of 100× penicillin/streptomycin. Make fresh aliquots each time. Store at 4 °C up to a month.
2. Laminin coating solution: Dilute 150 μL of 1 mg/mL laminin stock (Sigma) with 10 mL of PBS to make 15 μg/mL laminin solution.
3. Doxycycline: Make up 2 mg/mL stock. Use at 1:2000 in neural medium for 1 μg/mL.
4. Hygromycin B (50 mg/mL, Thermo).
5. AraC (Sigma): Make up 5 mM stock. Dilute with neural medium at 1:2000 for 2.5 μM.
6. Recombinant Human BDNF (PeproTech): Make up 10 μg/mL stock. Use at 1:1000 in neural medium for 10 ng/mL.

3 Methods

3.1 Polymerase Chain Reaction (PCR) and Gel Electrophoresis

1. General composition of PCR mixture (*see* **Note 3**):

PrimeSTAR Max Premix (2×)	5 μL
Forward primer (5 μM)	0.5 μL
Reverse primer (5 μM)	0.5 μL

(continued)

Template (plasmid DNA 0.1 ng/μL)	2 μL
Sterile distilled water	2 μL
Total	10 μL

2. General PCR condition (*see* **Note 4**):

98°C	10 s	30 cycles
55°C	5s	
72°C	5 s/kb	
72°C	5 min	
4°C	Hold	

3. Agarose gel: To make a 1% agarose gel, weigh out 1 g of agarose and transfer to 100 mL of 1× TAE buffer in a flask (*see* **Note 5**). Microwave for 1–2 min with occasional swirling until the agarose is completely dissolved. Cool down the agarose solution to about 50 °C. Add 5 μL of SyBR Safe DNA Gel Stain and mix well. Pour the solution into a gel tray and place a well comb in it. Transfer the solidified gel into a gel tank and fill the tank with 1× TAE buffer until the gel is covered.
4. Gel electrophoresis: Add loading buffer to DNA samples. Load a molecular weight ladder (Quick-Load Purple Ladder) and the samples into the gel wells. Run the gel at 120 V for 30–45 min. Visualize the DNA fragments with a UV transilluminator.

3.2 Transformation

1. Thaw a tube of NEB 5-alpha competent cells (C2987) on ice for 10 min (*see* **Note 6**).
2. Add 1–2 μL of plasmid to the cell mixture and mix gently by pipetting up and down. Place the tube on ice for 30 min (*see* **Note 7**).
3. Heat shock at 42 °C for 30 s and place back on ice for 5 min (*see* **Note 8**).
4. Add 950 μL of SOC medium to the mixture and incubate at 37 °C for an hour (*see* **Note 9**). Warm up ampicillin agar plates at 37 °C.
5. Spread 50–100 μL of the cell mixture onto a selection plate. Incubate the plate at 37 °C overnight.

3.3 Rapid Cracking

1. Aliquot 30 μL of the cracking buffer to 1.5 mL microcentrifuge tubes.
2. Randomly choose and label a few colonies with a marker.
3. Use sterile 10 μL pipette tips to pick a bit of each colony into each tube (*see* **Note 10**).
4. Heat the tubes at 70 °C for 30 min and then vortex for 30 s.
5. Load 10 ng of a control plasmid and 10 μL of each sample onto gel.
6. Run the gel and visualize the bands with a UV transilluminator (*see* **Note 11**).

3.4 Liquid Culture

1. Add 5 mL of LB medium and 5 μL of 100 mg/mL ampicillin into a 50 mL falcon tube.
2. Use a sterile inoculation loop to put a bit of selected colony into the LB-antibiotic mixture and swirl well.
3. Incubate broth culture at 37 °C, 180 rpm for 16–18 h in a shaking incubator.

3.5 Mini-prep (Plasmid Purification)

1. Transfer 1.5 mL of an overnight liquid culture to a 1.5 mL microcentrifuge tube. Centrifuge at 11,000 × *g* for 1 min. Discard the supernatant and remove as much of the liquid as possible.
2. Completely resuspend the cell pellet with 250 μL of Buffer A1 (added with RNase A) by pipetting up and down.
3. Add 250 μL of Buffer A2 and mix gently by inverting the tube 6–8 times (*see* **Note 12**). Incubate at room temperature for up to 5 min.
4. Add 300 μL of Buffer A3 and mix thoroughly by inverting the tube 6–8 times (*see* **Note 13**). Centrifuge at 11,000 × *g* for 5 min.
5. Prepare a NucleoSpin column in a collection tube. Transfer 750 μL of the supernatant onto the column. Centrifuge at 11,000 × *g* for 1 min. Discard the flow-through and place the column back to the collection tube.
6. Add 600 μL of Buffer A4 (supplemented with ethanol) and centrifuge at 11,000 × *g* for 1 min. Discard the flow-through and place the column back.
7. Centrifuge at 11,000 × *g* for 2 min. Discard the collection tube and place the column in a 1.5 mL microcentrifuge tube.
8. Add 50 μL of Buffer AE and incubate at room temperature for 1 min. Centrifuge at 11,000 × *g* for 1 min.
9. Measure the concentration and quality of purified plasmid by NanoDrop (*see* **Note 14**).

3.6 Restriction Digest

1. Set up the following reaction in a sterile 1.5 mL tube:

CutSmart buffer (10×)	1 μL
Plasmid DNA (0.2–0.5 μg)	× μL
Restriction enzyme	0.25 μL
Deionized water	8.75-× μL
Total	10 μL

2. Incubate the tube at 37 °C for 30–60 min.
3. Add 2 μL of Gel Loading Dye (6×) to the tube and load 10 μL of the mixture to gel. Load a molecular weight marker to the gel.
4. Run the gel and visualize the bands with a UV transilluminator.

3.7 Gibson Assembly Overall Outline

1. Design plasmid constructs and order primers with 25–30 bp homologous overlaps.
2. Generate DNA fragments by PCR.
3. Set up the assembly reaction on ice (Table 2) (*see* **Note 15**).
4. Pre-warm and incubate the assembly reaction in a thermocycler at 50 °C for 15 min when 2–3 fragments are being assembled or 1 h when 4–6 fragments are being assembled.
5. Digest the assembly product with DpnI enzyme at 37 °C for at least 1 h (*see* **Note 16**).
6. Transform NEB 5-alpha Competent cells with 2 μL of digested assembly product.
7. Perform rapid cracking to identify positive colonies with insert.
8. Pick positive colonies for broth culture.
9. Extract plasmid DNA by mini-prep.
10. Prepare samples for Sanger sequencing.

3.8 Introduce a LoxP Site to PB-GFP Vector After Puromycin Resistance Gene

1. Set up PCR to amplify DNA fragments for cloning (Table 3).
2. Run 5 μL of each sample on agarose gel to check the quality of PCR products. Estimate DNA concentration from gel by comparing its intensity with the bands in the ladder (*see* **Note 17**). Keep the rest for Gibson assembly.

Table 2
Instruction for Gibson assembly reaction preparation

	Recommended volume of reagents used for assembly		
	2–3 fragment assembly	4–6 fragment assembly	Positive control
Total amount of fragments	1 μL (0.05–0.25 pmols)	2 μL (0.1–0.5 pmols)	2.5 μL
Assembly master mix (2×)	2.5 μL	5 μL	2.5 μL
Deionized water	1.5 μL	3 μL	0
Total volume	5 μL	10 μL	5 μL

Table 3
Instruction for amplification of Gibson assembly fragments

Fragment name	Template	Primers	Size
PB 1	PB-GFP	S18.44/S18.17	3.1 kb
PB 2	PB-GFP	S18.45/S18.16	4.3 kb.

3. Set up the following assembly reaction in a PCR tube on ice:

PB 1	0.45 μL
PB 2	0.55 μL

NEBuilder HiFi DNA Assembly master mix	2.5 μL
Sterile distilled water	1.5 μL
Total	5 μL

4. Incubate the tube in a thermocycler at 50 °C for 15 min.
5. Add 0.6 μL of 10× CutSmart buffer and 0.4 μL of DpnI enzyme to the assembly product. Incubate at 37 °C for an hour.
6. Transform competent cells (C2987) with 2 μL of the digested assembly product.
7. Select and mark eight colonies for liquid culture (*see* **Note 18**).
8. Isolate plasmid DNA from the saturated broth culture.
9. Digest 250 ng of each sample with BamHI and run the gel (Fig. 1):

   Vector, linearized one band, 7.3 kb;

   PB + loxP, two bands, 5.2 kb and 2.1 kb.

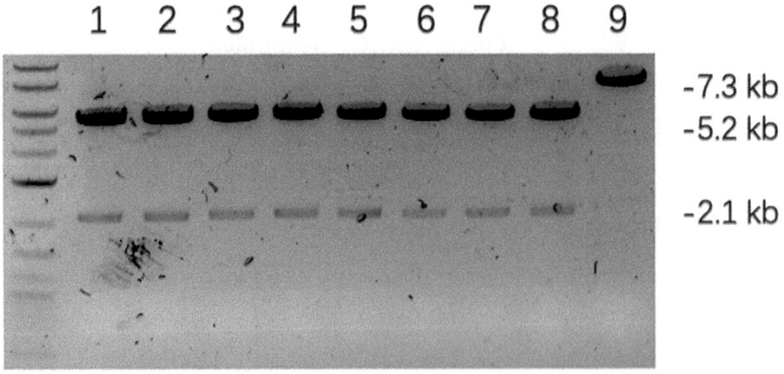

Fig. 1 Gel image of BamHI-digested plasmids for identifying samples with loxP insert. 1–8: purified plasmid samples; 9: negative control (PB-GFP plasmid). All samples are modified plasmids

3.9 Introduce CreER^{T2}-loxP System Under EF1α Promoter

1. Set up PCR to amplify DNA fragments for cloning (Table 4).
2. Run 5 μL of each sample on agarose gel to check the quality of PCR products. Estimate DNA concentration from gel by comparing its intensity with the bands in the ladder. Keep the rest for Gibson assembly.
3. Set up the following assembly reaction in a PCR tube on ice

PB-loxP 1	0.3 μL
PB-loxP 2	0.45 μL
CreERT	0.25 μL

NEBuilder HiFi DNA Assembly master mix	2.5 μL
Sterile distilled water	1.5 μL
Total	5 μL

4. Incubate the tube in a thermocycler at 50 °C for 30 min.
5. Add 0.6 μL of CutSmart buffer (10×) and 0.4 μL of DpnI enzyme to the assembly product. Incubate at 37 °C for an hour.
6. Select and mark eight colonies for rapid cracking. Use PB-GFP plasmid as the negative control (Fig. 2).
7. Pick five positive samples for broth culture.
8. Extract plasmid DNA by minipreps.
9. Digest 250 ng of each sample with BamHI and run the gel (Fig. 3):

   Vector, 5.2 kb and 2.1 kb fragments;

   PB + CreERT, 5.2 kb, 3.1 kb, and 1 kb.

Table 4
Instruction for amplification of Gibson assembly fragments

Fragment name	Template	Primers	Size
CreERT	CreP2ANgn2	S18.40/S18.43	2 kb
PB-loxP1	PB-loxP	S18.41/S18.16	2.8 kb
PB-loxP2	PB-loxP	S18.42/S18.17	4.6 kb.

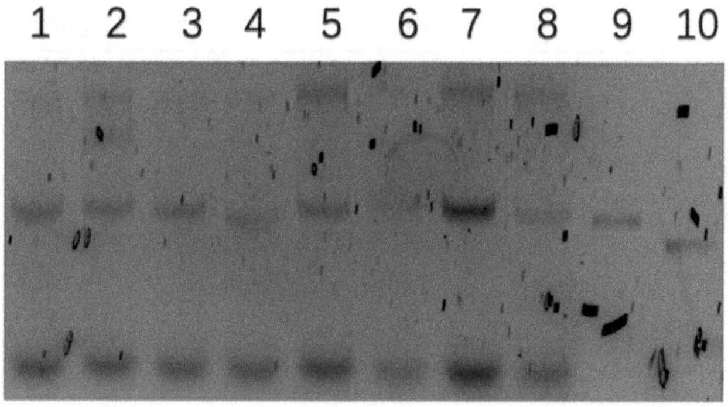

Fig. 2 Gel image of rapid cracking for identifying colonies with CreERT insert. 1–9: colony samples; 10: negative control (PB-GFP plasmid). Compared to the band of the vector, 7 samples may contain the assembled plasmid

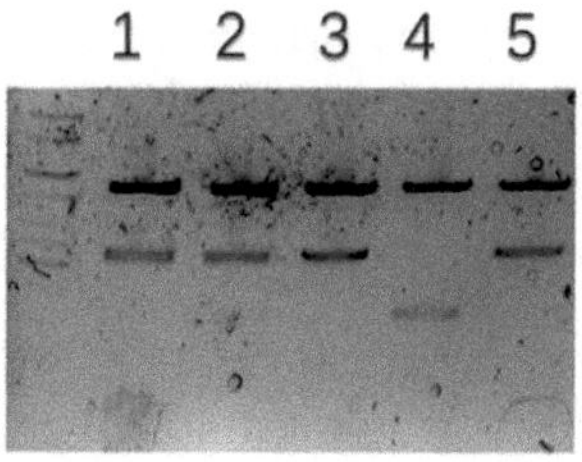

Fig. 3 Gel image of BamHI-digested plasmids. 1–5: purified plasmid samples. Four samples are assembled plasmid, whereas one sample is the vector

3.10 Introduce Tet-on System Under CMV Promoter

1. Set up PCR to amplify DNA fragments for cloning (Table 5).
2. Set up the following assembly reaction in a PCR tube on ice:

PB-CreERT 1	0.3 μL
PB-CreERT 2	0.7 μL
rtTA	0.3 μL

(continued)

Table 5
Instruction for amplification of Gibson assembly fragments

Fragment name	Template	Primers	Size
rtTA	FUW-M2rtTA	S18.34/S18.52	1.4 kb
Ngn2	TetO-Ngn2	S18.51/S18.47	1.3 kb
Hygro	Dlx2-hygro	S18.46/S18.39	2.3 kb
PB-CreERT 1	PB-CreERT	S18.35/S18.16	2.2 kb
PB-CreERT 2	PB-CreERT	S18.38/S18.17	7.2 kb

Ngn2	0.3 μL
Hygro	0.4 μL
NEBuilder HiFi DNA Assembly master mix	5 μL
Sterile distilled water	3 μL
Total	10 μL

3. Incubate the tube in a thermocycler at 50 °C for an hour.
4. Add 1.2 μL of 10× CutSmart buffer and 0.8 μL of DpnI enzyme to the assembly product. Incubate at 37 °C for an hour.
5. Transform competent cells (C2987) with 2 μL of each digested assembly product.
6. Select and mark nine colonies of each plate for rapid cracking. Use PB-CreERT-GFP plasmid as a negative control (Fig. 4).
7. Pick two positive colonies for broth culture.
8. Purify plasmid DNA by minipreps.
9. Digest 250 ng of each sample with BamHI and run the gel (Fig. 5):

   Vector, 5.2 kb, 3.1 kb, and 982 bp;

   PB-Ngn2, 5.2 kb, 3.3 kb, 3.1 kb, and 2.7 kb.

3.11 Sanger Sequencing Preparation

For purified DNA samples (in 12 μL), the recommended quantity of double-stranded plasmid is 600–1500 ng and primer quantity is 9.6 pmol (0.8 pmol/μL).

1. Prepare the samples in sterile 1.5 mL tubes. Add appropriate primer (forward or reverse only), plasmid and sterile MilliQ water to a total volume of 12 μL (*see* **Note 19**).
2. Wrap and send the samples for sequencing.

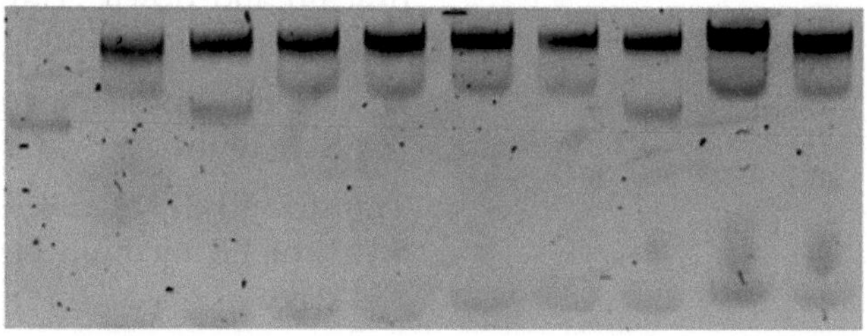

Fig. 4 Gel image of rapid cracking for identifying colonies with insert. 1: negative control (PB-CreERT-GFP plasmid); 2–10: colony samples. Seven colonies may contain assembled PB-Ngn2 plasmid

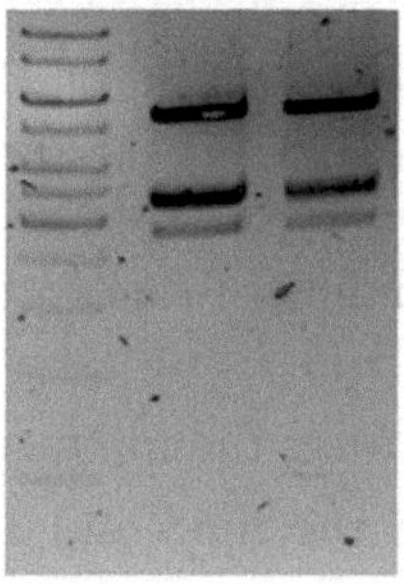

Fig. 5 Gel image of BamHI-digested plasmids. 1–2: purified plasmid samples. The samples chosen for minipreps are assembled plasmids

3.12 Thaw and Recover ES Cells

1. Thaw a cryovial of mouse ES cells in a 37 °C water bath with gentle rotation until a tiny bit of ice is left (*see* **Note 20**).
2. Spray the vial with 70% ethanol and transfer it to a biosafety cabinet.
3. Slowly add 1 mL of pre-warmed ES medium to the vial and transfer the cell suspension to a sterile 15 mL tube. Centrifuge the cells at 200 × *g* for 3 min.
4. Discard the supernatant and resuspend the cell pellet in 5 mL of ES medium.
5. Transfer the cell suspension to a T25 flask. Place the flask in a 37 °C, humidified incubator with 5% CO_2.

3.13 Sub-culture ES Cells

1. Warm up PBS, TrypLE, and ES medium in a 37 °C water bath.
2. Passage the cells when they reach 70–80% confluency. Remove old medium and wash the cells twice with 2 mL of PBS.
3. Dissociate the cells with 1 mL of TrypLE at 37 °C for 2–4 min.

4. Add 1 mL of ES medium and gently detach the cells by pipetting up and down. Transfer the cell suspension to a 15 mL tube. Centrifuge at 200 × *g* for 3 min.
5. Discard the supernatant and resuspend the cell pellet with 2 mL of ES medium.
6. Transfer 50 μL of cell suspension to a T25 flask added with 5 mL of ES medium (1:50 dilution). Passage the cells every 3–4 days.

3.14 Transfection

1. Seed cells onto a 24-well plate to be 70–90% confluent at the day of transfection.
2. Dilute 1.5 μL of Lipofectamine 3000 reagent (Thermo) in 25 μL of Opti-MEM Medium and mix well (*see* **Note 21**).
3. Dilute 1 μL of P3000 reagent in 25 μL of Opti-MEM.
4. Add 450 ng of the PB-Ngn2 plasmid and 150 ng of the Super PiggyBac Transposase expression vector to diluted P3000 solution (*see* **Note 22**).
5. Add diluted Lipofectamine 3000 reagent to the tube of diluted DNA (1:1 ratio) and mix well.
6. Incubate the mixture at room temperature for 10–15 min.
7. Add the lipid and DNA complex to the cells in a dropwise fashion.
8. Gently rock the plate and place back in the incubator.
9. After overnight incubation, visualize transfected cells with a fluorescent microscope to assess transfection efficiency.
10. Dissociate and replate the cells onto a 6-well plate.
11. Add 1 μg/mL puromycin to the cells 48 h after transfection.
12. Replace the medium supplemented with 1 μg/mL puromycin every 3 days.
13. After 5–7 days of antibiotic selection, split and plate stably transfected cells to a T25 flask (*see* **Note 23**).
14. Cryopreserve the stable cell line when they reach 80% confluency.

3.15 Neural Induction and Differentiation

1. Coat each well of a 24-well plate with 250 μL of 15 μg/mL laminin at 37 °C for at least 2 h or at 4 °C overnight.
2. On day 1, dissociate ES cells with 1 mL of TrypLE at 37 °C for 3–4 min. Add 1 mL of ES medium to detach the cells. Centrifuge at 200 × *g* for 3 min. Discard the supernatant and resuspend in 2 mL of ES medium. Count and calculate cell concentration using a hemocytometer. Plate cells at 50000/well of a 24-well plate in 500 μL of ES medium. Culture the cells in a 37 °C, 5% CO_2 incubator overnight.

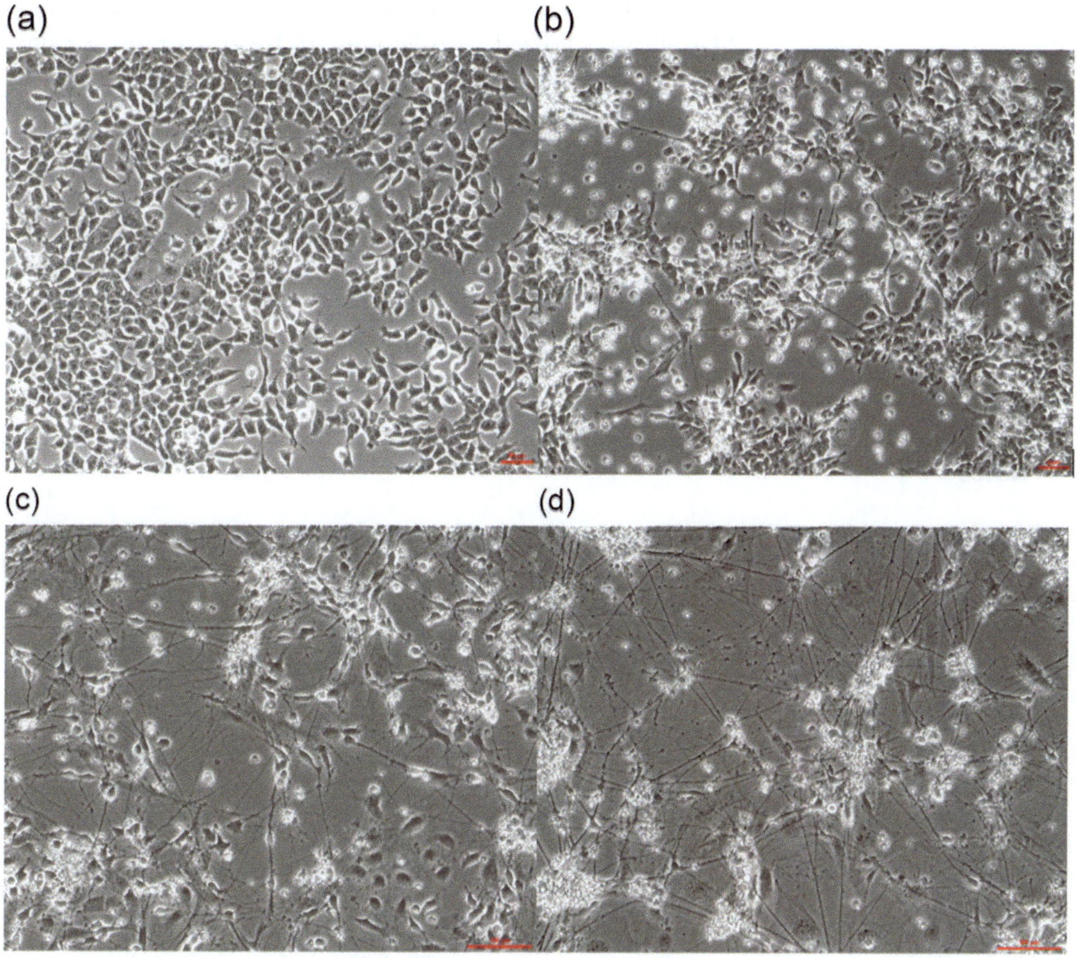

Fig. 6 Neural induction of mouse ES cells stably transfected with Ngn2 expression cassette. (**a**) day 1 post-induction. (**b**) day 4 post-induction. (**c**) day 7 post-induction. (**d**) day 14 post-induction. Scale bar: (**a**) and (**b**) 50 μm; (**c**) and (**d**) 100 μm

3. At day 0, remove ES medium and add 500 μL of N2B27 neural medium with 1 μg/mL doxycycline (*see* **Note 24**).
4. At day 1, change induction medium (N2B27 with 1 μg/mL doxycycline) and add 200 μg/mL hygromycin (Fig. 6a) (*see* **Note 25**).
5. At day 2 and day 3, change medium supplemented with doxycycline and hygromycin daily.
6. On day 4, split the cells (*see* **Note 26**). Remove the medium and wash the cells with 250 μL of PBS twice. Add 250 μL of TrypLE to dissociate the cells at 37 °C for 3–4 min. Add 750 μL of neural medium and gently detach the cells by pipetting. Transfer cell suspension to a 15 mL tube and centrifuge at 200 × *g* for 3 min. Discard the supernatant and resuspend cell

pellet in 1 mL of neural medium. Plate the cells onto wells of 24-well plate coated with laminin at 1:5–1:10 ratio, or onto relevant assay format (patch clamping, multielectrode array, immunostaining, etc.).

7. Cells are cultured in 500 μL of N2B27 medium supplemented with 1 μg/mL doxycycline and 10 ng/mL BDNF (*see* **Note 27**). Replace the medium every 2–3 days (Fig. 6b).
8. At day 6 or 7, treat the cells with 2.5 μM AraC for 48 h to kill proliferating non-neuronal cells (Fig. 6c).
9. On day 10, withdraw doxycycline and maintain the cells in neural medium with 10 ng/mL BDNF (Fig. 6d) (*see* **Note 28**).

4 Notes

1. Use a magnetic stir bar to help dissolve EDTA disodium salt. Keep adding small volumes of 5 M NaOH solution. The pH should be around 8 when EDTA completely dissolves.
2. If BCG is not available in the lab, add gel loading dye before loading the samples to gel.
3. The PCR mixture can be prepared at room temperature but keep all components on ice during sample preparation.
4. The PCR conditions need to be optimized based on template DNA, primers, and product size. Primer design should follow the basic rules in order to obtain the target products. Raise the annealing temperature to 58–63 °C if nonspecific bands are amplified. Increase the annealing time to 15 s or lower the annealing temperature to 50–53 °C if the target DNA fragment is not amplified.
5. The commonly used percentage of agarose gel is between 0.7% and 2%, depending on the band sizes to be separated. Use lower percentage of agarose (0.7–1.0%) for large DNA fragments and higher percentage (1.0–2.0%) for small fragments.
6. It is recommended to use chemical competent cells with efficiency higher than 1×10^9 cfu/μg. Electroporation is another method which can improve transformation efficiency. Use 1 μL of assembly product for electroporation and plate serial dilutions.
7. Recovery of E. coli in SOC outgrowth medium increases transformation efficiency. LB medium can also be used but the efficiency will be compromised.
8. Incubation on ice less than 30 min can result in loss in transformation efficiency.

9. Heat shock should be performed at exactly 42 °C for 30 s. Longer time will not increase transformation efficiency but may impact the efficiency.
10. Do not pick too large of a colony, otherwise the bands may be hard to distinguish on the gel.
11. Rapid cracking is a quick method for screening cloned inserts in plasmids. RNA smear is at the bottom, genomic DNA is at the top and plasmid DNA is in the middle. The samples with upper plasmid bands are positive clones containing the insert.
12. Avoid vortex which might shear host chromosomal DNA.
13. Do not vortex to avoid gDNA contamination.
14. The 260/280 ratio between 1.80 and 2.00 indicates a good DNA purity.
15. Use all fragments from PCR products in equimolar amounts. Gel purification or PCR clean-up of PCR products is not necessary. However, the total volume of unpurified PCR products should not exceed 20% in Gibson assembly reaction.
16. Lengthen the incubation time can digest the vectors more completely and then reduce background growth on agar plates. Heat-inactivation of DpnI enzyme prior to transformation is not necessary.
17. It is more accurate to quantify DNA on agarose gel by densitometry analysis software.
18. Since loxP sequence is only a tiny insert, rapid cracking is not able to distinguish the assembled plasmid. Add a restriction enzyme site within the insert to identify positive clones. Ensure that DpnI digest is complete to reduce the number of colonies for screening.
19. Design sequencing primers located around 50–100 bases upstream the target sequence.
20. Thaw the vial and dilute with culture medium quickly as DMSO is toxic to cells.
21. The ratio of DNA to lipofectamine should be optimized for different cell lines. The presence of serum affects lipid-based transfection by interfering the formation of DNA–lipid complexes. Hence, Opti-MEM reduced serum medium is recommended for use with lipofectamine reagent.
22. It is recommended to use a ratio of transposase to transposon vector between 1:2.5 and 1:5 for transfection.
23. Clonal selection or FACS sorting can be performed to obtain a homogenous cell population for neural differentiation.
24. The presence of doxycycline initiates the expression of transcription factor in the inducible transgene cassette.

25. The hygromycin resistance gene is co-expressed with transcription factor linked by 2A self-cleaving peptides. Therefore, induced cells can be selected with hygromycin B.
26. Split the cells at early differentiation stage can improve cell viability. Older neurons with more mature morphology tend to have low viability after dissociation.
27. As a neurotransmitter modulator, BDNF can support the survival and growth of neurons. Co-culture with glial cells helps the formation of synapses between neurons, which are essential for signal transmission.
28. Endogenous Ngn2 expression has been activated at day 10 post-induction, so doxycycline can be withdrawn to terminate transgene expression. Differentiated neurons will exhibit mature morphology and functional activity.

Acknowledgments

This work was supported by a Tuition Fee Scholarship from Swinburne University of Technology.

References

1. Zhao S, Jiang E, Chen S, Gu Y, Shangguan AJ, Lv T, Luo L, Yu Z (2016) PiggyBac transposon vectors: the tools of the human gene encoding. Transl Lung Cancer Res 5(1):120–125. https://doi.org/10.3978/j.issn.2218-6751.2016.01.05
2. Lu X, Huang W (2014) PiggyBac mediated multiplex gene transfer in mouse embryonic stem cell. PLoS One 9(12):e115072. https://doi.org/10.1371/journal.pone.0115072
3. Gibson DG, Young L, Chuang R-Y, Venter JC, Hutchison CA, Smith HO (2009) Enzymatic assembly of DNA molecules up to several hundred kilobases. Nat Methods 6(5):343–345. https://doi.org/10.1038/nmeth.1318
4. Benoit RM, Ostermeier C, Geiser M, Li JSZ, Widmer H, Auer M (2016) Seamless insert-plasmid assembly at high efficiency and low cost. PLoS One 11(4):e0153158. https://doi.org/10.1371/journal.pone.0153158
5. Romito A, Cobellis G (2016) Pluripotent stem cells: current understanding and future directions. Stem Cells Int 2016:20. https://doi.org/10.1155/2016/9451492
6. Oh Y, Jang J (2019) Directed differentiation of pluripotent stem cells by transcription factors. Mol Cells 42(3):200–209. https://doi.org/10.14348/molcells.2019.2439
7. Thoma E, Wischmeyer E, Offen N, Maurus K, Schartl M, Wagner T (2012) Ectopic expression of neurogenin 2 alone is sufficient to induce differentiation of embryonic stem cells into mature neurons. PLoS One 7(6):e38651. https://doi.org/10.1371/journal.pone.0038651
8. Zhang Y, Pak C, Han Y, Ahlenius H, Zhang Z, Chanda S, Marro S, Patzke C, Acuna C, Covy J, Xu W, Yang N, Danko T, Chen L, Wernig M, Südhof Thomas C (2013) Rapid single-step induction of functional neurons from human pluripotent stem cells. Neuron 78(5):785–798. https://doi.org/10.1016/j.neuron.2013.05.029

Chapter 4

Modifying Bacterial Artificial Chromosomes for Extended Genome Modification

Hannah Auch, Nikolai Klymiuk, and Petra Runa-Vochozkova

Abstract

Bacterial artificial chromosomes have been used extensively for the exploration of mammalian genomes. Although novel approaches made their initial function expendable, the available BAC libraries are a precious source for life science. Their comprising of extended genomic regions provides an ideal basis for creating a large targeting vector. Here, we describe the identification of suitable BACs from their libraries and their verification prior to manipulation. Further, protocols for modifying BAC, confirming the desired modification and the preparation of transfection into mammalian cells are given.

Key words Instructions: Bacterial artificial chromosomes (BAC), Bacterial recombination, BAC libraries, BAC fingerprinting, BAC Sanger sequencing, Genome modification

1 Introduction

Bacterial artificial chromosomes (BACs) are huge plasmids comprising pieces of 150–300 kb from vertebrate genomes [1]. BACs have been initially developed for the generation of genomic libraries in the early ages of full genome exploration. Such libraries have been produced by partial digestion of genomic DNA with restriction enzymes and cloning of these fragments into a plasmid backbone of approximately 20 kb. The BACs thus comprised overlapping inserts and systematic restriction enzyme digestions of all BACs from a library allowed their ordered assembling and finally the construction of a chromosome. Reference genomes, such as for human, have then been generated by shot-gun sequencing of BACs with minimal overlap [2–4]. Next-generation sequencing approaches were based on shot-gun sequencing of the whole genome [5, 6] avoiding the intermediate step of BAC libraries, but the assembly of these short sequence reads required the previously defined genomic structure. Present-day sequencing approaches generate extremely large contigs [7, 8] facilitating de novo genome

Paul John Verma, Huseyin Sumer and Jun Liu (eds.), *Applications of Genome Modulation and Editing*, Methods in Molecular Biology, vol. 2495, https://doi.org/10.1007/978-1-0716-2301-5_4,

exploration and making the once precious BAC libraries dispensable. BAC libraries (Table 1) are, however, still accessible and by carrying large genomic regions they provide a source for advanced experimental genetic modification strategies in life science.

In contrast to classical targeting or gene editing by CRISPR/Cas and short single-stranded oligo-deoxy-nucleotides (ssODN), the large genomic segments in BACs allow either the introduction of huge modifications into a genome or provide extended homologous arms (HA) for homologous recombination (HR), which is often helpful if well-characterized target genomes such as in stem cells are missing. The availability of different bacterial recombination systems such as the RecET system [9, 10] or λ Red system [11, 12] promotes the effective manipulation of BACs in *E. coli* and facilitates an unrestricted portfolio of shaping BACs for genetic modification in vertebrate cells.

We have gained extensive experience in tailoring porcine BACs for generating pig models for biomedical research, but the protocol below describes a very general procedure of preparing a BAC as a targeting vector for the site-specific introduction of exogenous DNA into the mammalian genome. For the complexity and multiplicity of a possible modification, however, we cannot refer to the design and construction of the desired modification. Rather, we generally speak about a "modification of interest" (MOI) which shall be representative for any approach on reporter genes, fusion genes, gene mutations, etc. Importantly, for genome manipulation of vertebrate cells, BAC vectors can be combined with gene-editing tools to promote site-directed HR. For gene editing in porcine and bovine somatic cells and their preparation for somatic cell nuclear transfer, we would like to refer to [13].

2 Materials

2.1 Design and Construct a Desired Modification of Interest

1. Vector pL451/pL452 (Addgene).
2. Electrocompetent *E. coli* strain DH10B.
3. Kanamycin (storage concentration 25 mg/mL, working concentration 1000× diluted).
4. Restriction enzymes with an appropriate buffer.
5. T4 ligase and ligation buffer.
6. Calf Intestinal Alkaline Phosphatase (CIAP).
7. General equipment for agarose gel electrophoresis: Universal agarose powder (BioSell), Bromophenol blue, 1 kb DNA ladder (GeneRuler 1 kb DNA Ladder), GelRed, 50× TAE buffer (2 M Tris base, 50 mM EDTA, 1 M acetic acid) storage solution and 1× working solution, agarose gel electrophoresis system, Imager for gel documentation UVP GelStudio Plus (Analytic Jena).

Table 1
BAC libraries

	Library	Source	Digest	Average insert size	Coverage	Vector[a]	Constructed by	Distributor
Human	CHORI-17	Embryo	*Eco*RI	197 kb	11.3×	pBACGK1.1	BACPAC[b]	BACPAC[b]
	CalTech-D	Sperm	*Hin*DIII	129 kb	17×	pBeloBAC11	CalTech[c]	Invitrogen[d]
	RPCI-11	Blood	*Eco*RI, *Mbo*I	178 kb	32.2×	pBACe3.6, pTARBAC1	[17]	BACPAC[b]
Porcine	CHORI-242	Blood	*Mbo*I	173 kb	11.4×	pTARBAC1.3	BACPAC[b]	BACPAC[b]
	SBAB	Fibroblasts	*Hin*DIII	135 kb	5×	pBeloBAC11	[18]	INRA[e]
	RPCI-44	Blood	*Eco*RI	165 kb	10.2×	pTARBAC2	[19]	BACPAC[b]
Bovine	CHORI-240	Blood	*Mbo*I	167 kb	10.7×	pTARBAC1.3	BACPAC[b]	BACPAC[b]
	RPCI-42	Blood	*Eco*RI	164 kb	11.9×	pBACe3.6	BACPCA[b]	BACPAC[b]
	BtINRA	Fibroblasts	*Hin*DIII	120 kb	4×	pBeloBAC11	[20]	INRA[e]
Sheep	CHORI-243	Blood	*Eco*RI	184 kb	5.4–14×[f]	pTARBAC2.1	[21]	BACPAC[b]

[a] In all backbones, resistance is provided against cam, the integration site is flanked by SP6 and T7 primer-binding sites, NotI and AscI sites for linearization are provided.
[b] Pieter de Jong's laboratory at BACPAC Resources, Children's Hospital Oakland Research Institute (https://bacpacresources.org/)
[c] California Institute of Technology
[d] Invitrogen is shipping the BAC clones as a glycerol stock of bacteria (https://www.thermofisher.com/de/en/home/life-science/cloning/clone-collections.html)
[e] The BAC-YAC Resource Center of the Animal Genetics Department of the INRA (http://dga.jouy.inra.fr/grafra/)
[f] Different information is given in the original publication and an update at https://www.sheephapmap.org/bes.php

2.2 Verification of BAC Clone Genomic Location

2.2.1 Long-Term Storage of the Received BAC Clones

1. BAC clone in *E. coli* strain DH10B (BACPAC Resources at Children's Hospital Oakland Research Institute).
2. LB-medium: 2% tryptone, 0.5% yeast extract, 10 mM NaCl, ddH_2O to 1000 mL, autoclaved, then supplemented with chloramphenicol (cam, storage concentration 12.5 mg/mL in EtOH, working concentration 1000× diluted).
3. 60% glycerol, sterile filtrated (pores 0.22 μm).
4. Culture tubes 12 mL.
5. Cryo vials 1.5 mL.
6. Shaking incubator GFL 3031 with orbital motion.

2.2.2 Isolation of BAC Clones by Basic Alkaline Lysis Plasmid Miniprep

1. LB-medium: 2% tryptone, 0.5% yeast extract, 10 mM NaCl, ddH_2O to 1000 mL, autoclaved, then supplemented with cam (storage concentration 12.5 mg/mL in EtOH, working concentration 1000× diluted).
2. STE: 10 mM Tris–HCl pH 8.0, 100 mM NaCl, 1 mM EDTA/NaOH pH 8.0.
3. Plasmid A: 50 mM Glucose, 25 mM Tris–HCl pH 8.0, 10 mM EDTA/NaOH pH 8.0.
4. Plasmid B (always prepare freshly): 0.1 M NaOH, 0.5% SDS.
5. Plasmid C: 3 M KOAc pH 4.8 with 9 M HOAc.
6. RNase A (20 mg/mL).
7. PCiA: phenol, chloroform, isoamyl alcohol (25:24:1), store at 4 °C max. 2 months.
8. Isopropanol (iPrOH).
9. Ethanol (EtOH).
10. T-Buffer: 10 mM Tris–HCl, pH 8.0.
11. Culture tubes 12 mL.
12. Shaking incubator GFL 3031 with orbital motion.
13. Centrifuge with swing rotor (Eppendorf, 5910R, 4× Universal).
14. Nanodrop (SimpliNano spectrophotometer, Biochrom).

2.2.3 BAC Clones Fingerprinting

1. General equipment for agarose gel electrophoresis (Midi gel chamber: electrodes in distance of approx. 19 cm and gel box 12 × 14 cm with 12 wells comb).
2. Restriction enzymes with an appropriate buffer.
3. Thermoblock or incubator for 37 °C.

2.2.4 BAC-End Sequencing

1. PEG-$MgCl_2$: 40% PEG 8000, 30 mM $MgCl_2$.
2. Commercially synthesized primers.
3. BigDye Terminator v3.1 Sequencing Kit.

4. 125 mM EDTA pH 8.0.
5. EtOH.
6. Thermal cycler.
7. Refrigerated centrifuge (Eppendorf, 5910R, FA-48 × 2).

2.3 Modification of BAC Clone by Bacterial Recombineering

2.3.1 Making Electrocompetent Cells

1. LB-medium: 2% tryptone, 0.5% yeast extract, 10 mM NaCl, ddH_2O to 1000 mL, autoclaved, then supplemented with cam (storage concentration 12.5 mg/mL in EtOH, working concentration 1000× diluted).
2. *E. coli* SW strain (SW102, SW105, SW106 from National Cancer Institute, Frederick, USA).
3. 10% glycerol, sterile filtrated (pores 0.22 μm).
4. ddH_2O.
5. 500 mL Erlenmeyer flask.
6. Glass cuvettes.
7. 50 mL falcon tubes with round bottom.
8. Shaking Incubator GFL 3031 with orbital motion.
9. Photometer GeneQuant Pro (Amersham Biosciences).
10. Centrifuge with fixed angle rotor (Eppendorf, 5910R, rotor FA-6 × 50).

2.3.2 Transfer BAC Clone into Bacteria

1. SOB medium: 2% tryptone, 0.5% yeast extract, 10 mM NaCl, 2.5 mM KCl, 10 mM $MgCl_2$, 10 mM $MgSO_4$, ddH_2O to 1000 mL, autoclaved.
2. Petri dishes coated with LB-agar supplemented with appropriate antibiotic.
3. Electroporation cuvettes, gap width 1 mm.
4. Incubator for 32 °C.
5. Electroporator (Eppendorf Eporator).

2.3.3 Preparation of Modification as a Linearized dsDNA Fragment

1. Restriction enzyme with appropriate buffer.
2. Low melting agarose powder (Low Melting Point Agarose, Thermo Scientific).
3. General equipment for agarose gel electrophoresis.
4. Agarase.
5. PCiA: phenol, chloroform, isoamyl alcohol (25:24:1), store at 4 °C max. 2 months.
6. 3 M NaAOc, pH 5.2.
7. EtOH.
8. T-Buffer: 10 mM Tris–HCl, pH 8.0.
9. DNA extraction kit (Double Pure Combi Kit, BioSell).

10. Nanodrop (SimpliNano spectrophotometer, Biochrom).
11. Refrigerated centrifuge (Eppendorf, 5910R, FA-48 × 2).
12. Thermoblock.

2.3.4 BAC Recombineering

See Subheadings 2.3.1. and 2.3.2.

2.3.5 Arabinose-Induced FLPE and Cre Recombination

1. LB-medium: 2% tryptone, 0.5% yeast extract, 10 mM NaCl, ddH_2O to 1000 mL, autoclaved, then supplemented with cam (storage concentration 12.5 mg/mL in EtOH, working concentration 1000× diluted).
2. Arabinose.
3. Culture Tubes 12 mL.
4. Glass cuvettes.
5. Shaking Incubator GFL 3031 with orbital motion.
6. Photometer GeneQuant Pro (Amersham Biosciences).

2.4 Verification of BAC Recombinant

2.4.1 Confirmation of Recombinants by End-Point PCR

1. Commercially synthesized primers.
2. 2 M dNTPs.
3. Herculase II Fusion DNA Polymerase Kit.
4. T-Buffer: 10 mM Tris–HCl, pH 8.0.
5. Thermal cycler.
6. General equipment for agarose gel electrophoresis system.

2.4.2 Confirmation of Recombinants by BAC Fingerprinting

See Subheading 2.2.3

2.4.3 Confirmation of Recombinants by Sanger Sequencing

See Subheading 2.2.4

2.5 Preparing of BAC Targeting Vector for Nucleofection into Cells

1. LB-medium: 2% tryptone, 0.5% yeast extract, 10 mM NaCl, ddH_2O to 1000 mL, autoclaved, then supplemented with appropriate antibiotic.
2. BAC DNA isolation kit (Large Construct Kit, QIAgen).
3. 3 M NaOAc, pH 5.2.
4. PCiA.
5. iPrOH.
6. EtOH.
7. TE Buffer: 10 mM Tris–HCl pH 8.0, 1 mM EDTA/NaOH pH 8.0.

8. 500 mL Erlenmeyer flask.
9. Refrigerated centrifuge with fixed angle rotors (Eppendorf, 5910R, FA-6 × 50 and FA-48 × 2).
10. Shaking Incubator GFL 3031 with orbital motion.
11. Nanodrop (SimpliNano spectrophotometer, Biochrom).

3 Methods

3.1 Design and Construct a Desired Modification of Interest (MOI)

In principle, any modification of a BAC can be defined as the introduction of an MOI into the genomic segment by bacterial recombination (Fig. 1b). Bacterial recombination without positive selection is theoretically possible, but rather ineffective. As most approaches in mammalian cells require positive selection as well, the usage of a combinatorial positive selection cassette is recommended (*see* **Note 1**). The opportunity of bacterial recombination and huge genomic regions in BACs provide an unlimited portfolio of potential modifications. For limitations of space and the focus on BAC modification protocols, we can only give very superficial advice on designing and constructing the MOI.

1. Design the modification and assemble it in silico (*see* **Note 2**). The modification vector in principle comprises the MOI itself, resistance cassette, 5′ and 3′ HAs complementary to an integration site in BAC clone (*see* **Note 3**), and additional features such a restriction sites or lox/FRT sites (*see* illustration in Fig. 1a, b).
2. Make available the necessary components (*see* **Note 4**).
3. Assemble the final modification plasmid from all designed parts, either obtained from gene synthesis or plasmid cloning.
4. Verify the modification plasmid (*see* **Note 5**).

3.2 Finding BAC Clones Covering the Target Region

Exploring a BAC map is necessary for identifying appropriate BAC clones covering the desired target region. While in the past distinct sources were provided for searching BAC maps, many BAC libraries are available meanwhile in the National Center for Biotechnology Information (NCBI) database (*https://www.ncbi.nlm.nih.gov/*). For this reason, we describe the usage of this flexible resource, while other or similar procedures might work on alternative online or off-line BAC map viewer programs. Depending on the coverage of the BAC library, a number of clones can be selected for a target region of interest. For matters of time and efficacy, we recommend purchasing 3 different BAC clones for a given target region and select one of them upon comprehensive verification for the MOI.

1. Choose the Genome Data Viewer at the NCBI web page (*https://www.ncbi.nlm.nih.gov/genome/gdv/*).

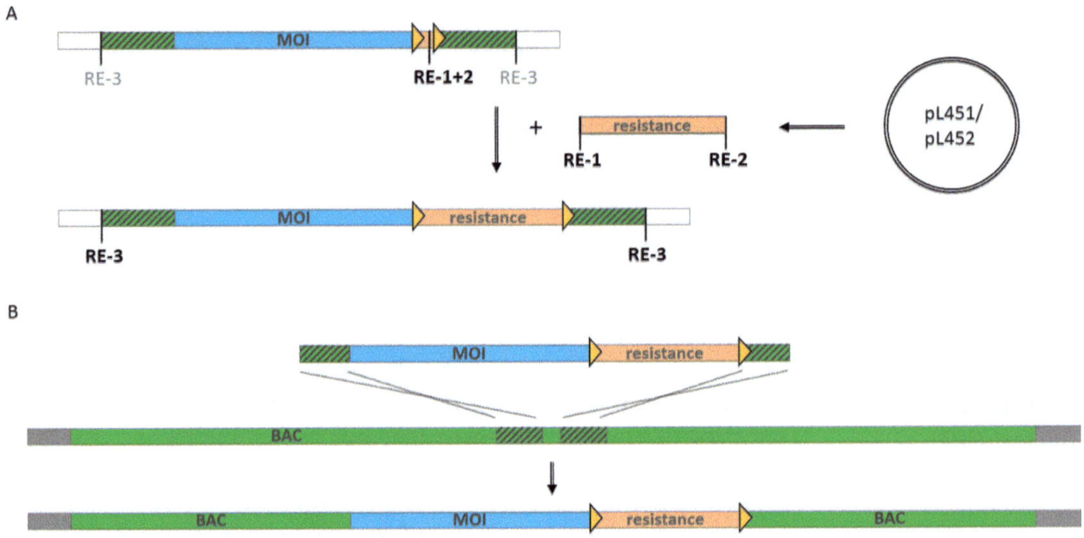

Fig. 1 Schematic illustration of BAC clone modification. (**a**) Assembly of modification vector carrying a desired modification of interest (MOI) in one cloning step. The MOI, homology arms (striped boxes) complementary to target genomic region in BAC clone, lox/FRT sites (arrowheads), and sites for restriction enzymes are synthesized in vitro into standard plasmid (white boxes). The resistance cassette is excised from pL451 or pL452 vector and cloned between lox/FRT sites in pUC vector by restriction enzymes, RE1 and RE2. The sites for restriction enzyme (RE-3) creating the modification fragment used for bacterial recombination into the BAC clone are placed at the ends of modification. (**b**) Introduction of the MOI into the BAC clone. The modification fragment carrying MOI and a resistance cassette with lox/FRT sites (arrowheads) is recombined into the BAC clone via short homology arms (striped boxes)

2. Choose your desired species in the "search organism" window and your target region in the "search in the genome" field (*see* **Note 6**) and browse the genome.
3. Upon direction to the desired target region, you can zoom out or into the genomic location with the scale bar or move the shown region by the arrows in the control bar.
4. In the same control bar, there is a "Tracks" button with which you can "Configure Tracks" in a newly opened window. By choosing "Genomic Clones" you may select the preferred libraries and confirm by clicking "Configure".
5. The Genome Data Viewer now contains the BAC clones covering your target region of interest and shows their approximate size (Fig. 2a).

3.3 Verification of BAC Clone Genomic Location

Our preferred source for BAC clones is BACPAC Genomics (Emeryville, CA, USA), formerly BACPAC Resources at Children's Hospital Oakland Research Institute, as they provide BAC clones from most of the libraries explorable at NCBI (Table 1). BAC clones are provided in the *E. coli* strain DH10B as an LB-agar stab culture, facilitating shipment at room temperature

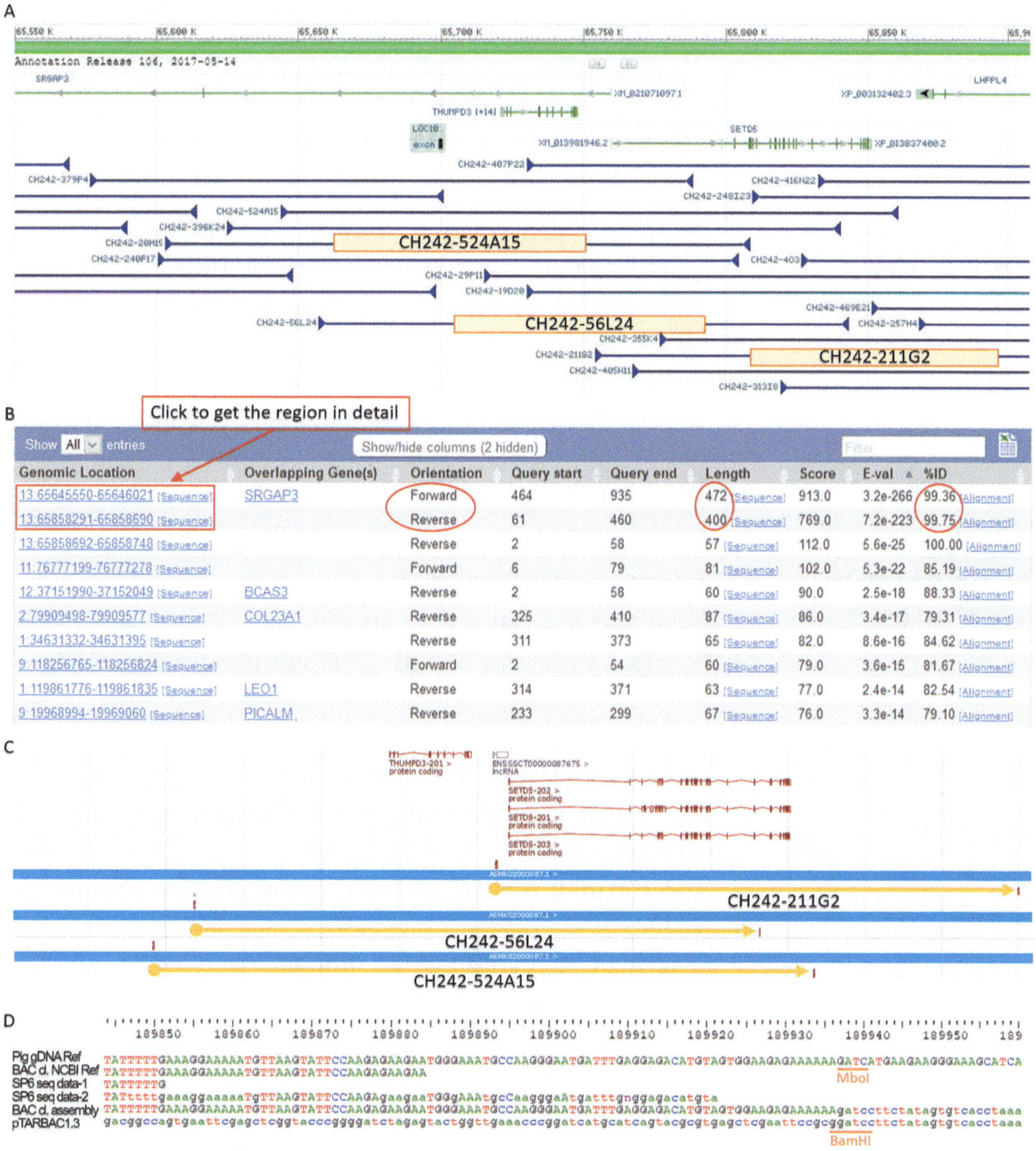

Genomic Location	Overlapping Gene(s)	Orientation	Query start	Query end	Length	Score	E-val	%ID
13:65645550-65646021 [Sequence]	SRGAP3	Forward	464	935	472 [Sequence]	913.0	3.2e-266	99.36 [Alignment]
13:65858291-65858690 [Sequence]		Reverse	61	460	400 [Sequence]	769.0	7.2e-223	99.75 [Alignment]
13:65858692-65858748 [Sequence]		Reverse	2	58	57 [Sequence]	112.0	5.6e-25	100.00 [Alignment]
11:76777199-76777278 [Sequence]		Forward	6	79	81 [Sequence]	102.0	5.3e-22	85.19 [Alignment]
12:37151990-37152049 [Sequence]	BCAS3	Reverse	2	58	60 [Sequence]	90.0	2.5e-18	88.33 [Alignment]
2:79909498-79909577 [Sequence]	COL23A1	Reverse	325	410	87 [Sequence]	86.0	3.4e-17	79.31 [Alignment]
1:34631332-34631395 [Sequence]		Reverse	311	373	65 [Sequence]	82.0	8.6e-16	84.62 [Alignment]
9:118256765-118256824 [Sequence]		Forward	2	54	60 [Sequence]	79.0	3.6e-15	81.67 [Alignment]
1:119861776-119861835 [Sequence]	LEO1	Reverse	314	371	63 [Sequence]	77.0	2.4e-14	82.54 [Alignment]
9:19968994-19969060 [Sequence]	PICALM	Reverse	233	299	67 [Sequence]	76.0	3.3e-14	79.10 [Alignment]

Fig. 2 Finding a suitable BAC clone for targeting a pig rosa locus. (**a**) Visualization of BAC clones from Chori-242 Library located in the region from 65.55 Mb to 65.95 Mb on Chr13 in NCBI database. Chosen BAC clones potentially carrying the rosa locus are marked in boxes. (**b**) Searching location of BAC clone CH242-524A15 based on BES data via Blast in Ensemble. The important information of both sequence ends (from T7 and SP6 primer) such a genomic location, their orientation and the procentual identity (ID) are marked by red frame. The table is extracted from Ensemble. (**c**) Location of BAC clones in porcine genome (Sus scrofa 11.1) via Blast of BES in Ensemble. BAC clone ends are represented as ● for T7 primer sequence and → as the SP6 primer sequence end. Location of upstream (thumpd3) and downstream (setd5) genes is shown above in the window of region from 65.6 Mb to 65.93 Mb on Chr13. (**d**) Assembly of BAC clone CH242-56L24 based on porcine reference genomic sequence, BES from SP6 and T7 primers and sequence of BAC vector (pTARBAC1.3). The SP6 end is displayed in alignment. The restriction sites used for construction of BAC clone are underlined

(RT) around the globe. Incubation temperature is always at 37 °C, the antibiotic resistance encoded on the BAC backbone is, to the best of our knowledge, always cam. For this, we simplify the general term "the appropriate antibiotic" with "cam". Upon arrival, each BAC clone should be prepared for long-term storage before it is verified for its identity.

3.3.1 Long-Term Storage of the Received BAC Clones

1. From each stab culture, create a single cells streak on an LB-cam agar plate and incubate plates overnight (o/N).
2. Pick 4 single bacterial clones and inoculate them into a 2.5 mL LB-cam medium.
3. Shake at 180 rpm up to 16 h.
4. Transfer 900 μL o/N-inoculum of each clone into a cryo vial, add 300 μL of 60% glycerol, mix by slowly pipetting up and down, and store at −80 °C.

3.3.2 Isolation of BAC Clones by Basic Alkaline Lysis Plasmid Miniprep

1. Use 5 mL o/N-bacterial culture containing BAC clone and growing in LB-cam medium.
2. Centrifuge at 1150 × *g* for 10 min.
3. Resuspend pellet in 750 μL STE and transfer into a 1.5 mL reaction tube.
4. Centrifuge at 4600 × *g* for 5 min.
5. Resuspend pellet in 200 μL Plasmid A.
6. Add 400 μL Plasmid B, mix by inverting six times, incubate on ice for 5 min.
7. Add 300 μL Plasmid C, mix by inverting six times, incubate on ice for 3 min.
8. Centrifuge at 16,400 × *g* for 10 min.
9. Transfer the supernatant to a new 1.5 mL reaction tube.
10. Add 4 μL RNase A (20 mg/mL), mix by inverting, and incubate at 37 °C for 45 min.
11. Add 300 μL of PCiA (*see* **Note 7**), mix by shaking for 2 min, and centrifuge at 16,400 × *g* for 2.5 min.
12. Transfer the upper, aqueous layer to a new 1.5 mL reaction tube.
13. Add 650 μL iPrOH, mix by shaking for 2 min, and centrifuge at 16,400 × *g* for 10 min.
14. Remove the supernatant, add 700 μL of 70% EtOH, and incubate at 4 °C o/N.
15. Centrifuge at 16,400 × *g* for 2.5 min and remove the supernatant.

16. Air-dry the pellet for 3 min and resolve in 100 μL T-Buffer. Resolving can be supported by incubation in a water bath at 42 °C for 1 h with occasional tapping.
17. Measure concentration (*see* **Note 8**).

3.3.3 BAC Clones Fingerprinting

1. Prepare a total volume of 40 μL per reaction:

   1 μL of restriction enzyme (*see* **Note 9**).

   4 μL of the appropriate buffer.

   25–40 μg of isolated DNA (*see* **Note 8**).
2. Digest o/N at 37 °C.
3. Separate the DNA fragments in 0.5–0.6 % agarose gel by gel electrophoresis. Run the gel electrophoresis for 1 h at 10 V, then for another 6 h at 60 V or longer if necessary to separate fragments of similar size (*see* **Notes 10** and **11**).

3.3.4 BAC-End Sequencing

The positioning of clones in the BAC maps is approximate, based on their assembly by restriction digest fingerprinting. In the case a sequence has been deposited for a given clone to any database, it often comprises an assembly of unordered and incomplete pieces (Fig. 2c, d). For a precise localization of the purchased clones we, therefore, routinely sequence the ends of BACs from both sides with the Sanger method, using primers binding near the cloning sites in the backbones. Then, we BLAST the respective reference genome with the obtained sequence. End sequencing requires high purity of the BAC, but with the following protocol, we commonly achieve >300 bp.

1. Mix 50–150 μg of isolated DNA from miniprep (3.3.2) in a volume of 50 μL (*see* **Note 8**) with 25 μL of PEG-$MgCl_2$ by sufficient mixing by pipetting up and down.
2. After 10 min incubation at RT, spin down the DNA at 16400 × *g* for 20 min and wash the pellet with 70% EtOH o/N. Resolve DNA in 10–20 μL T-Buffer.
3. Mix sequencing reaction: 1–5 μL DNA (take a minimum of 1 μg BAC DNA per 10 μL reaction, better 2–3 μg DNA per reaction), 5 pmol primer (*see* **Note 12**), 1 μL BigDye 3.1, 4 μL buffer and fill in dH_2O up to 10 μL.
4. Program: 95 °C for 5 min, 95 °C for 30 sec, 50 °C for 10 sec, 60 °C for 4 min, step 2–4 repeat 50×, 4 °C for 10 min.
5. Purify DNA with EtOH precipitation: Add 2.5 μL of 125 mM EDTA to each sequencing reaction, mix with 30 μL 100% EtOH by pipetting up and down several times. Incubate for 15 min on ice, then centrifuge 30 min at 16,650 × *g* at 4 °C. Remove supernatant and add 150 μL of 70% EtOH. Vortex, centrifuge 2.5 min at 15,900 × *g*. Remove carefully absolutely all supernatant and air-dry pellet for 6 min. Resolve in 30 μL dH_2O for separation on capillary electrophoresis.

3.3.5 Determination of BAC-End Sequences in the Reference Genome

1. Reference genomes are available at several sites on an open-source basis. We prefer and recommend using Ensembl (http://www.ensembl.org/index.html), but other sources should work as well. At ensembl.org, the command line has a "BLAST/BLAT" button tool, directing you to the genome search side. Click the "New job" button, paste your sequence (*see* **Note 13**), and select the appropriate database at "add/remove" species (*see* **Note 14**). Use the default settings for the other parameters and "Run".
2. BLAT will provide an output list for the best matches. Get the list of complementary sequences located on the genome in a new window (Fig. 2b). The data can be re-assorted for a number of parameters. We specifically consider:
   (a) genomic location—the location of the BAC-ends should at least approximately correlate to the chromosomal position of the BACs in the NCBI Genome Viewer Database.
   (b) orientation—both end sequences must be in opposite directions.
   (c) length and %ID—the end sequences are assumed to be (almost) identical to the genome in their entire length. Single nucleotide polymorphism or short insertion/deletion might, however, decrease the % ID below 100%. In the case the identity falls below 90%, you are confronted with an artificial matching.
3. Clicking the link with genomic localization directs you to a new window showing the genomic localization of the BAC-end sequences. The "Region in detail" window illustrates the adjacent region of the BAC locus, confirming the match. By using the genomic localization of both BAC-end sequences, you can specifically zoom into the "localization" covered by the BAC (Fig. 2c).
4. The entire BAC sequence can be extracted from the reference genome by clicking the blue "contig" bar and choosing the "export primary assembly sequence/features" option. When a new window pops up you "Select location" by indicating the chromosome number and the terminal positions of the BAC-ends. Clicking "Next" provides the choice of the "Text" option for providing the entire BAC sequence in a .fasta format.
5. Copy-pasting the sequence into an appropriate viewer facilitates the comprehensive analysis of the region covered by the BAC (*see* **Note 15**).

3.4 Modification of BAC Clone by Bacterial Recombineering

Our bacterial recombination protocol is based on the "recombineering" system, initially described by [14] and modified by [15]. The BAC clone selected has to be transferred into a SW *E. coli* strain derivate and single cell clones are then prepared for recombination of MOI (*see* **Note 16**).

3.4.1 Making Electrocompetent Cells

1. Inoculate a 1 mL o/N-culture with the desired SW strain, incubate it at 180 rpm at 32 °C.
2. Transfer 1 mL of o/N-culture in 50 mL LB-medium in a 500 mL Erlenmeyer flask. Continue with incubation at 32 °C and shaking at 180 rpm until $OD_{600} = 0.6–0.8$.
3. Cool cell suspension with occasional shaking on ice for 10 min.
4. Transfer cell suspension into 50 mL-tube and centrifuge them for 10 min at 2200 × *g* at 4 °C. Continue with additional centrifugations at 4 °C to wash cells. Besides, always keep cells on ice (*see* **Note 17**).
5. Resuspend cell pellet in 25 mL ice-cold ddH_2O and centrifuge for 10 min at 2800 × *g*.
6. Resuspend cell pellet in 12.5 mL ice-cold ddH_2O and centrifuge for 10 min at 3200 × *g*.
7. Resuspend cell pellet in 5 mL ice-cold 10% glycerol and centrifuge for 10 min at 4200 × *g*.
8. Resuspend in 0.5 mL ice-cold 10% glycerol.
9. Prepare aliquots of 80 μL. Cells can be used directly or stored at −80 °C.

3.4.2 Transfer BAC Clone into Recombineering-Competent *E. coli* Strain

1. Prepare materials: pre-cool cuvettes and cell aliquots on ice. Pre-warm 1 mL SOB aliquots in 1.5 mL reaction tubes at 32 °C.
2. Prepare reaction tube with BAC DNA amount proportionate to isolation method. BAC integrity should have been checked before electroporation by restriction digest.
3. Mix 80 μL of electrocompetent cells (as prepared in 3.4.1) with 1–6 μL DNA containing 3–6 μg by pipetting 2–3× up and down and transfer it into the cuvette.
4. Dry the cuvette on the outside with a paper wipe (*see* **Note 18**), put it into an electroporator, and apply a 1.75 kV pulse. Make notes of the actual voltage and time applied (*see* **Note 19**).
5. Add 1 mL SOB to wash cells from the cuvette and transfer cells back to the reaction tube (*see* **Note 20**).
6. Incubate the cells at 32 °C for 1 h and clean the cuvette (*see* **Note 21**).
7. Centrifuge cells for 5 min at 2350 × *g*.

8. Remove most of the supernatant (leave 100–200 μL in the tube).
9. Resuspend cells and plate on appropriate antibiotic LB-agar plates.
10. Incubate plates for 24–48 h (*see* **Note 22**).

3.4.3 Preparation of Modification as a Linearized dsDNA Fragment

1. Excise the constructed and verified modification vector (3.1) from the plasmid backbone by appropriate restriction enzyme (s) (*see* **Notes 23** and **24**).

2.1 For purification of modification fragments <10 kb:

(a) Separate modification fragment from other elements of the plasmid on 0.6–0.7% agarose.

(b) Excise the modification fragment from the gel and extract DNA from agarose by conventional column-based DNA kit and elute in TE Buffer (*see* **Note 25**).

2.2 For purification of modification fragments >10 kb:

(a) Separate restriction enzyme digest on 1% low melting agarose.

(b) Excise modification fragment, transfer it into a 1.5 mL reaction tube, weigh and completely resolve agarose, normally 12 min at 70 °C is sufficient.

(c) Equilibrate the sample to 42 °C for 5 min, add agarose (1 U/100 mg gel) and digest for 30 min at 42 °C.

(d) Centrifuge undigested carbohydrates at 15,000 × *g* for 10 min.

(e) Add 1/10 volume 3 M NaOAc 5.2 pH and extract with 200 μL PCiA by slowly inverting the reaction tube for 2 min.

(f) Centrifuge for 2 min at 15,000 × *g* and transfer the upper aqueous phase into the new reaction tube.

(g) Repeat the PCiA extraction twice or until the interphase is clear.

(h) Add 2.5× volumes of 100% EtOH, mix gently, and incubate for 30 min at −80 °C.

(i) Centrifuge at 15,000 × *g* for 30 min at 4 °C and wash the DNA pellet with 70% EtOH o/N.

(j) Resolve the DNA in 10 μL T-Buffer.

3. Determine the concentration of the excised fragment and run a small aliquot on agarose gel for verification of its integrity.
4. Use modification fragment directly for bacterial recombineering into BAC clone (*see* Subheading 3.4.4.) or store DNA fragment at −20 °C.

3.4.4 BAC Recombineering

This step brings together the desired BAC and the modification fragment in the recombineering competent SW cells. The protocol therefore largely refers to previously conducted steps, with some essential variations.

1. Prepare o/N-culture with an SW-clone carrying the desired BAC, and inoculate 100 mL culture and grow it to $OD_{600} = 0.4–0.5$ according to Subheading 3.4.1.
2. Transfer 100 mL of culture to a 42 °C water bath and keep it for 15 min with constant shaking to induce the expression of λ-Red proteins.
3. Cool on ice for 10 min with occasional shaking and prepare electrocompetent cells according to Subheading 3.4.1, **step 3**.
4. For a 5 kb modification fragment, electroporate 100 ng excised vector DNA from Subheading 3.4.3 into 80 μL recombineering-electrocompetent cells according to Subheading 3.4.2, with the exception that recovery time in SOB at 32 °C after electroporation needs to be 2 h instead of 1 h to guarantee proficient recombination (*see* **Note 26**).
5. Plate the cells on agar carrying the appropriate antibiotic (*see* **Note 27**) and grow them for a maximum of 48 h. Prepare back-up streaks on a plate with the same antibiotic (*see* **Note 28**) to promote analysis of defined colonies.

3.4.5 Arabinose-Induced FLPE and Cre Recombination

In addition to the heat-induced bacterial recombination, the SW105 and SW106 strains provide the opportunity of inducible FLPE and Cre recombinases. This might be used for exchanging or deleting a resistance cassette, in the case it is flanked by appropriate FTR or lox sites. Exchange of resistance cassettes can be carried out according to 3.4.4, using a linearized fragment carrying an alternative resistance cassette, flanked by FTR or lox sites according to the cassette in the BAC. Deletion of a selection cassette can be done according to a simplified protocol.

1. Inoculate 3 mL SOB-cam medium with bacterial cells containing BAC in which the region should be removed.
2. Reach the $OD_{600} = 0.2–0.3$. Measure the cell culture density by spectrophotometer.
3. Add 10 μL arabinose (100 mg/1 mL) per 1 mL inoculum.
4. Incubate the culture for 1 h with shaking and transfer bacteria on LB-agar plates by making a streak of single cells or plating a small volume of 5 μL culture diluted in LB-medium on LB-cam.
5. Pick single colonies (can be four colonies per inoculum as the Cre recombination is very effective) to create a back-up LB-agar plate and 2.5 mL o/N-culture for checking them by relevant restriction digest.

3.5 Verification of BAC Recombinant

Multiple approaches can be followed to verify the correct recombination of the modification fragment into the BAC. We commonly follow a combination of end-point PCR, BAC fingerprinting, and Sanger sequencing. A full BAC exploration by next-generation sequencing is optional but still costly and time-consuming.

3.5.1 End-Point PCR

1. Prepare 100 μL of T-Buffer in PCR tubes.
2. Transfer a visible amount of bacterial colony/clone into the PCR tube (*see* **Note 29**). Make sure to wash the bacteria off thoroughly.
3. Disrupt bacteria by 10 min at 95 °C, 5 min at 4 °C
4. Centrifuge at 2000 × *g* for 10 min to pellet cell debris.
5. For standard end-point PCR reaction, use 2 μL of the supernatant as DNA template and primer pairs flanking both HAs. *See* schematic illustration in Fig. 3b.

3.5.2 BAC Clone Fingerprinting

According to Subheading 3.3.3, any enzyme can be chosen for checking the general integrity of the BAC. As the modification normally affects only a rather small region of the BAC, only a few enzymes will indicate whether the modification has been correctly integrated.

1. Perform in silico digestion of the original BAC sequence and the BAC sequence comprising the desired modification for enzymes producing 30–40 bands.
2. Compare digestion patterns and choose an enzyme that produces a significant difference between the two constellations (*see* **Note 30**).
3. Perform fingerprinting restriction enzyme digest and run it on an agarose gel, according to Subheading 3.3.3 (Fig. 3a).

3.5.3 Sanger Sequencing

The procedure can be carried out according to Subheading 3.3.4, albeit with primers located within the BAC. We recommend verifying the sequences across the HA and across the essential components of the modification (Fig. 3b).

3.6 Preparing of BAC Targeting Vector for Nucleofection into Cells

Once when the BAC clone has been verified to contain all modifications, it can be prepared for transfection into vertebrate cells. Sufficient amounts and purity of BAC vectors can be produced by using a commercially available Large Construct Kit (QIAgen). For homologous recombination in somatic cells, the BAC needs linearization and co-transfection with CRISPR/Cas components.

1. Use a large construct kit and follow the manufacturers' instructions. Resolve the pellet in an appropriate amount of TE Buffer, for example, 100 μL in the case of 250 mL o/N-culture. Following the protocol, the concentration of circular BAC is

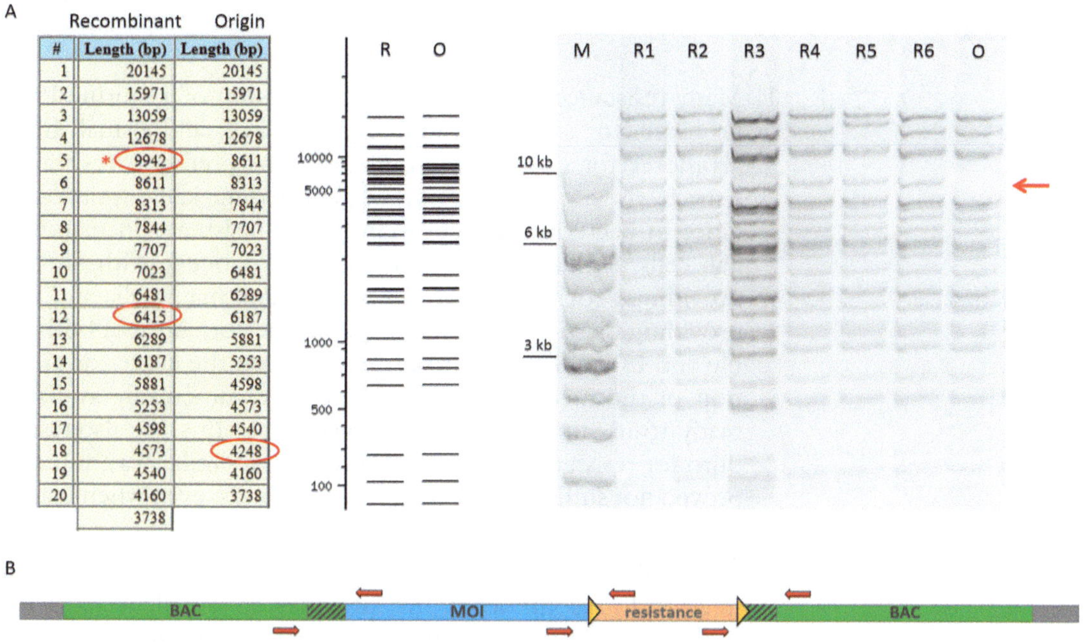

#	Recombinant Length (bp)	Origin Length (bp)
1	20145	20145
2	15971	15971
3	13059	13059
4	12678	12678
5	* 9942	8611
6	8611	8313
7	8313	7844
8	7844	7707
9	7707	7023
10	7023	6481
11	6481	6289
12	6415	6187
13	6289	5881
14	6187	5253
15	5881	4598
16	5253	4573
17	4598	4540
18	4573	4248
19	4540	4160
20	4160	3738
	3738	

Fig. 3 Verification of modified BAC recombinants. (**a**) BAC clones fingerprinting. The digest patterns of the modified BAC recombinant (R) and origin BAC clone (O) are created in silico and compared to find the distinguishing fragments (right, the fragments are circled). The digestion patterns are also visualized by simulation of separation in agarose gel (the middle) displaying the significance of distinguishing fragments. The fragment of 9942 bp (star) mainly significant for BAC modification is missing in the digest pattern of origin BAC clone (left). This fragment is also detected after DNA fragments separation of digested origin BAC clone (O) and modified BAC recombinants (R1-R6) by agarose gel electrophoresis (right). The other two distinguishing fragments are so close to neighbor fragments that they are not separated on agarose gel. (**b**) Position of primers verifying the correct location of modification of interest (MOI) in modified BAC recombinants by end-point PCR and Sanger sequencing. The primers (arrows) are placed at positions to amplify and sequence the homology arms (striped boxes) and the MOI-resistance cassette border in modified BAC recombinants

mostly 200–600 ng/μL. Linearize 20–25 μg of BAC with an appropriate enzyme o/N at 37 °C in a total volume of 200 μL (*see* **Note 31**).

2. Add 20 μL 3 M NaOAc 5.2 pH and extract with 200 μL PCiA by slowly inverting the reaction tube.
3. Centrifuge at 16,400 × *g* for 2 min and transfer the upper aqueous phase into a new tube (*see* **Note 32**).
4. Add 400 μL of 100% EtOH, mix gently, and incubate at −80 °C for 30 min.
5. Centrifuge at 16,400 × *g* for 30 min at 4 °C. Carefully remove supernatant and wash DNA pellet in 70% EtOH at 4 °C o/N.
6. Centrifuge at 16,400 × *g* for 2 min, carefully remove the supernatant, air-dry for 2–3 min, and resolve in 11 μL TE buffer (*see* **Note 33**).
7. Determine concentration and mix with linearized BAC with CRISPR/Cas components (*see* **Note 34**).

4 Notes

1. Many resistance cassettes use a combination of a murine PGK promoter for expression in mammalian cells and a small EM7 promoter for expression in *E. coli*. Our preferred resistance gene is the acetyltransferase neo, providing kanamycin resistance in bacteria and neomycin/G418-resistance in mammalian cells. Similarly, we had sufficient experience with Sh ble, providing zeocin-resistance in bacteria and mammalia and bsd providing blasticidin resistance. As the performance of both Sh ble and bsd in *E. coli* critically depends on the exact composition of the medium, we make use of commercially available ready-to-use media or agar-media compositions. We also tested puro for resistance against puromycin in a similar way, but this proved not sufficient in bacteria in our hands, even when ready-to-use media were used.
2. We found it helpful to truly assemble the modification at a nucleotide resolution as this provides an excellent basis for analysis by restriction enzyme digest, designing end-point PCR, Sanger sequencing, estimation of splice sites, and open reading frames, etc. BioEdit [16] is our preferred program.
3. Common bacterial recombination protocols claim 50 bp of homologies as sufficient. We presume that this suggestion was based on the ability to create such elements from ssODN in a very cost-efficient manner. We mostly let HAs of 300 bp commercially synthesize, as gene synthesis has become very cost-efficient as well. Gene synthesis also allows us to generate both HAs in the same plasmid, separated by one or several sites for restriction enzymes. This facilitates the integration of MOI and resistance cassette. Further, this strategy provides the opportunity to place further restriction enzyme sites at the terminal ends of the HAs for the final release of the modification vector from the backbone (*see* Subheading 3.4.3).
4. Given the effective gene synthesis services available, the entire modification vector might be commercially synthesized. This would simplify the procedure, but to our assumption, the integration of available components such as resistance cassettes of some 2 kb size or larger from the existing plasmid by cloning is still cheaper than its full synthesis for each new project.
5. Ideally, the entire modification plasmid should be verified by Sanger sequencing. Larger parts that have been acquired from one source, either an established plasmid or a verified gene synthesis, however, we normally do not sequence completely. Rather we make use of several fingerprinting restriction enzyme digests and the verification of the terminal ends of such cassettes by Sanger sequencing.

6. We assume that mostly a gene is the desired target region. As most mammalian genes are properly annotated in the meanwhile, giving the gene name should direct you to the desired locus. In the case the gene name is not known in the species of interest or an intergenic region is the target region, we recommend searching for neighboring genes and to explore the target region by zooming or moving with the control bar. The search for these neighbor sites can be done in the species of interest or, as most loci appear conserved among mammals, in another species; for sure, the information density on genome annotation is highest in human. The current nomenclature of gene names can be explored by using the "gene" database at NCBI (https://www.ncbi.nlm.nih.gov/gene), which recognizes also older names or aliases.
7. Phenol is nowadays often seen critical for its harmful properties. Following alternative protocols (https://bacpacresources.org/protocols.htm), its usage can be avoided, but the DNA is then more impure, which affects resolving and electroporation into bacteria. The sample quality does, however, not interfere with restriction digest and visualization of band patterns by gel electrophoresis. We use the simplified protocol regularly for fingerprinting of BAC clones, but in this case, we resolve the DNA pellet isolated from only 2.5 mL o/N-culture in 100 μL T-Buffer.
8. In a common photometer, we normally detect 1000–2500 ng/μL from a 5 mL o/N-culture resolved in 100 μL. This is, however, totally misleading as the method co-precipitates disrupted chromosomal DNA of the *E. coli* genome to a large extent. A more realistic assumption is that you harvest some 1–2 μg of BAC DNA (<25 ng/μL in 100 μL). Alternative methods might avoid this complication, but they are more complicated to conduct. As co-precipitation of *E. coli* DNA does, in principle, not affect the subsequent steps, we stick to the incorrect but measurable values and talk about "isolated DNA" rather than "BAC DNA."
9. We found it appropriate to digest BACs in a way that 30–40 bands appear, correlating with a mean band size of approximately 3–6 kb. This fits with many of the commonly used enzymes such as *Hin*DIII, *Spe*I, *Hpa*I, *Eco*RI, and *Xba*I. Our preferred choice for initial fingerprinting is mostly *Eco*RI.
10. Fingerprinting of BACs was initially performed with pulsed-field electrophoresis for creating BAC maps. We found it appropriate to separate band sizes from 3 to 50 kb within this much simpler setting. Care needs to be taken that combs are cleaned with detergent before placing them in the agarose to avoid disruption of the slot surface with debris. Usage of fresh electrophoresis buffer is recommended as well.

11. Visualization of DNA bands with EtBr performed outstandingly well when adding it to agarose that has been cooled to 60–65 °C after boiling. Similarly, the less harmful GelRed can be used, but care must be taken that GelRed has to be used at a concentration of 2.5 μL staining dye to 100 mL agarose gel. GelRed has to be kept in the fridge and cannot be mixed directly into the sample. Otherwise, bands from a BAC digest will run at a different speed and the patterns will be diffuse.
12. On the commonly used BAC backbone, the insertion site of the genomic region is flanked by primer binding sites for T7 (5′-GTAATACGACTCACTATAGGG-3′) and SP6 (5′-ATT TAGGTGACACTATAGAA) (Table 1).
13. The BLAST at Ensembl allows the parallel search for several sequences, for example, the 5′-end and the 3′-end of a BAC. This can be done either by choosing "add more sequences" or, very proficient in our hands, by inserting the 5′-end and 3′-end sequence in the same window, separated by "NNNN". Treating both sides as a single sequence helps if one of the end sequences is completely within a highly repetitive segment. In this case, the connection of the sequence to the complementary end will prioritize the identification of the correct locus in the output list. Theoretically, both end sequences might match highly repetitive elements in the genome, but in practice, this did never happen in our hands.
14. For most of the species, multiple genomes are meanwhile available, but quality often differs substantially. For comparison to other databases such as NCBI, we recommend choosing the respective reference genome. This is, by now, GRCH38 for human, GRCm39 for mouse, ARS-UCD1.2 for cattle, Sscrofa11.1 for pig, Oar_rambouillet_v1.0 for sheep.
15. We normally use BioEdit for identifying the BAC-end sequences and the restriction enzyme sites used for cloning the fragment into the BAC backbone (Fig. 2d). Further, the position of genes and their exons can be defined. For confirming the restriction enzyme digest pattern from the fingerprinting analysis, the BAC sequence should be combined with the sequence of the vector backbone and examined for the used restriction enzyme sites (http://nc2.neb.com/NEBcutter2/index.php).
16. The most recent generation of recombineering provides three main *E. coli* strains: SW102, SW105, and SW106. While all of them provide a heat-inducible λ-Red recombination system comprising of the λ-encoded exo, bet, and gam components under a temperature-sensitive repressor. For this cells have to be propagated only at 32 °C, except when the λ-Red system is switched at 42 °C. Strains SW105 and SW106 contain FLPE or

Cre genes in addition. In both cases, the genes can be induced by arabinose.

17. Avoid repetitive centrifugation of cell suspensions. In the case you have more than 50 mL o/N-culture, separate it onto several tubes. Otherwise, the competent cells are of a low quality which might impair electroporation efficacy.
18. Precisely remove any traces of liquid as this might result in a short-cut and therefore loss of the reaction.
19. In our Eppendorf electroporator (Eporator), we usually get a variation of voltage between 1.65 and 1.69 kV, and a pulse time of 4.3–4.9 m. When checking the quality of new batches of electrocompetent cells by applying a pulse to an empty cell aliquot, a pulse time > 5.0 m indicates good quality of cells. When using electrocompetent cells of lower quality, clones might still arise, albeit at decreased numbers.
20. The recovery of cells in SOB after electroporation is a critical step. It is said that each additional 30 s of keeping cells in the original suspension might decrease viability by half. Moreover, the usage of pre-warmed medium and placing the cells at 32 °C is seen as an additional heat-shock, supporting the transfer of BAC DNA into the cells. Speed is therefore essential and we normally do not make use of special tips or Pasteur pipets to transfer the cells out of the cuvette. Rather, we spill 1 mL of pre-warmed SOB into the cuvette, pipet up and down 2–3 times, and with the same tip transfer it back to the reaction tube and the reaction tube into the incubator. The advantage of fast processing should overcome the detrimental aspects of leaving small volumes of cell suspension in the cuvette. For convenient and fast work, we placed the electroporator next to the incubator.
21. Cuvettes can be reused several times until visible damage. Washing of the cuvette dH_2O 10 times is sufficient, but after the end of an electroporation session, we additionally sterilize it for 30 min under UV light.
22. The efficacy of BAC transfer into SW strains strongly depends on the quality of BAC DNA. We only get few colonies when using the simplified protocol of basic alkaline lysis, yield is increased to 20–30 bacterial colonies if the described phenol/chloroform extraction step is included. More than 100 clones can be obtained when BACs are isolated by commercially available time-consuming and costly column-based kits.
23. For normal-sized modifications (approx. 5 kb), we digest 6 μg plasmid from column-based plasmid isolations or 15 μg DNA from plasmid minipreps. Without columns). When modifications of 10 kb are prepared, we use around 40 μg plasmid DNA.

24. In this case, the modification fragment is of similar size as the plasmid backbone, a clear separation of the two bands might be challenging. We found it helpful to use additional restriction enzymes in the digest that cut the backbone into two (or more) fragments. Importantly, separation of bands on agarose is impaired, when DNA concentration is high in the restriction enzyme digestion reaction. It is often helpful to increase the volume of the digest before loading onto the gel, preferentially by filling one or several slots almost completely. Further, the usage of GelRed as DNA stain dye often results in incomplete separation of bands. We normally load GelRed directly into the sample.
25. To get the maximum DNA amount from the purification column, elution should be carried out with pre-warmed (i.e., 70 °C) elution buffer, incubate membrane with elution buffer for 5 min, and elute membrane 2× with the elution buffer. If necessary, concentrate the DNA by EtOH precipitation in a smaller volume. Add 1/10 volume 3 M NaOAc and 2.5 volumes 100% EtOH. Invert a few times and incubate for 30 min at −80 °C. Centrifuge (16,400 × *g*, 30 min, 4 °C) and wash with 70% EtOH o/N.
26. The amount of DNA can be adapted, according to fragment length. We successfully used DNA amounts of 60–1000 ng. It is important, however, to keep the DNA volume lower than 10% of the cell suspension volume of 80 μL.
27. For the growing cells that underwent bacterial recombination, it is essential to select for the resistance encoded by the modification vector, for example, kanamycin, blasticidin, and zeocin.
28. After selection for and verification of the modification, selection can be switched back to cam. In contrast to older protocols that claim this impossible, we do recover cells for 2 h in LB without antibiotic and then plate them on cam. It is particularly recommended to keep bacteria under zeocin selection only temporarily.
29. End-point PCR can be immediately driven on material from the original clone, but it is necessary to prepare a back-up streak on a second plate to regenerate enough material for further analysis.
30. We found it appropriate to choose an enzyme that produces a band of 6–12 kb that is appearing only in the original or the modified BAC. For sure this band should be clearly separated from the other fragments of the digestion pattern. Smaller bands are often difficult to visualize, in the case of larger bands, it is often difficult to estimate the band size. Ideally, two distinct digestion approaches are selected and compared.

31. The appropriate enzyme should produce extended HAs to your MOI. We found it convenient to cut the final BAC targeting vector with *Not*I, whose sites flank the genomic segment in the BAC clone of the most frequent libraries in NCBI and appear very rarely in the genome (Table 1). The second option can be AscI, which, unlike NotI, only opens the vector. However, you must always check that the enzyme you have selected does not affect your MOI.
32. The transferring of the aqueous phase needs to be done with absolute caution to not have the remainder of phenol in the DNA sample. It could be added next extraction with chloroform to remove phenol residue, but the DNA loss can be too big after two extractions. Unfortunately, we have not seen an evident improvement in cell response to nucleofection.
33. Following the protocol, the concentration of linearized BAC is mostly 1000–1500 ng/μL. For further processing, we recommend using fresh BAC DNA (1–2 days old). Longer storage in the fridge or freezing can harm the quality of DNA samples. For longer storage, it is better to keep DNA as a pellet in 70% EtOH.
34. Linearized BAC can be combined with distinct constellations of CRISPR/Cas, either applied as ribonucleoprotein, RNA, or plasmids expressing gRNA and Cas9. For us, it appeared most flexible to use separate plasmids comprising Cas9 and gRNA elements [17], with the latter being commercially synthesized under a human U6 promoter. A common composition of BAC with CRISPR/Cas plasmid for nucleofection into somatic cells is 250 ng/μL of Cas9 plasmid, 250 ng/μL of gRNA-expressing plasmid, and 500 ng/μL of linearized BAC in 5 μL reaction.

References

1. Shizuya H et al (1992) Cloning and stable maintenance of 300-kilobase-pair fragments of human DNA in Escherichia coli using an F-factor-based vector. Proc Natl Acad Sci U S A 89(18):8794–8797
2. Lander ES et al (2001) Initial sequencing and analysis of the human genome. Nature 409(6822):860–921
3. Archibald AL et al (2010) Pig genome sequence--analysis and publication strategy. BMC Genomics 11:438
4. Bovine Genome S et al (2009) The genome sequence of taurine cattle: a window to ruminant biology and evolution. Science 324(5926):522–528
5. Wheeler DA et al (2008) The complete genome of an individual by massively parallel DNA sequencing. Nature 452(7189): 872–876
6. Genomes Project, C et al (2010) A map of human genome variation from population-scale sequencing. Nature 467(7319): 1061–1073
7. Jain M et al (2018) Nanopore sequencing and assembly of a human genome with ultra-long reads. Nat Biotechnol 36(4):338–345
8. Gordon D et al (2016) Long-read sequence assembly of the gorilla genome. Science 352(6281):aae0344

9. Zhang Y et al (1998) A new logic for DNA engineering using recombination in Escherichia coli. Nat Genet 20(2):123–128
10. Muyrers JP et al (1999) Rapid modification of bacterial artificial chromosomes by ET-recombination. Nucleic Acids Res 27(6): 1555–1557
11. Murphy KC (1998) Use of bacteriophage lambda recombination functions to promote gene replacement in Escherichia coli. J Bacteriol 180(8):2063–2071
12. Yu D et al (2000) An efficient recombination system for chromosome engineering in Escherichia coli. Proc Natl Acad Sci U S A 97(11): 5978–5983
13. Vochozkova P et al (2019) Gene editing in primary cells of cattle and pig. Methods Mol Biol 1961:271–289
14. Lee EC et al (2001) A highly efficient Escherichia coli-based chromosome engineering system adapted for recombinogenic targeting and subcloning of BAC DNA. Genomics 73(1): 56–65
15. Warming S et al (2005) Simple and highly efficient BAC recombineering using galK selection. Nucleic Acids Res 33(4):e36
16. Hall TA (1999) BioEdit: a user-friendly biological sequence alignment editor and analysis program for Windows 95/98/NT, Nucleic acids symposium series. Information Retrieval Ltd., London, pp c1979–c2000
17. Osoegawa K et al (2001) A bacterial artificial chromosome library for sequencing the complete human genome. Genome Res 11(3): 483–496
18. Rogel-Gaillard C et al (1999) Construction of a swine BAC library: application to the characterization and mapping of porcine type C endoviral elements. Cytogenet Cell Genet 85(3–4):205–211
19. Fahrenkrug SC et al (2001) A porcine BAC library with tenfold genome coverage: a resource for physical and genetic map integration. Mamm Genome 12(6):472–474
20. Eggen A et al (2001) Construction and characterization of a bovine BAC library with four genome-equivalent coverage. Genet Sel Evol 33(5):543–548
21. Dalrymple BP et al (2007) Using comparative genomics to reorder the human genome sequence into a virtual sheep genome. Genome Biol 8(7):R152

Chapter 5

Immortalised Cas9-expressing Cell lines for Gene interrogation

Luis F. Malaver-Ortega and Joseph Rosenbluh

Abstract

The ability of modifying the genome of multiple species, precisely and without or minimal off-targeted effects, have opened numerous opportunities for the biotechnology industry. In this chapter, we describe an easy to establish, robust, and practical pipeline that can be used to generate immortalized cell lines, from different tissues, to capture cell linage context and validate the tools required for genome editing and genetic modification. This pipeline serves as a reference for similar approaches for gene interrogation in other species.

Key words Genome editing, CRISPR/Cas9, Cell immortalization, Cas9 assay

1 Introduction

The development of CRISPR/Cas9 for genome editing has enabled multiple applications in biotechnology. Systematic interrogation of gene Knock Out (KO)-induced phenotype requires a robust and reliable cell line that expresses constitutively Cas9 enzyme, and it is not limited in the number of passages that can be maintained. Immortalization of primary cells requires human telomerase reverse transcriptase (hTERT) [1, 2]. However, many primary cells require in addition to hTERT inhibition of other cellular pathways such as TP53 and the RB pathways [1]. SV40 large T antigen derived from the polyomavirus SV40 inhibits both of these pathways [3] and enables immortalization of many type of primary cell lines [4–7]. In this chapter, we outline a basic pipeline that can be used to validate sgRNA guides, in Cas9-expressing immortalized human cell lines as a first step on the validation of genome edits. For design and delivery of guide RNA (sgRNA), the reader can refer to [8].

Paul John Verma, Huseyin Sumer and Jun Liu (eds.), *Applications of Genome Modulation and Editing*,
Methods in Molecular Biology, vol. 2495, https://doi.org/10.1007/978-1-0716-2301-5_5,

2 Materials

2.1 Lentivectors (See Note 1)

1. pRRLsin-SV40 T antigen-IRES-mCherry [7] (Addgene #58993) lentiviral vector for the expression of SV40 large T antigen and mCherry red fluorescent protein.
2. pLV-hTERT-IRES-hygro [9] (Addgene #85140) Lentiviral vector for expression of hTERT.
3. Lenti-Cas9-2A-Blast [10] (Addgene #73310) Lentiviral vector for expression of human codon-optimized S. pyogenes Cas9 and blasticidin resistance.
4. pXPR_011 [11] (Addgene #59702) Lentiviral vector for expression of EGFP and an sgRNA against EGFP. Used to test Cas9 activity.

2.2 Envelop and Packaging Plasmids (See Note 1)

1. pMD2.G (Addgene #12259) Envelop.
2. psPAX2 (Addgene #12260) Packaging.

2.3 Packaging Cell Line: 293FT Cell Line (Invitrogen) (See Note 2)

2.3.1 Culture Media

1. Adult Fibroblast Basic Medium (BM) and 293FT medium: 10% Fetal Bovine Serum, 2 mM GlutaMAX supplement, 10 μM Non-Essential Amino Acids (NEAA), 25 units/mL of Penicillin, and 25 μg/mL of Streptomycin in Dulbecco's Modified Eagle Medium (DMEM).
2. Virus collection medium: 30% Fetal Bovine Serum, 2 mM GlutaMAX supplement, 10 μM NEAA, 25 units/mL of Penicillin, and 25 μg/mL of Streptomycin in DMEM.

2.4 Equipment

1. Centrifuge (*see* **Note 3**).

3 Methods

This protocol involves the generation of viral vectors. It is necessary to consult and follow the local and institutional guidelines applicable to the use of recombinant viruses and animal experimentation. This protocol can be used as a general guide for establishing immortalized cell lines from diverse tissues in human and other species. The following sections describe the steps require to produce the lentivectors; transduction of the cells; and selection of the transduced cells, in order to establish the proposed pipeline as described in Fig. 1 (*see* **Note 4**). Isolation of primary fibroblast has been described previously [12].

3.1 Lentiviral Production

1. (Day 1) Seed six million 293FT cells per 10 cm dish in 10 mL of 293FT medium. Place dishes in an incubator and culture for 24 h.

Pipeline

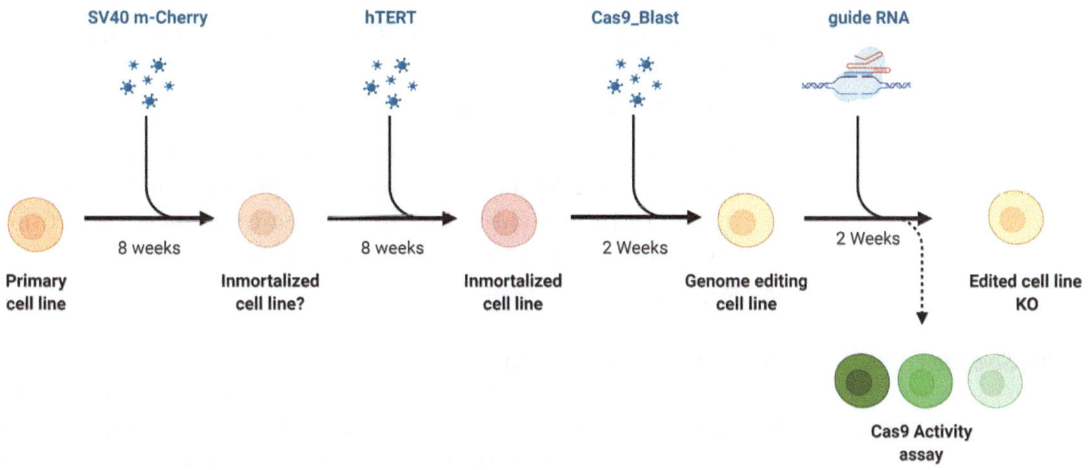

Fig. 1 Pipeline for generation of immortalized Cas9 cell lines

2. (Day 2) Prepare transfection mix (quantity for one dish): mix 1.4 μg VSV-G plasmid, 14 μg psPAX2 plasmid, and 12 μg lentivector (*see* **Note 5**) in 85 μL of Opti-MEM.
3. Into a different tube, dispense 103 μL Fugene6 in 1.5 mL of Opti-MEM.
4. Incubate maximum 5 min at room temperature.
5. Combine DNA mix with Fugene6/Opti-MEM mix.
6. Incubate 20 min at room temperature.
7. Dispense 1585 μL of transfection mix onto the cells (*see* **Note 6**). Rock dish back and forth to mix, then return the dish to the incubator.
8. (Day 3) Aspirate and discard 9 mL of the medium (*see* **Note 7**). Immediately add 15 mL of virus collection medium. Return to the incubator.
9. (Day 4) Aspirate 15 mL of virus-containing medium and transfer to a clean 50 mL tube. Immediately add 15 mL of virus collection medium. Return dish to incubator and store collected virus-containing medium at 4 °C.
10. (Day 5) Aspirate remaining virus-containing medium and combine with the virus-containing medium collected the day before. Mix well, aliquot into 2 mL tubes and store at −80 °C.

3.2 Lentiviral Infection

1. For each lentivector to be used, adding 100,000 freshly trypsinized targeted cells in 2 mL of BM medium with 8 μg/mL of polybrene in each well of a 12-well tissue culture plate. Prepare 4 wells.

2. Thaw the virus-containing medium at 37 °C and add 50, 500, and 1000 μL of the medium into 3 wells, respectively. The fourth well is used as negative control for selection.
3. Centrifuge the plate at 1064 RFC for 30 min at room temperature (*see* **Note 8**).
4. Place the plate back into an incubator.
5. 24 h post transduction, remove medium from all wells and replace with BM medium containing predetermined concentration of proper antibiotic. In the case of mCherry selection, use BM medium only without antibiotic.
6. Return the plate to an incubator.

3.3 Selection

Depending on the lentivectors, antibiotic selection or mCherry expression can be used for selection of positive transduced cells. Antibiotic concentration and maintenance concentration need to be determined by dose response experiments prior and for each batch of primary cell isolation. (*see* **Note 9**)

1. From 48 h post transduction, monitor for cell death or mCherry expression under a fluorescent microscope.
2. Death of untransduced cells will be apparent from around 5 days post transduction. Surviving cells should be clearly identifiable if the transduction was successful. Expand the cells in culture medium with the antibiotic.
3. For mCherry expressing cell selection, two rounds of FACS of the positive cell populations are recommended.

3.4 Cas9 Activity Assay

Once the cells have been transduced with lentivirus containing the bicistronic construct pXPR-011 encoding the sgRNA GFP and GFP protein, Cas9 activity can be determined.

1. After day 1 post selection with puromycin, quantify GFP production cells by FACS for at least 1 week.
2. The wild-type, parental cell line will keep a high expression of GFP protein over time. In contrast, Cas9-expressing cell lines show an early peak of GFP expression in the following days after transduction but levels drop to zero after a week post transduction (Fig. 2) (*see* **Note 10**).

4 Notes

1. pRRLsin-SV40 T antigen-IRES-mCherry was a gift from Snorri Thorgeirsson (Addgene plasmid # 58993; http://n2t.net/addgene:58993; RRID:Addgene_58993); pLV-hTERT-IRES-hygro was a gift from Tobias Meyer (Addgene plasmid # 85140; http://n2t.net/addgene:85140; RRID:

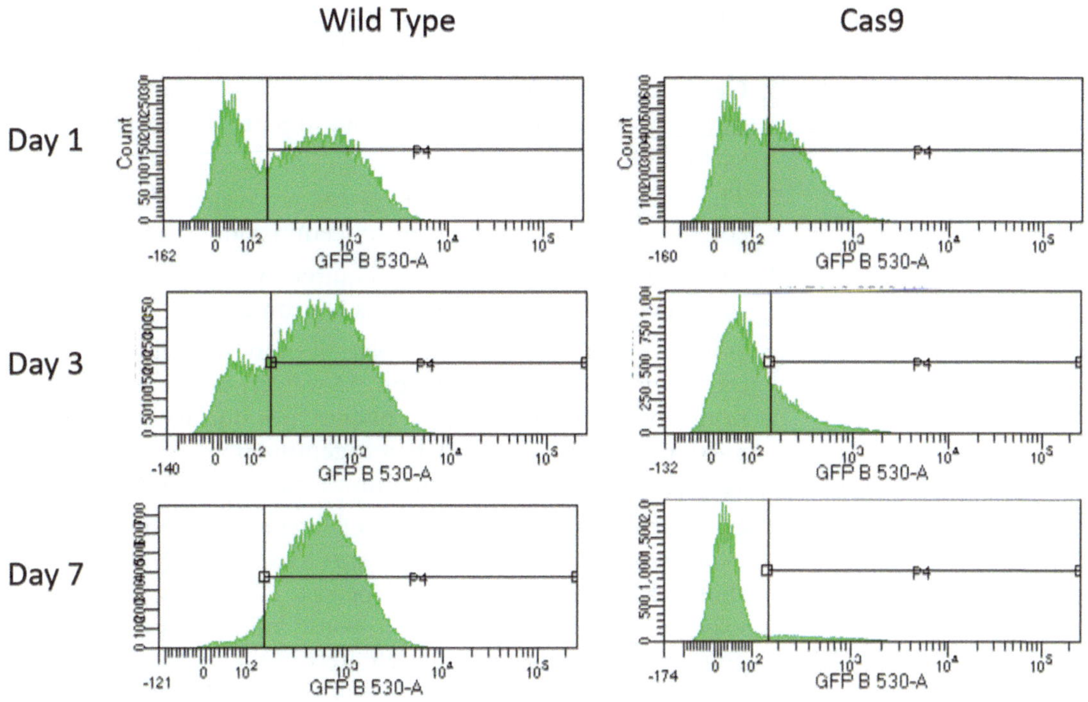

Fig. 2 Cas9 activity by FACS

Addgene_85140); pXPR_011 was a gift from John Doench & David Root (Addgene plasmid # 59702; http://n2t.net/addgene:59702; RRID:Addgene_59702); Lenti-Cas9-2A-Blast was a gift from Jason Moffat (Addgene plasmid # 73310; http://n2t.net/addgene:73310; RRID: Addgene_73310) psPAX2 and pMD2.G were a gift from Didier Trono (Addgene #12260; http://n2t.net/addgene:12260; RRID:Addgene_12260) and (Addgene #12259; http://n2t.net/addgene:12259; RRID: Addgene_12259), respectively.

2. Maintain cells under 70% confluency. Higher densities although do not affect cell proliferation has a harmful impact in virus yield.
3. Eppendorf Centrifuge 5810 rotor s-4-104 or a similar model.
4. Times could vary but as a general observation selection based in mCherry, or Hygromycin takes around 8 weeks depending on the capacity of the cells to recover from FACS sorting, sensitivity to Hygromycin and doubling time. Blasticidin and Puromycin are strong selection markers that usually take less time, around 2 weeks.
5. Lentivector is referred to pRRLsin-SV40 T antigen-IRES-mCherry, pLV-hTERT-IRES-hygro, Lenti-Cas9-2A-Blast,

and pXPR_011 for the four steps of transduction to establish immortalized Cas9-expressing cells shown in Fig. 1.

6. Gently deliver drops of transfection mix over the cells, taking care not to dislodge cells (hold pipette horizontally, not vertically).
7. This medium will contain some virus at a very low titters that if conserved will dilute the final concentration.
8. Ensure adherence to biosafety requirements (e.g., aerosol tight caps open only inside of the class II cabinet).
9. It is possible to attempt the isolation of individual clones from the original pooled population. However, one has to be cautious about the risk of skew phenotypes due to the genomic make out of the clone. A heterogenous population compensates to some extent this sometimes-undesirable effect.
10. Cas9 activity may vary between cell lines and in sometimes GFP expressed protein could take more than a week to clear. In those cases when GFP signal persist a longer assessment of expression by FACS is required. Some residual GFP expression could nevertheless persist indefinitely in the no-clonal cell line. Values below 10% GFP positive of the population are acceptable. A comprehensive method for GFP analysis can be found elsewhere [13].

References

1. Maqsood MI, Matin MM, Bahrami AR, Ghasroldasht MM (2013) Immortality of cell lines: challenges and advantages of establishment. Cell Biol Int 37(10):1038–1045. https://doi.org/10.1002/cbin.10137
2. Klingelhutz AJ, Barber SA, Smith PP, Dyer K, McDougall JK (1994) Restoration of telomeres in human papillomavirus-immortalized human anogenital epithelial cells. Mol Cell Biol 14(2):961–969. https://doi.org/10.1128/mcb.14.2.961-969.1994
3. Shay JW, Wright WE, Werbin H (1991) Defining the molecular mechanisms of human cell immortalization. Biochim Biophys Acta 1072(1):1–7. https://doi.org/10.1016/0304-419x(91)90003-4
4. Yoshioka M, Takenouchi T, Kitani H, Okada H, Yamanaka N (2016) Establishment of SV40 large T antigen-immortalized bovine liver sinusoidal cell lines and their immunological responses to deoxynivalenol and lipopolysaccharide. Cell Biol Int 40(12):1372–1379. https://doi.org/10.1002/cbin.10682
5. Takenouchi T, Iwamaru Y, Sato M, Yokoyama T, Shinagawa M, Kitani H (2007) Establishment and characterization of SV40 large T antigen-immortalized cell lines derived from fetal bovine brain tissues after prolonged cryopreservation. Cell Biol Int 31(1):57–64. https://doi.org/10.1016/j.cellbi.2006.09.006
6. Takenouchi T, Iwamaru Y, Sato M, Yokoyama T, Kitani H (2009) Establishment of an SV40 large T antigen-immortalized bovine brain cell line and its neuronal differentiation by dibutyryl-cyclic AMP. Cell Biol Int 33(2):187–191. https://doi.org/10.1016/j.cellbi.2008.11.001
7. Holczbauer A, Factor VM, Andersen JB, Marquardt JU, Kleiner DE, Raggi C, Kitade M, Seo D, Akita H, Durkin ME, Thorgeirsson SS (2013) Modeling pathogenesis of primary liver cancer in lineage-specific mouse cell types. Gastroenterology 145(1):221–231. https://doi.org/10.1053/j.gastro.2013.03.013
8. Nageshwaran S, Chavez A, Cher Yeo N, Guo X, Lance-Byrne A, Tung A, Collins JJ, Church GM (2018) CRISPR guide RNA cloning for mammalian systems. J Vis Exp 140:57998. https://doi.org/10.3791/57998

9. Hayer A, Shao L, Chung M, Joubert L-M, Yang HW, Tsai F-C, Bisaria A, Betzig E, Meyer T (2016) Engulfed cadherin fingers are polarized junctional structures between collectively migrating endothelial cells. Nat Cell Biol 18(12):1311–1323. https://doi.org/10.1038/ncb3438
10. Hart T, Chandrashekhar M, Aregger M, Steinhart Z, Brown KR, MacLeod G, Mis M, Zimmermann M, Fradet-Turcotte A, Sun S, Mero P, Dirks P, Sidhu S, Roth FP, Rissland OS, Durocher D, Angers S, Moffat J (2015) High-resolution CRISPR screens reveal fitness genes and genotype-specific cancer liabilities. Cell 163(6): 1515–1526. https://doi.org/10.1016/j.cell.2015.11.015
11. Doench JG, Hartenian E, Graham DB, Tothova Z, Hegde M, Smith I, Sullender M, Ebert BL, Xavier RJ, Root DE (2014) Rational design of highly active sgRNAs for CRISPR-Cas9-mediated gene inactivation. Nat Biotechnol 32(12):1262–1267. https://doi.org/10.1038/nbt.3026
12. Kisiel MA, Klar AS (2019) Isolation and culture of human dermal fibroblasts. Methods Mol Biol 1993:71–78. https://doi.org/10.1007/978-1-4939-9473-1_6
13. Kain SR (2000) Flow cytometric analysis of GFP expression in mammalian cells. In: Diamond RA, Demaggio S (eds) Living color: protocols in flow cytometry and cell sorting. Springer, Berlin, Heidelberg, pp 199–226. https://doi.org/10.1007/978-3-642-57049-0_19

Chapter 6

Targeting the AAVS1 Site by CRISPR/Cas9 with an Inducible Transgene Cassette for the Neuronal Differentiation of Human Pluripotent Stem Cells

Jinchao Gu, Ben Rollo, Huseyin Sumer, and Brett Cromer

Abstract

CRISPR/Cas9 system is a powerful genome-editing technology for studying genetics and cell biology. Safe harbor sites are ideal genomic locations for transgene integration with minimal interference in cellular functions. Gene targeting of the AAVS1 locus enables stable transgene expression without phenotypic effects in host cells. Here, we describe the strategy for targeting the AAVS1 site with an inducible Neurogenin-2 (Ngn2) donor template by CRISPR/Cas9 in hiPSCs, which facilitates generation of an inducible cell line that can rapidly and homogenously differentiate into excitatory neurons.

Key words CRISPR/Cas9, AAVS1, Ngn2, Stem cells, Neuronal differentiation

1 Introduction

Transposition and viral transduction are widely used for genomic integration of transgene elements in human pluripotent stem cells (hPSCs) including human embryonic stem cells (hESCs) and human-induced pluripotent stem cells (hiPSCs). However, these traditional methods may result in variable transgene expression or disruption of cellular functions through insertional mutagenesis. Safe harbor sites are genomic locations in which an exogenous construct can be inserted for consistent level of transgene expression without adverse effects on cell function [1]. The three commonly used human safe harbor loci are AAVS1, hROSA26, and CCR5 sites [1]. The AAVS1 site is located within intron 1 of the protein phosphate 1, regulatory subunit 12C (PPP1R12C) on human chromosome 19, which is one of the major integration hotspots of adeno-associated virus (AAV) [2]. As one of the most widely targeted safe harbor loci in the human genome, insertion into the AAVS1 locus allows robust and persistent transgene expression without causing discernible phenotypic effects

Paul John Verma, Huseyin Sumer and Jun Liu (eds.), *Applications of Genome Modulation and Editing*, Methods in Molecular Biology, vol. 2495, https://doi.org/10.1007/978-1-0716-2301-5_6,

[2]. Gene targeting strategies such as ZFNs, TALENs, and CRISPR have been successfully utilized to modify AAVS1 site in many cell types including hPSCs. To date, systematic AAVS1 targeting materials have been developed and standardized for transgene integration in hPSCs using TALENs or CRISPR/Cas9. Donor plasmids share a common design containing homology arms and selection markers such as antibiotic resistance or fluorescent reporter genes. Furthermore, the probability of off-target integration can be reduced by employing promoterless selection markers with a splice acceptor site [2]. In this system, the selection marker is expressed only if the donor construct is inserted within an intron.

While there are a lot of AAVS1 targeting vectors, an inducible construct enables versatile conditional transgene expression in cells. This has been achieved using TALEN-mediated homologous recombination at the AAVS1 site using a donor vector designed to establish transgenic hPSC lines that regulate transgene activity in response to doxycycline (dox) treatment [3]. Pluripotent cells with a modified AAVS1 locus maintain their ability to differentiate into all cell types derived from the three germ layers. This allows insertion of inducible transcription factor (TF) genes to control differentiation fate, such as neuronal differentiation.

For example, an inducible mouse Ngn2 transgene cassette has been stably harbored at the AAVS1 site in hiPSCs by TALENs [4]. With dox treatment, these cells rapidly differentiated into functional cortical glutamatergic neurons. By taking the advantage of a safe harbor site, this isogenic Ngn2 cell line was able to generate a uniform neuronal population with minimum variability. Similarly, the human Ngn2-coding sequence was cloned into an AAVS1 targeting construct for site-specific transgene insertion in hiPSCs by TALENs [5]. A similar approach was used for AAVS1 insertion of codon-optimized human-coding sequences for TFs Ascl1 and Dlx2 to drive inducible differentiation to inhibitory neurons [6]. This construct was targeted into both AAVS1 alleles in hiPSC clones for equivalent expression of TFs. Dox induction of the edited cell lines yielded almost pure GABAergic neural population. Thus, genome editing of safe harbor sites facilitates reproducible TF-induced neurogenesis with low variability, which can be leveraged for drug development and disease modeling of human brain disorders.

In this methods chapter, we describe the strategy for targeting the AAVS1 site in hiPSCs with CRISPR/Cas9. We use an inducible Ngn2 donor template which allows for the rapid and homogenous differentiation of hiPSCs into excitatory neurons.

2 Materials

2.1 Plasmids

1. pSpCas9(BB)-2A-GFP (PX458 Addgene #48138).
2. AAVS1-hNgn2 (kind gift from Michael Peitz, [5]).
3. pCE-mp53DD (Addgene #41856),

2.2 Oligos

1. Oligonucleotide PCR primers (Table 1).

2.3 Molecular Cloning

1. Luria-Bertani (LB) broth.
2. LB agar plates with 100 μg/mL Ampicillin.
3. Tris acetate EDTA (TAE) buffer.
4. Restriction enzymes BbsI and EcoRI (NEB) and NEBuffer 2.1.
5. T4 polynucleotide kinase (T4 PNK, NEB), T4 PNK Buffer (NEB), and 10 mM ATP.
6. T4 DNA Ligase (Promega) and 2× Rapid Ligation Buffer (Promega).
7. PrimeSTAR Max DNA Polymerase (2×, Takara Bio).
8. Phire Animal Tissue Direct PCR Kit (Thermo Fisher).
9. Thermocycler.
10. SyBR Safe DNA Gel Stain (10,000× in DMSO, Thermo Fisher).
11. Gel tray, comb, tank, and electrophoresis equipment.
12. Gel loading dye and Quick-Load Purple 1 kb Plus DNA Ladder 100 bp–10 kb (NEB).
13. UV transilluminator.
14. NEB 5-alpha competent E. coli (C2987).
15. Super optimal broth with catabolite repression (SOC medium, NEB).
16. Shaking incubator.
17. Macherey-Nagel NucleoSpin Plasmid Mini Kit.
18. Table top centrifuge.
19. NanoDrop Spectrophotometer.

2.4 hiPSC Culture

1. The iPSC line was generated from human neonatal foreskin fibroblasts ATCC PCS-201-010.
2. Essential 8 Medium (Thermo Fisher): Mix 1 mL of Essential 8 Supplement (50×) with 49 mL of Essential 8 Basal Medium. Store at 4 °C (*see* **Note 1**).

Table 1
Oligos

Primer name	Sequence (5′-3′)
AAVS1 fwd	CACCGGGGCCACTAGGGACAGGAT
AAVS1 rvs	AAACATCCTGTCCCTAGTGGCCCC
U6 sequencing fwd	GAGGGCCTATTTCCCATGATTCC
AAVS1 5′ Arm fwd	TGCTTTCTTTGCCTGGACAC
AAVS1 3′ Arm rvs	GGTTCTGGCAAGGAGAGAGA
AAVS1 puro fwd	CCATAGCTCAGGTCTGGTCTAT
AAVS1 puro rvs	AGGAAGAGAAGAGGTCAGAAGC

3. Dulbecco's PBS (DPBS) without Calcium and Magnesium ions (Thermo Fisher) (*see* **Note 2**).
4. Vitronectin coating reagent: Dilute Vitronectin, truncated recombinant human (VTN-N 0.5 mg/mL, Thermo Fisher) at 1:100 with DPBS for 5 μg/mL.
5. EDTA cell dissociation reagent: Dilute UltraPure 0.5 M EDTA, pH 8.0 (Thermo Fisher) at 1:1000 with DPBS for 0.5 mM.
6. StemPro Accutase Cell Dissociation Reagent (Thermo Fisher).
7. Y-27632 ROCK inhibitor: Make up 10 mM stock solution of Y-27632 dihydrochloride (Sigma) with DPBS. Store at −20 °C. Use at 1:1000 in culture medium for 10 μM final concentration.
8. DMSO.
9. Ethanol 70% v/v in H_2O.
10. Cell culture plastics, pipettes, and pipette tips.
11. Class II biosafety cabinet.
12. Table top centrifuge.
13. CO_2 incubator (37 °C, 5% CO_2).
14. Water bath (37 °C).
15. Hemocytometer.
16. Trypan blue staining solution.
17. Bright-field microscope.
18. Cryopreservation box.

2.5 Transfection

1. Opti-MEM Reduced Serum Medium (Gibco).
2. Lipofectamine STEM Transfection Reagent (Thermo Fisher).

3. Fluorescent microscope.
4. Puromycin (stock at 10 mg/mL, Thermo Fisher).

2.6 Neuronal Differentiation

1. N2B27 neural medium: To make 50 mL medium, mix 23.75 mL of DMEM/F12 medium, 23.75 mL of neurobasal medium, 500 μL of 50× B27 supplement, 250 μL of 100× N2 supplement, 250 μL of 100× GlutaMAX, 500 μL of 100× non-essential amino acids, 45 μL of β-mercaptoethanol, 500 μL of 100× penicillin/streptomycin. Make fresh aliquots each time. Store at 4 °C for up to 1 month.
2. Laminin coating solution: Dilute 150 μL of 1 mg/mL laminin stock (L2020, Sigma) with 10 mL of PBS to make 15 μg/mL laminin solution.
3. Doxycycline (D9891, Sigma): Make up 2 mg/mL stock. Use at 1:2000 in neural medium for 1 μg/mL.
4. AraC (C1768, Sigma): Make up 5 mM stock. Dilute with neural medium at 1:2000 for 2.5 μM.
5. Recombinant Human BDNF (PeproTech): Make up 10 μg/mL stock. Use at 1:1000 in neural medium for 10 ng/mL.

3 Methods

3.1 Oligo Annealing and Phosphorylation to Produce AAVS1 Targeting gRNA

1. Set up the following reaction in a PCR tube (*see* **Note 3**):

AAVS1 fwd (100 μM)	1 μL
AAVS1 rvs (100 μM)	1 μL
10× T4 PNK Buffer	1 μL
10 mM ATP	1 μL
ddH2O	5.5 μL
T4 PNK	0.5 μL
Total	10 μL

2. Incubate in a thermocycler using the following parameters (*see* **Note 4**)

37 °C	30 min
95 °C	5 min and then ramp down to 25 °C at 0.1 °C/s
4 °C	Hold

3. Dilute the phosphorylated and annealed oligo duplex at 1:200 in ddH_2O.

3.2 Oligo Cloning into Backbone Vector to Produce PX458-AAVS1

1. Set up the following single-step reaction of digestion and ligation in a PCR tube (*see* **Note 5**):

PX458 plasmid (50 ng total)	× μL
diluted oligo duplex	1 μL
2× Rapid ligation Buffer	5 μL
ddH_2O	2.5-× μL
BbsI	0.5 μL
T4 DNA ligase	1 μL
Total	10 μL

2. Incubate the tube in a thermocycler using the following program:

37°C 5 min ⎤
23°C 5 min ⎦ 6 cycles
4°C Hold

3. Transform C2987 competent cells with 2 μL of ligation product.
4. Inoculate two colonies for liquid culture.
5. Extract plasmid DNA using a miniprep kit.
6. Screen for correct construct insertion by digesting plasmid samples with BbsI and EcoRI (*see* **Note 6**):

NEBuffer 2.1 (10×)	1 μL
Plasmid DNA (200 ng)	× μL
ddH_2O	8.5-× μL
BbsI	0.25 μL
EcoRI	0.25 μL
Total	10 μL

Incubate at 37 °C for 1 h. Perform gel electrophoresis (Fig. 1).

7. Prepare PCR mixture to confirm constructs:

PrimeSTAR Max Premix (2×)	5 μL
U6 sequencing primer (5 μM)	0.5 μL
AAVS1 reverse primer (5 μM)	0.5 μL
ddH_2O	2 μL

(continued)

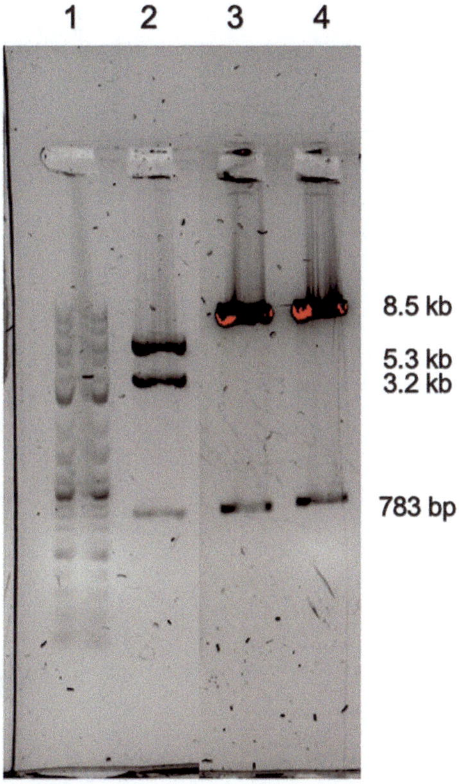

Fig. 1 Gel image of double digestion with BbsI and EcoRI. The correct construct was only cut by EcoRI, which generated the 8.5 kb and 783 bp bands.1: DNA Ladder; 2: PX458 vector; 3–4: purified plasmid samples

Template (plasmid DNA 0.1 ng/μL)	2 μL
Total	10 μL

Set up thermocycler program:

98°C	10 s	30 cycles
59°C	5 s	
72°C	5 s/kb	
72°C	5 min	
4°C	Hold	

Perform gel electrophoresis (Fig. 2).

8. Confirm construct by Sanger sequencing using U6 primer.

3.3 Thaw and Recover hiPSCs

1. Prior to thawing cells, coat one well of a 6-well plate with 1 mL of 5 μg/mL Vitronectin at room temperature for 1 h or at 4 °C overnight.

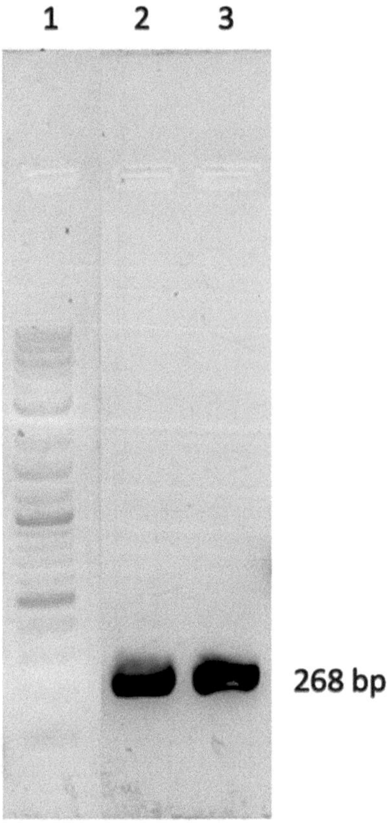

Fig. 2 Gel image of PCR amplifications using U6 universal primer and AAVS1 reverse primer. The correct construct was expected to produce a 268 bp band. 1: DNA ladder; 2–3: ligated plasmid templates

2. Thaw a cryovial of hiPSCs in a 37 °C water bath with gentle rotation until almost completely thawed (*see* **Note** 7).
3. Decontaminate the vial with 70% ethanol and transfer it to a biosafety cabinet.
4. Slowly add 1 mL of pre-warmed Essential 8 medium to the vial and transfer the cell suspension to a sterile 15 mL tube. Centrifuge the cells at 200 × *g* for 3 min.
5. Discard the supernatant and resuspend the cell pellet in 2 mL of growth medium supplemented with 10 μM Y-27632 (*see* **Note 8**).
6. Transfer the cell suspension to one vitronectin-coated well of a 6-well plate. Place the flask in a 37 °C, humidified incubator with 5% CO_2.

3.4 Sub-culture hiPSCs

1. Prior to passaging, coat one well of a 6-well plate with 1 mL of 5 μg/mL Vitronectin at 37 °C for 1 h.

2. Pre-warm Essential 8 medium at room temperature until it is no longer cold to touch. Warm up DPBS and 0.5 mM EDTA in a 37 °C water bath.
3. Passage the cells when they reach 70–80% confluency. Remove old medium and rinse the cells with 1 mL of DPBS.
4. Dissociate cells with 1 mL 0.5 mM EDTA at 37 °C for 2–4 min.
5. Remove EDTA when cells begin to separate and round up and holes appear in hiPSC colonies.
6. Add 2 mL of growth medium to detach the cells by gentle trituration (*see* **Note 9**). Transfer the cell suspension to a 15 mL tube.
7. Add 200 μL of cell suspension to one well of a 6-well dish pre-coated with vitronectin (*see* **Note 10**).
8. Disperse the cells evenly across the culture surface by gently rocking the plate. Place the plate into the 37 °C, 5% CO_2 incubator, and incubate the cells overnight.
9. Replace spent culture medium daily until cells are ready for passaging (*see* **Note 11**).

3.5 Transfection and Selection

1. Grow hiPSC in one well of a 6-well plate to 80% confluency for transfection.
2. Aspirate old medium and rinse the cells with 1 mL of DPBS.
3. Add 500 μL of Accutase and incubate at 37 °C for 5 min or until most hiPSC colonies have detached.
4. Add 1.5 mL of culture medium to neutralize and dilute Accutase. Pipette the solution several times down to the surface of the well to break apart cell colonies. Transfer the single cell suspension to a conical centrifuge tube.
5. Centrifuge at 200 × *g* for 3 min.
6. Discard the supernatant and resuspend cell pellet in 2 mL of culture medium.
7. Count the cells and adjust cell density to 2×10^5/mL in Essential 8 medium added with 10 μM Y-27632 (*see* **Note 12**).
8. Transfer 500 μL of cell suspension to one vitronectin-coated well of a 24-well plate and place it in the incubator.
9. Allow the cells to adhere for 4–6 h before adding the transfection solution (*see* **Note 13**).
10. Prepare the transfection complex for CRISPR knock-in. In tube A, add 2 μL of Lipofectamine STEM reagent to 25 μL of Opti-MEM and mix well. In tube B, add 200 ng of PX458-AAVS1, 200 ng of pCE-mp53DD, and 200 ng of AAVS1-hNgn2 to 25 μL of Opti-MEM and mix well (*see* **Note 14**).

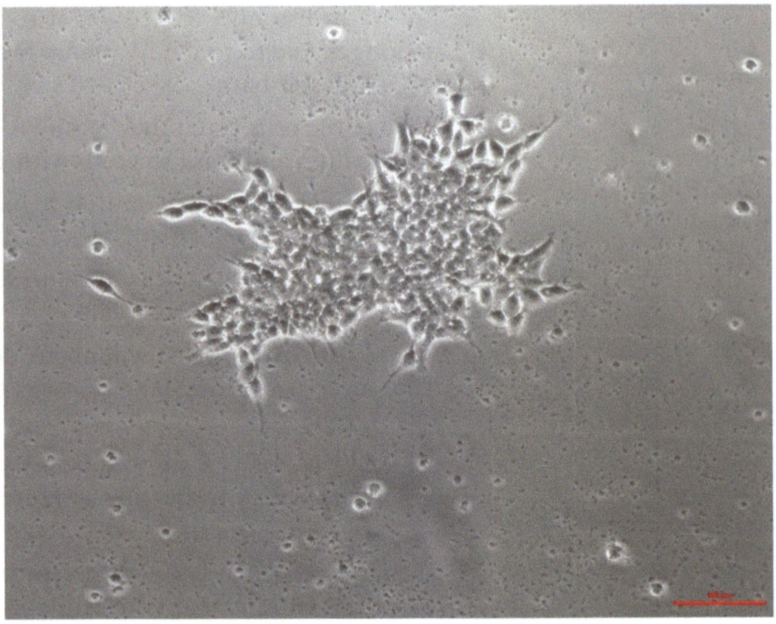

Fig. 3 A stable colony after puromycin selection for 7 days. Y-27632 was added to support cell growth and expansion

Combine the contents of tube A and B. Mix well and incubate at room temperature for 10 min.

11. Add the lipid/DNA complexes dropwise evenly across the surface of a single well. Return the multiwell plate to the incubator and culture overnight.
12. The following day assess transfection efficiency by fluorescent microscopy. Replace the medium with fresh growth medium.
13. Start 0.2 μg/mL puromycin selection 48 h after transfection. Change selection medium daily (*see* **Note 15**).
14. Pick individual puromycin-resistant colonies with a pipette using a stereo microscope. Add individually selected clones to a 24-well plate and culture.
15. Passage the cells to a 6-well plate for expansion when stable colonies reach 80% confluency (Fig. 3).
16. Cryopreserve the stable cell line with Essential 8 medium +10% DMSO.

3.6 Genotyping

1. During sub-culture, centrifuge the rest of the dissociated cells at 200 × *g* for 3 min. Discard the supernatant and collect the cell pellet for genotyping using Phire Direct PCR kit.
2. Dilute the cell pellet in 20 μL of Dilution Buffer. Add 0.5 μL of Release Additive and mix by vortexing the tube briefly. Transfer the sample to a 1.5 mL tube.

3. Spin down the solution at 11,000 × *g* for 1 min. Incubate the reaction at room temperature for 2–5 min.
4. Heat the sample at 98 °C for 2 min.
5. Centrifuge at 11,000 × *g* for 1 min. Transfer the supernatant to a new tube.
6. Set up the following PCR mixture:

Phire PCR Buffer 2×	10 μL
Forward primer (5 μM)	2 μL
Reverse primer (5 μM)	2 μL
Phire DNA polymerase	0.4 μL
ddH$_2$O	4.6 μL
Supernatant	1 μL
Total	20 μL

Use AAVS1 5′ and 3′ Arm primers for Reaction 1 and use AAVS1 puro primers for Reaction 2 [7] (*see* **Note 16**).

7. Perform PCR reactions in a thermocycler Only 1 cycle for initial denaturation (98 °C 5 min).

98°C	5 min	
98°C	5 s	30 cycles
60°C	5 s	
72°C	20 s	
72°C	1 min	
4°C	Hold	

8. Run the PCR products on an agarose gel and visualize with UV transilluminator (Fig. 4).

3.7 Neural Induction

1. Coat each well of a 24-well plate with 250 μL of 15 μg/mL laminin at 37 °C for at least 2 h or at 4 °C overnight.
2. On day -1, after routine passaging, centrifuge the remaining cell suspension at 200 × *g* for 3 min. Discard the supernatant and resuspend the cell pellet with 300 μL of Accutase to break the cell clumps into single cells. Incubate at 37 °C for 5 min with gentle trituration. Add 1.7 mL of culture medium to neutralize Accutase and centrifuge at 200 × *g* for 3 min. Remove supernatant and resuspend in 2 mL of culture medium. Count and calculate cell concentration using a hemocytometer. Seed cells at 5 × 10^4/well of a 24-well plate in 500 μL of Essential 8 medium with 10 μM Y-27632. Place the multiwell plate in the incubator overnight.

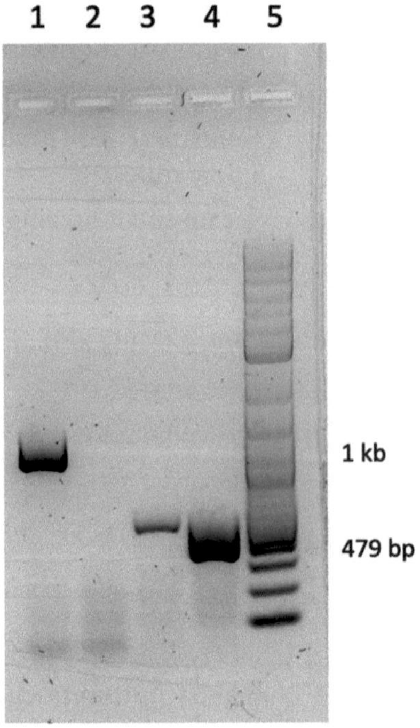

Fig. 4 Gel image of genotyping using Phire Direct PCR Kit. The 1 kb band (1) with AAVS1 puro primers verified successful knock-in of Ngn2 donor construct and the presence of a 479 bp band (4) with AAVS1 5′ and 3′ Arm primers indicated unmodified alleles. It was expected to generate a band about 7 kb (3) from modified alleles. However, the Phire DNA polymerase was inefficient to amplify such a big fragment from genome. 1,3: Targeted cell line; 2, 4: Untransfected iPSCs; 5: DNA ladder

3. At day 0, remove the medium and add 500 μL of N2B27 neural medium with 1 μg/mL doxycycline and 5 μM Y-27632 (Fig. 5) (*see* **Note 17**).
4. At day 1, 2, and 3, replace neural induction medium daily.
5. On day 4, split the cells using Accutase (Fig. 5) (*see* **Note 18**). Remove the medium and wash the cells with 250 μL of PBS twice. Add 200 μL of Accutase to dissociate the cells at 37 °C for 5 min. Add 800 μL of neural medium and gently detach the cells by pipetting. Transfer the cell suspension to a conical tube and centrifuge at 200 × *g* for 3 min. Discard the supernatant and resuspend the cell pellet in 1 mL of neural medium. Plate the cells at 1:5 ratio in wells of a new 24-well plate pre-coated with laminin, or onto relevant assay format (patch clamping, multielectrode array, immunostaining, etc.).
6. Cells are cultured in 500 μL of N2B27 medium supplemented with 1 μg/mL doxycycline, 5 μM Y-27632, and 10 ng/mL BDNF (*see* **Note 19**). Replace the medium without Y-27632 every 2–3 days.

Day 0 Day 4

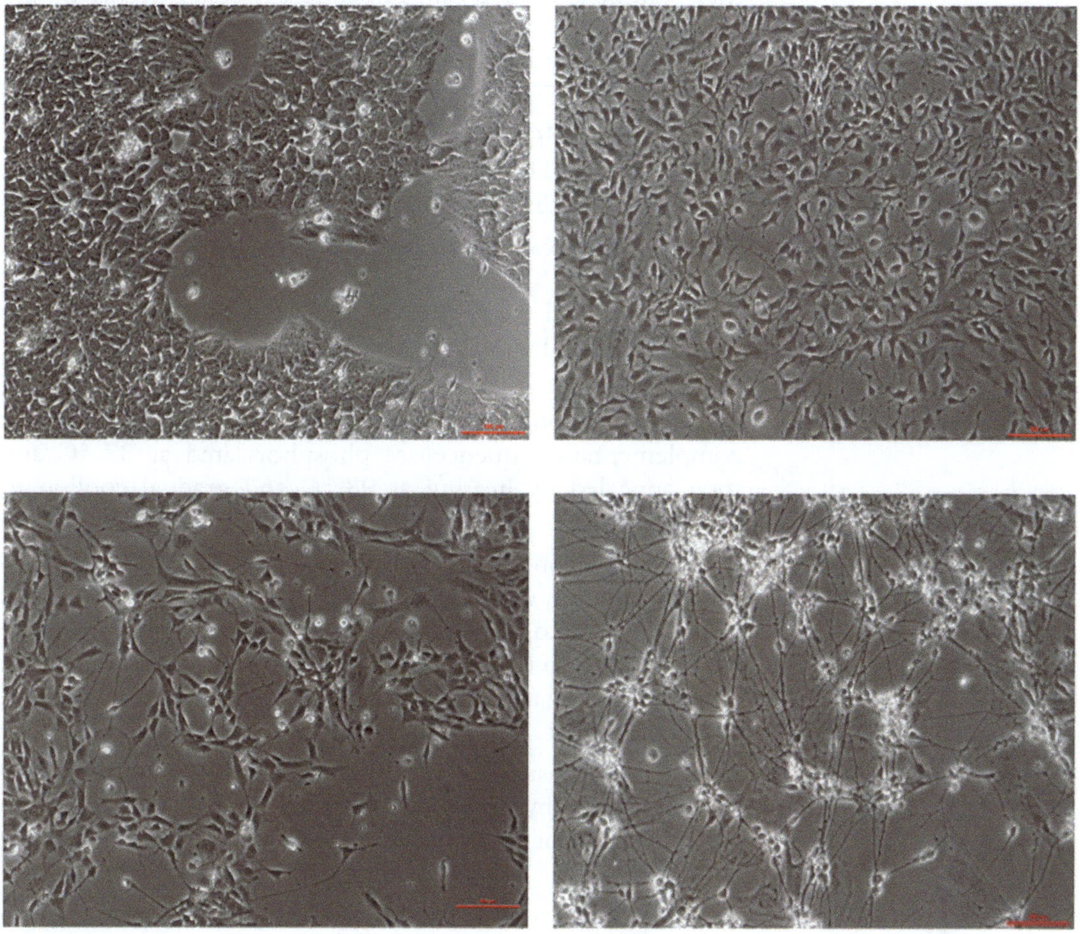

Day 7 Day 11

Fig. 5 Images showing a neural induction of AAVS1 targeted Ngn2 hiPSCs. Cells differentiated into neurons with complex neuronal morphology in prolonged culture. Scale bars: 100 μm

7. At day 6 or 7, treat the cells with 2.5 μM AraC for 48 h to kill proliferating non-neuronal cells (Fig. 5).
8. On day 10, withdraw doxycycline and maintain the cells in neural medium with 10 ng/mL BDNF (Fig. 5) (*see* **Note 20**).

4 Notes

1. As the growth factors in Essential 8 supplement (50×) are unstable at 37 °C, the frozen supplement should be thawed at room temperature. Complete Essential 8 medium can be stored at 4 °C for up to a month. It is recommended to make

up 50 mL complete medium each time. The thawed supplement can be aliquoted into Eppendorf tubes (1 mL each) and stored at −20 °C. The complete medium should be warmed at room temperature. Warming medium at 37 °C may denature the growth factors.

2. The presence of Calcium and Magnesium promotes cell adhesion; therefore, DPBS without Ca^{2+} and Mg^{2+} should be used for rinsing and dissociating cells.
3. T4 PNK requires ATP to fulfill its activity, but ATP was not added in T4 PNK Buffer. The reaction needs to be supplemented with 1 mM ATP. Alternatively, 1× T4 Ligation Buffer which contains 1 mM ATP (NEB) can be used in place of T4 PNK buffer.
4. Ligation needs oligos with a 5′ phosphate. The oligos with complementary sequence are phosphorylated at 37 °C and then annealed by heating at 98 °C and gradual cooling to 25 °C.
5. The original protocol used BbsI-digested and gel-purified plasmid for ligation, which took a few more steps. As BbsI cleaves outside of its recognition sequence, plasmids ligated with guide sequence can no longer be cut by BbsI. After several cycles of digestion and ligation, there will be no plasmid containing BbsI sites left.
6. Since ligated plasmids do not have BbsI recognition sites, they can only be cut by EcoRI to generate fragments different from the PX458 vector. This is one reliable method to screen positive clones.
7. DMSO is toxic to cells. The thawed cells should be diluted with culture medium and centrifuged to remove DMSO as soon as possible.
8. Addition of ROCK inhibitors for the first 24 h after thawing can improve cell viability.
9. Break up cell clumps with gentle trituration for two or three times. Avoid extra pipetting and scraping cells left behind from the well.
10. For maintaining established cell lines, the split ratio can be adjusted between 1:3 and 1:12 depending on sub-culture frequency.
11. Due to the thermal instability of growth factors in Essential 8 medium, cells require daily feeding on fresh growth medium.
12. Singularization of hiPSCs can cause dissociation-induced apoptosis. Addition of ROCK inhibitor significantly improves cell viability after cell dissociation by Accutase.

13. Transfection efficiency can be increased when cells are seeded as single cells. The cells will adhere to the laminin-coated surface a few hours after seeding.
14. Recent studies reported that p53 toxicity could lead to double-strand break-mediated cell death in pluripotent cells, decreasing genome-editing efficiency [8–10]. The episomal p53 dominant-negative plasmid can transiently inhibit p53 function, so viability of transfected cells is improved.
15. ROCK inhibitors can be added to improve cell survival under drug selection if transfection efficiency is very low.
16. Anneal for 5 s at the Tm of the lower Tm primer when the primers are shorter than 20 nucleotides. Anneal at Tm + 3 °C of the lower Tm primer when the primers are longer than 20 nucleotides.
17. The treatment of doxycycline induces Ngn2 expression in the transgene cassette at the AAVS1 locus. Y-27632 improves survival and attachment of hiPSCs at the initiation of neuronal differentiation.
18. Split the induced neurons not later than day 6. Older more mature neurons with branched networks are usually difficult to detach and result in a reduced survival rate.
19. BDNF is a neural growth factor that supports neuronal survival and growth. The induced neurons can also be co-cultured with glial/astrocyte cells to enhance synapse formation for coordinated electrical signal transmission.
20. When compared to viral transduction or piggyBac transposition methods, the rate of conversion from pluripotent cells to neurons was slightly delayed (i.e., 5 days slower) in the AAVS1 targeted cell line.

Acknowledgments

This work was supported by a Tuition Fee Scholarship from Swinburne University of Technology.

References

1. Pellenz S, Phelps M, Tang W, Hovde BT, Sinit RB, Fu W, Li H, Chen E, Monnat RJ (2018) New human chromosomal safe harbor sites for genome engineering with CRISPR/Cas9, TAL effector and homing endonucleases. bioRxiv:396390. https://doi.org/10.1101/396390
2. Oceguera-Yanez F, Kim S-I, Matsumoto T, Tan GW, Xiang L, Hatani T, Kondo T, Ikeya M, Yoshida Y, Inoue H, Woltjen K (2016) Engineering the AAVS1 locus for consistent and scalable transgene expression in human iPSCs and their differentiated derivatives. Methods 101:43–55. https://doi.org/10.1016/j.ymeth.2015.12.012
3. Qian K, Huang C-L, Chen H, Blackbourn Iv LW, Chen Y, Cao J, Yao L, Sauvey C, Du Z, Zhang S-C (2014) A simple and efficient

system for regulating gene expression in human pluripotent stem cells and derivatives. Stem Cells 32(5):1230–1238. https://doi.org/10.1002/stem.1653

4. Wang C, Ward ME, Chen R, Liu K, Tracy TE, Chen X, Xie M, Sohn PD, Ludwig C, Meyer-Franke A, Karch CM, Ding S, Gan L (2017) Scalable production of iPSC-derived human neurons to identify tau-lowering compounds by high-content screening. Stem Cell Rep 9(4):1221–1233. https://doi.org/10.1016/j.stemcr.2017.08.019
5. Meijer M, Rehbach K, Brunner JW, Classen JA, Lammertse HCA, van Linge LA, Schut D, Krutenko T, Hebisch M, Cornelisse LN, Sullivan PF, Peitz M, Toonen RF, Brüstle O, Verhage M (2019) A single-cell model for synaptic transmission and plasticity in human iPSC-derived neurons. Cell Rep 27(7):2199–2211.e2196. https://doi.org/10.1016/j.celrep.2019.04.058
6. Rhee HJ, Shaib AH, Rehbach K, Lee C, Seif P, Thomas C, Gideons E, Guenther A, Krutenko T, Hebisch M, Peitz M, Brose N, Brustle O, Rhee JS (2019) An autaptic culture system for standardized analyses of iPSC-derived human neurons. Cell Rep 27(7):2212–2228.e2217. https://doi.org/10.1016/j.celrep.2019.04.059
7. Jiang Y, Zhou Y, Bao X, Chen C, Randolph LN, Du J, Lian XL (2018) An ultrasensitive calcium reporter system via CRISPR-Cas9-mediated genome editing in human pluripotent stem cells. iScience 9:27–35. https://doi.org/10.1016/j.isci.2018.10.007
8. Ihry RJ, Worringer KA, Salick MR, Frias E, Ho D, Theriault K, Kommineni S, Chen J, Sondey M, Ye C, Randhawa R, Kulkarni T, Yang Z, McAllister G, Russ C, Reece-Hoyes J, Forrester W, Hoffman GR, Dolmetsch R, Kaykas A (2017) P53 toxicity is a hurdle to CRISPR/CAS9 screening and engineering in human pluripotent stem cells. bioRxiv:168443. https://doi.org/10.1101/168443
9. Ihry RJ, Worringer KA, Salick MR, Frias E, Ho D, Theriault K, Kommineni S, Chen J, Sondey M, Ye C, Randhawa R, Kulkarni T, Yang Z, McAllister G, Russ C, Reece-Hoyes J, Forrester W, Hoffman GR, Dolmetsch R, Kaykas A (2018) p53 inhibits CRISPR–Cas9 engineering in human pluripotent stem cells. Nat Med 24(7):939–946. https://doi.org/10.1038/s41591-018-0050-6
10. Haapaniemi E, Botla S, Persson J, Schmierer B, Taipale J (2018) CRISPR–Cas9 genome editing induces a p53-mediated DNA damage response. Nat Med 24(7):927–930. https://doi.org/10.1038/s41591-018-0049-z

Chapter 7

Microinjection of Zygotes for CRISPR/Cas9-Mediated Insertion of Transgenes into the Murine Rosa26 Safe Harbor

Fabien Delerue and Lars M. Ittner

Abstract

Genetically modified (GM) mice are widely used in biomedical research because they can address complex questions in an in-vivo setting that could not otherwise be addressed in-vitro. Microinjection of zygotes remains the most common technique to generate GM animals to date. Here, we describe the targeted insertion (knock-in) of transgenes by microinjection of 1-cell or 2-cell stage embryos into the murine Rosa26 safe harbor.

Key words Microinjection, Zygotes, Pronuclei, Transgene, Rosa26, Reimplantation, Genotyping

1 Introduction

Genetically modified (GM) animals are the most powerful tools to understand the mechanisms underlying physiological processes and their pathological counterparts. They are also invaluable tools to search for disease modifiers and to develop and test novel treatment strategies. Beside medical research, GM animals are generated to improve animal welfare and for production of biological products.

Cutting-edge technologies such as engineered endonucleases (e.g. CRISPR/Cas9) designed to target and eventually cleave-specific DNA sequences have emerged as alternative methods to accelerate the process of genome editing and apply it to virtually any living species [1]. In the meantime, assisted reproductive techniques and gene transfer technologies have also extensively been refined, facilitating the generation of GM animals. As such, electroporation of fertilized oocytes recently became an alternative to microinjection [2, 3]. Nevertheless, the main limitation of electroporation lies within the size of the exogenous DNA that can be brought inside the zygotes. To date, there has been no report of a successful transformation of double-stranded DNA (e.g., plasmid or transgene) of intact zygotes, and the largest insertion reported

Paul John Verma, Huseyin Sumer and Jun Liu (eds.), *Applications of Genome Modulation and Editing*,
Methods in Molecular Biology, vol. 2495, https://doi.org/10.1007/978-1-0716-2301-5_7,

so far by electroporation was obtained using a single-stranded transgene of 1 kb [4]. This is likely due to the Zona Pellucida preventing large transgenes to be shuttled inside the embryos. Therefore, by physically penetrating all protective membranes (i.e. Zona Pellucida, oolemma, nuclear envelope, and pronuclear membrane) microinjection allows for direct delivery of exogenous DNA of virtually any size into the pronuclei. Consequently, random integration of transgenes of few kilobases in size to generate sophisticated tools such as inducible reporter mice can be achieved by microinjection [5].

Yet, random integration of transgenes is limited by position effects. As most of the genome is "silent," a transgene integrated randomly into the genome is likely to not be expressed [6]. Overcoming this caveat requires the generation of several transgenic lines which then need to be further analyzed to check for genuine expression of the transgene in-vivo. This time-consuming and labor-intensive effort can be addressed by targeted integration of transgenes to genomic "safe harbors." Safe harbors are defined as genomic regions able to accommodate the predictable expression of newly integrated DNA without adverse effects on the host organism [7]. In mice, several safe harbors have been discovered, including the TIGRE (also known as Igs7) [8], Hipp11 (also known as Igs2) [9], Col1a1 [10], HPRT1 [11] and Rosa26 loci [12]. The Rosa26 locus is the most commonly used safe harbor in mice, with over a thousand transgenic lines generated in this permissive locus.

Although genomic integration of circular transgenes is achievable, linearization is highly recommended as it increases the success rate of both random [13] and targeted integration [14]. We have successfully generated several Rosa26 knock-in (KI) mouse lines by microinjection of zygotes.

2 Materials

Prepare all microinjection mixes using DNAse-free and RNAse-free embryo-tested ultrapure water. Manipulate all consumables and equipment wearing sterile single-use gloves and apply purifying agent to eliminate potential DNase or RNase contamination from glassware and plastic surfaces. Prepare all reagents at room temperature and store at −20 °C.

Diligently follow all waste disposal regulations when disposing waste materials.

2.1 Injection Mix

1. Microinjection buffer: 8 mM Tris–HCl and 0.15 mM EDTA. Weight 63.04 mg of Tris–Hcl and 2.79 mg of EDTA and dilute in 50 mL of DNAse/RNAse-free water, sterile filter with a 0.22 μm syringe filter, and adjust the pH to 7.4.

2. 1% agarose gel: add 1.5 g of agarose to 150 mL pre-heated Tris Acetate EDTA (1× TAE) buffer stained with safe DNA stain and cool down to initiate polymerization.
3. Electrophoresis set up.
4. Linearized transgene: the transgene is visualized on an ultraviolet (UV) transilluminator and excised with a clean scalpel before purification using a gel extraction kit following the manufacturer's recommendations.
5. Heat block.
6. Spectrophotometer.
7. Endonuclease (e.g., spCas9): dilute the protein (or mRNA) at 1 μg/μL in microinjection buffer and aliquot in PCR tubes before freezing.
8. Single guide RNA (sgRNA): 5′-ACTCCAGTCTTTCTAGAA GA-3′ (*see* **Note 1**).

Dilute the sgRNA at 1 μg/μL in microinjection buffer and aliquot in PCR tubes before freezing.

2.2 Microinjection

1. Injection needle: pull borosilicate capillaries with internal filament (1 mm outer diameter) on a pipette puller with the recommended parameters: Pressure: 500—Heat: ramp value +5—Pull: 180—Velocity: 50—Time: 180 (*see* **Note 2**).
2. Holding needle: recommended 35° tip angle, 15 μm inner diameter, 100 μm outer diameter, 1 mm flange, smooth rounded tip.
3. Manual microinjector.
4. Loading microcapillaries: 0.5–20 μL, 100 mm, sterile.
5. Microinjection chamber: add one drop (50 μL) of HEPES-buffered medium to the center of the depression of a glass slide and overlay with paraffin oil.
6. Micromanipulation set up: micromanipulators (12–12,500 μm range) mounted on a differential interference contrast (DIC) inverted microscope with ×10 and ×40 optics. Manual microinjector (*see* **Note 3**). Pneumatic injector: recommended injection pressure 500 hPA, compensation pressure 50 hPa (*see* **Note 4**). Optional: piezo drill for 2-cell injections (*see* **Note 5**).

2.3 Embryo Culture

1. Sterile 3.5 cm plastic petri dishes.
2. Embryo tested, sterile-filtered paraffin oil.
3. Sterile embryo-grade handling (HEPES-buffered) medium (i.e., M2 medium) and culture medium (i.e., KSOM medium).
4. 5% O_2, 6% CO_2 humidified incubator.

5. Mouth pipette.
6. Hyaluronidase (10 mg/mL diluted in KSOM medium).

2.4 Embryo Production and Reimplantation (See Note 6)

1. A colony of vasectomized mice (15–20 males).
2. A colony of stud mice (8 males).
3. A colony of egg-donor mice (8 females).
4. A colony of recipient mice (~30 females).
5. Superovulation hormones (*see* **Note 7**):
   (a) Pregnant mare's serum gonadotropin (PMSG) diluted to 5 IU/100 μL in sterile Phosphate Buffer Saline (1× PBS).
   (b) Human chorionic gonadotropin (hCG) diluted to 5 IU/100 μL in sterile Phosphate Buffer Saline (1× PBS).
6. Dissecting tools (Dumont #5 forceps, Iris scissors, Vessel clamp, Micro-scissors, Wound clips, Clips applier, absorbable surgical sutures).
7. Isoflurane anesthetic machine.
8. Heating mat.
9. Injectable analgesics (*see* **Note 8**).
10. Dissecting microscope.

2.5 Screening of the Progeny

1. Alkaline lysis reagent (25 mM NaOH, 0.2 mM EDTA, pH = 12): dissolve 50 mg of NaOH and 3.72 mg of EDTA in 50 mL of DNAse/RNAse-free water (no pH adjustment needed).
2. Neutralizing reagent (40 mM Tris–HCl, pH = 5): dissolve 315 mg of Tris–HC in 50 mL of DNAse/RNAse-free water (no pH adjustment needed).
3. Heat block.
4. Centrifuge (with cooling option).
5. Thermocycler.
6. PCR reagents (i.e. Taq polymerase and buffers).
7. Primers for PCR genotyping.
   E.g., Transgene-specific PCR:
   - Fwd 5′-GCCTGAAGAACGAGATCAGC-3′
   - Rev 5′-AGAGCTGGGTGGTGTCTTTG-3′
   - 5′ Flanking PCR:
   - Fwd 5′-CCTAAAGAAGAGGCTGTGCTTTGG-3′
   - Rev 5′-GGGCCATTTACCGTAAGTTATGT-3′
   - 3′ Flanking PCR:

- Fwd 5′-GCCTGAAGAACGAGATCAGC-3′
- Rev 5′-CCTGAAGAAGCTTGGCAAAA-3′

8. 1% agarose gel: add 1.5 g of agarose to 150 mL pre-heated Tris Acetate EDTA (1× TAE) buffer stained with safe DNA stain and cool down to initiate polymerization.
9. Electrophoresis set up.
10. Gel imaging system.

3 Methods

All procedures must be carried out in accordance with the applicable local and national gene technology and ethical regulations.

3.1 Preparation of the Injection Mix (Fig. 1)

1. To perform targeted integration by Homologous Directed Repair (HDR), the donor transgene must contain homology arms of at least 800 bp (*see* **Note 9**), with sequences found on either side of the original *XbaI* site of the murine Rosa26 locus [15]. The transgene of interest (e.g. pCMV-LSL-MAPT-SV40pA, 2342 bp) is linearized by digestion with appropriate restriction enzyme(s) (*see* **Note 10**). Incubate few micrograms of plasmid at appropriate temperature until complete digestion is reached in a heat block. Run the linearized transgene on an agarose gel, visualize on the UV transilluminator, and gel extract using the manufacturer's recommendations. Final elution should be done in microinjection buffer and concentration should be checked with a spectrophotometer (*see* **Note 11**).
2. Mix the reagents in microinjection buffer at the following concentrations:
   - Linearized donor transgene: 10 ng/μL.
   - Endonuclease (protein or mRNA): 10 ng/μL.
   - Single guide RNA: 10 ng/μL.
3. Freeze down the aliquots until use. Keep on ice during microinjection. Do not refreeze potential leftover, discard after use.

3.2 Embryos Production (Fig. 2a)

1. Fertilized zygotes are obtained by superovulation of female mice (*see* **Note 6**). Inject intraperitoneally (i.p.) 5 IU of PMSG per mouse, followed by 5 IU i.p. hCG 46–48 h later.
2. Immediately mate overnight with males (*see* **Note 12**). At the same time, mate recipient females with vasectomized males.
3. The following morning, euthanize the females and collect the Cumulus-Oocyte-Complexes (COCs) in pre-warmed KSOM medium in a 3.5 cm plastic petri dish. Check the recipient females for oestrus and select the pseudopregnant females displaying a white vaginal plug for future reimplantation.

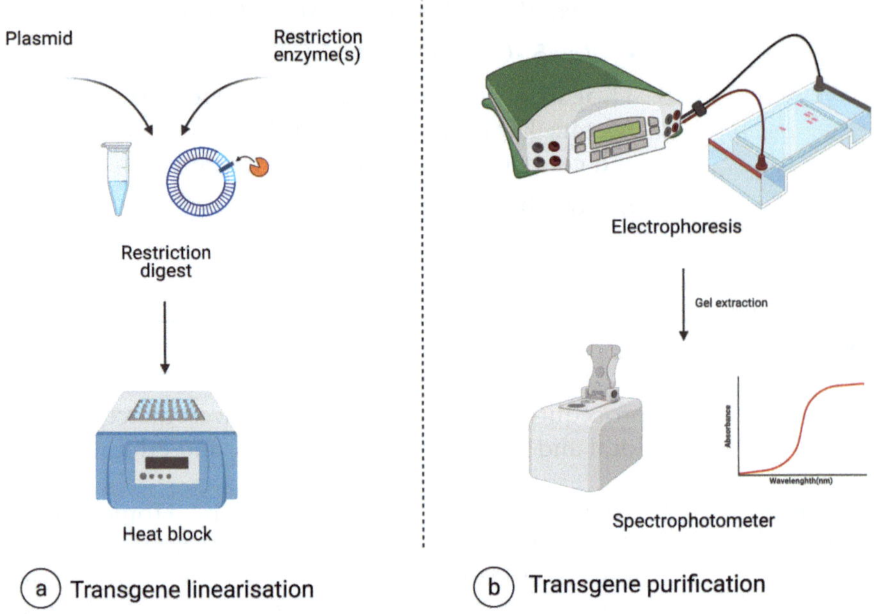

Fig. 1 Preparation of the injection mix. (**a**) The transgene is linearized by incubation with appropriate restriction enzyme(s). Incubation time and buffer composition (particularly for double digest) are determined following the manufacturer's recommendations. (**b**) The digested plasmid is run on an agarose gel and the linearized transgene is purified by gel extraction and eluted in microinjection buffer. Quality control is performed using a spectrophotometer before the transgene is mixed with ribonucleoprotein complexes for microinjection

4. Add 2–3 μL of hyaluronidase to the dish and let the cumulus cells detach from the zygotes. With a mouth pipette, transfer the purified zygotes into a drop of KSOM medium (overlaid with paraffin oil) and place the dish in the incubator.

3.3 Microinjection (Fig. 2b, c)

1. Connect the holding needle to the micromanipulator linked to the manual microinjector.
2. Back fill the injection needle with 3 μL of the injection mix using a loading microcapillary and connect it to the micromanipulator linked to the pneumatic injector.
3. Load around 50 zygotes into the injection chamber using the mouth pipette.
4. Place the injection chamber on the stage of the inverted microscope and position the needles in the injection chamber, in focus with the zygotes.
5. Secure one zygote to the holding needle using the manual microinjector.
6. Insert the tip of the injection needle into one of the two pronuclei (*see* **Note 13**).
7. Inject the mix with the pneumatic injector until the pronucleus visibly swells.

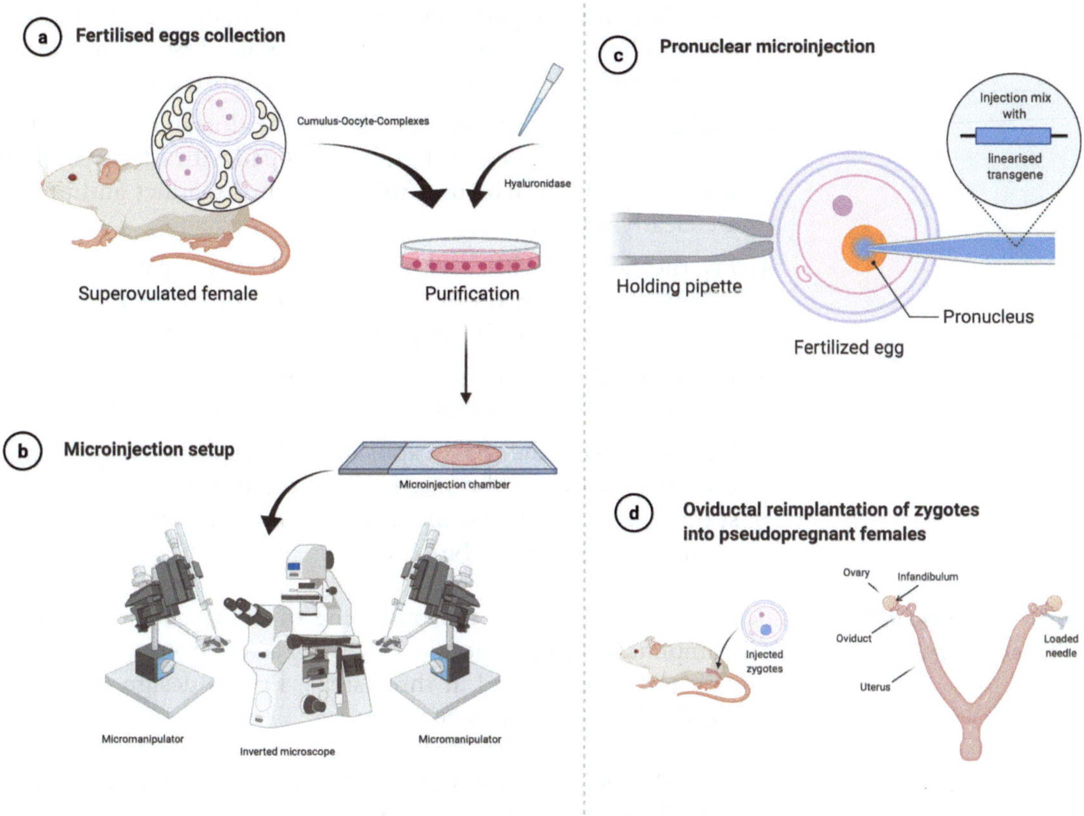

Fig. 2 Assisted reproductive technology (ART) workflow associated with the production of Rosa26-targeted KI mice. (**a**) Collection of fertilized zygotes from superovulated females is performed by incubation of cumulus-oocyte-complexes (COCs) with hyaluronidase. (**b**) Purified zygotes are transferred to the microinjection chamber for injection. (**c**) Pronuclear injection of the mix for targeted integration is visualized when the pronucleus (in focus with the holding and injection needles) swells. (**d**) Injected eggs are reimplanted through the wall of the oviduct in selected recipient females

8. Withdraw rapidly the injection needle from the zygote.
9. Move the zygote in a defined part of the depression slide (*see* **Note 14**).
10. Repeat the procedure for all zygotes.
11. Once all zygotes are injected, transfer them back into the culture dish (i.e. KSOM) using the mouth pipette and put the dish back in the incubator.

3.4 Oviductal Reimplantation (Fig. 2d) (See Note 15)

The entire procedure is performed using the dissecting microscope. Once all zygotes have been injected and cultured in the incubator for 30 min, observe them and discard all lysed zygotes.

1. Load 18–25 injected zygotes into the mouth pipette, minimizing the volume of culture medium carried over.

2. Anesthetize a plugged pseudopregnant female with the isoflurane machine, inject it with an appropriate dose of analgesics before placing it on the heating mat (*see* **Note 16**).
3. Make a small skin incision parallel to the dorsal midline of the mouse, cut the muscle and grab the fat pad with a forceps, then gently pull the ovary out until its attached oviduct and uterus are clearly visible.
4. Fix the fat pad with a vessel clamp. Using a pair of microscissors make an incision into the wall of the oviduct few millimeters upstream of the ampulla containing the COCs.
5. Introduce the mouth pipette into the oviduct and expel the zygotes until an air bubble is visible inside the ampulla.
6. Gently remove the glass capillary and place the reproductive tract back into the abdomen. Close the incision with surgical sutures, then apply wound clips. Monitor the mouse closely until full recovery (*see* **Note 17**).

3.5 Screening of the Progeny (Fig. 3)

Successful integration of the transgene of interest is screened by PCR amplification of genomic DNA extracted from tail (or ear) biopsies and is performed in three steps. Primer design (*see* **Note 18**) for PCR genotyping is critical.

3.5.1 DNA Extraction

1. Lyse the tissue biopsies at 95 °C in a heat block for 1 h in 100 μL of alkaline lysis reagent.
2. Add 100 μL of neutralizing reagent.
3. Centrifuge for 5 min at 12,000 × *g* at 4 °C.
4. Store the samples at 4 °C for up to 6 months.

Note that genomic DNA extraction can also be performed using proteinase-K based kits, following the manufacturer's instructions.

3.5.2 PCR Genotyping

1. Transgene-specific PCR:

   First, a transgene-specific PCR is performed to identify all pups carrying the transgene. The PCR amplicon is usually around 500 bp and both Forward and Reverse primers sit on the transgene (Fig. 3b). As such, this PCR will detect both targeted (i.e. by HDR) and random integration events.

   Next, correct integration of the transgene on either side of the double-strand break is checked by flanking PCRs at both 5′ and 3′ ends of the integration site.
2. 5′ Flanking PCR:

   For the 5′ flanking PCR (Fig. 3c), the Forward primer must bind the endogenous Rosa26 locus outside of the Left Homology Arm (LHA), while the Reverse primer needs to sit on the transgene.

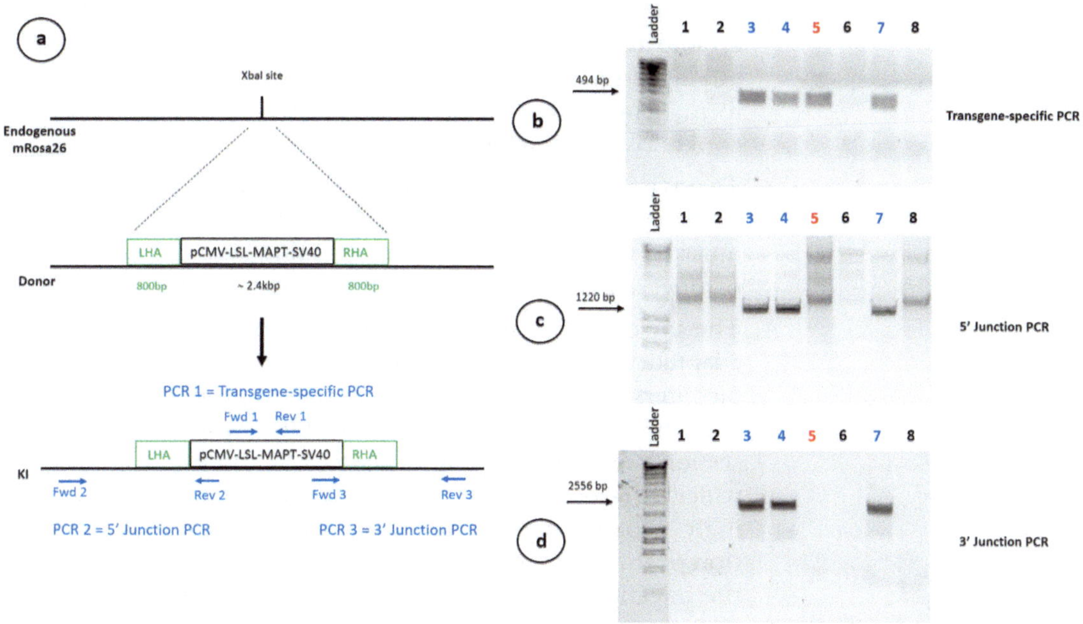

Fig. 3 Genotyping strategy for the screening of targeted insertion events. (**a**) Schematic of the design and positioning of the primers for the 3-step PCR genotyping of the offspring. (**b**) The transgene-specific PCR identifies all pups carrying a transgene integrated in their genome. (**c**) The 5′ junction PCR confirms that homologous directed repair (HDR) occurred properly at the left homology arm. (**d**) The 3′ junction PCR confirms that HDR occurred properly at the right homology arm. Pups #3, 4, and 7 are properly targeted, while pup #5 carry a random integration (i.e. the transgene-specific PCR is positive, however both junction PCRs are negative). Pups #1, 2, 6, and 8 do not carry any transgene (i.e. wildtype)

3. 3′ Flanking PCR:

   Conversely, for the 3′ flanking PCR (Fig. 3d), the Forward primer must sit on the transgene, while the Reverse primer must bind the endogenous Rosa26 locus outside of the Right Homology Arm (RHA).

Founders carrying a properly targeted transgene should present with a band on all three PCRs. Such Founders should ultimately be sequenced across the integration site by next-generation sequencing (i.e. targeted deep sequencing) to confirm that no undesired mutation (e.g. integration of the backbone) occurred in the vicinity of the integration site, and that no recombination events occurred within the transgene (e.g. concatemerization) before integration.

4 Notes

1. Other sgRNAs targeting the original *XbaI* site can be used although the specific sgRNA described herein has been shown to be very efficient in mice [16].

2. Each new batch of glass capillaries needs to first be run through a ramp test to determine their melting point before routine pulling is set up. Parameters should be adapted to the desired form, length, and shape of the injection needle. Refer the cookbook of the puller.
3. Two types of manual microinjectors can be used: oil-based and air-based. Should the oil-based microinjector be selected, it must be filled with paraffin oil, avoiding any air bubble in the oil line, as per the manufacturer's recommendations.
4. The injection pressure and compensation pressure should be adapted to the size of the opening of the tip to allow few picoliters to be delivered during the pronuclear injection while preventing backflow in the needle. As the microinjection session progresses, the tip of the needle can either clog or get wider. Pressure should then be increased or decreased accordingly. Clogged needles can be flushed using the "Clean" function of the injector. Injection needles should be replaced when the injection flow can no longer be controlled, or the lysing rate of the zygotes is too high.
5. The use of a piezo drill to penetrate the zygote is recommended if injecting at the 2-cell stage [17]. The entire procedure remains otherwise identical.
6. The choice of the mouse strain for both eggs donors and recipient (pseudopregnant) females is critical. Strains such as FVB or B6D2F1 hybrids are ideal when learning microinjection because their eggs are resistant to potential damages induced during the whole process. C57BL/6 is the most common strain used in biomedical research, however C57BL/6 eggs are sensitive to the microinjection process. Therefore, it is recommended to only attempt microinjection in C57BL/6 zygotes once experienced. Likewise, albino strains such as CD-1 or Swiss are good recipients as they care for their litters and are usually not prone to cannibalism.

   A genetically modified mouse line can be transferred from one background to the other by backcrossing it over several generations although this process is very lengthy and marker-assisted selection breeding protocols (i.e. "speed congenics") can help accelerate the process [18].

   The age of the mice is also a critical factor. Recipients need to be at least 8-week-old, whereas donor females cannot be used at age 5-8 weeks. The preferred age for donor females is 3-4 week-old [19].
7. Inhibin antiserum combined with equine chorionic gonadotropin (IASe) can be used in lieu of PMSG when the donor strain does not yield enough eggs [20]. However, IASe cannot

be used with young C57BL/6 females as the yield is too high and the sperm cannot fertilized the zygotes by natural mating (in vitro fertilization is then required) [21].

8. Injectable analgesics may be classified as restricted substances and may require licenses to obtain, use, and discard them according to applicable regulations.

9. Sequences for Homologous Directed Repair (HDR) are:
   - Left Homology Arm (LHA):

     5′GCGCCTGTCAGTTAACGGCAGCCGGAGTGCGCA
     GCCGCCGGCAGCCTCGCTCTGCCCACTGGGT
     GGGGCGGGAGGTAGGTGGGGTGAGGCGAGCT
     GGACGTGCGGGCGCGGTCGGCCTCTGGCGGG
     GCGGGGGAGGGGAGGGAGGGTCAGCGAAAGT
     AGCTCGCGCGCGAGCGGCCGCCCACCCTCCC
     CTTCCTCTGGGGGAGTCGTTTTACCCGCCGCC
     GGCCGGGCCTCGTCGTCTGATTGGCTCTCGG
     GGCCCAGAAAACTGGCCCTTGCCATTGGCTCGT
     GTTCGTGCAAGTTGAGTCCATCCGCCGGCCAGC
     GGGGGCGGCGAGGAGGCGCTCCCAGGTTCCGG
     CCCTCCCCTCGGCCCCGCGCCGCAGAGTCTGG
     CCGCGCGCCCCTGCGCAACGTGGCAGGAAGCG
     CGCGCTGGGGGCGGGGACGGGCAGTAGGGCTG
     AGCGGCTGCGGGGCGGGTGCAAGCACGTTTCC
     GACTTGAGTTGCCTCAAGAGGGGCGTGCTGAG
     CCAGACCTCCATCGCGCACTCCGGGGAGTGGA
     GGGAAGGAGCGAGGGCTCAGTTGGGCTGTTTT
     GGAGGCAGGAAGCACTTGCTCTCCCAAAGTCG
     CTCTGAGTTGTTATCAGTAAGGGAGCTGCAGTG
     GAGTAGGCGGGGAGAAGGCCGCACCCTTCTCC
     GGAGGGGGGAGGGGAGTGTTGCAATACCTTTC
     TGGGAGTTCTCTGCTGCCTCCTGGCTTCTGAG
     GACCGCCCTGGGCCTGGGAGAATCCCTTCCCC
     CTCTTCCCTCGTGATCTGCAACTCCAGTCTTTC
     T-3′

   - Right Homology Arm (RHA):

     5′ AGAAGATGGGCGGGAGTCTTCTGGGCAGGCTTA
     AAGGCTAACCTGGTGTGTGGGCGTTGTCCTGC
     AGGGGAATTGAACAGGTGTAAAATTGGAGGGA
     CAAGACTTCCCACAGATTTTCGGTTTTGTCGG
     GAAGTTTTTTAATAGGGGCAAATAAGGAAAATG
     GGAGGATAGGTAGTCATCTGGGGTTTTATGCA
     GCAAAACTACAGGTTATTATTGCTTGTGATCCG
     CCTCGGAGTATTTTCCATCGAGGTAGATTAAAG
     ACATGCTCACCCGAGTTTTATACTCTCCTGCTT
     GAGATCCTTACTACAGTATGAAATTACAGTGTC
     GCGAGTTAGACTATGTAAGCAGAATTTTAATCA

TTTTTAAAGAGCCCAGTACTTCATATCCATTTC
TCCCGCTCCTTCTGCAGCCTTATCAAAAGGTAT
TTTAGAACACTCATTTTAGCCCCATTTTCATTTA
TTATACTGGCTTATCCAACCCCTAGACAGAGCA
TTGGCATTTTCCCTTTCCTGATCTTAGAAGTCT
GATGACTCATGAAACCAGACAGATTAGTTACAT
ACACCACAAATCGAGGCTGTAGCTGGGGCCTC
AACACTGCAGTTCTTTTATAACTCCTTAGTACA
CTTTTTGTTGATCCTTTGCCTTGATCCTTAATT
TTCAGTGTCTATCACCTCTCCCGTCAGGTGGT
GTTCCACATTTGGGCCTATTCTCAGTCCAGGG
AGTTTTACAACAATAGATGTATTGAGAATCCAA
CCTAAAGCTTAACTTTCCACTCCCATGAATGCC
TCTCTCCTTTTTCT-3′

10. When using restriction enzyme(s) to linearize the transgene, always check that the transgene does not contain the corresponding restriction site(s) in its sequence to preserve its integrity.
11. The quality of the transgene preparation should also be checked using a spectrophotometer, as it is critical for subsequent microinjection. Check that the A260/A280 ratio is around 1.8 (i.e. no contamination by proteins) and the A260/A230 ratio is at least 2.0 (i.e. no contamination by organic solvents).
12. Males should be kept single-housed and mated 1:1 (or 1:2 maximum) with females. Recipient females not in estrus on the day of microinjection can be re-mated in subsequent sessions.
13. Both male and female pronuclei can be injected independently or in combination [22].
14. Two options could be chosen from and need to be tested empirically: either keeping the non-injected eggs below the needles and move the injected zygotes above the needles, or vice versa. Zygotes that do not display two visible pronuclei (unfertilized or parthenogenetic) should not be injected and should be discarded.
15. Infundibular reimplantation can be performed (in lieu of oviductal reimplantation) by tearing the bursa and inserting the mouth pipette in the infundibulum. However, oviductal reimplantation is less technically challenging and easier to perform [23].
16. Monitor the loss of toe pinch reflex to ensure the mouse is properly anesthetized before starting the surgery.
17. Recipient females can be pooled in a cage as they may assist each other with fostering the pups.

18. Manual primer design should be avoided. Instead, free online tools such as Primer-BLAST allow for an easy design of PCR primers, while ensuring that the selected primers are unique in (or excluded from) the murine genome.

References

1. Delerue F, Ittner LM (2015) Genome editing in mice using CRISPR/Cas9: achievements and prospects. Clon Transgen. https://doi.org/10.4172/2168-9849.1000135
2. Qin W, Dion SL, Kutny PM et al (2015) Efficient CRISPR/cas9-mediated genome editing in mice by zygote electroporation of nuclease. Genetics 200(2):423–430. https://doi.org/10.1534/genetics.115.176594
3. Kaneko T, Mashimo T (2015) Simple genome editing of rodent intact embryos by electroporation. PLoS One 10(11):e0142755. https://doi.org/10.1371/journal.pone.0142755
4. Miyasaka Y, Uno Y, Yoshimi K et al (2018) CLICK: one-step generation of conditional knockout mice. BMC Genomics 19(1):1–8. https://doi.org/10.1186/s12864-018-4713-y
5. Delerue F, White M, Ittner LM (2014) Inducible, tightly regulated and non-leaky neuronal gene expression in mice. Transgenic Res. https://doi.org/10.1007/s11248-013-9767-7
6. Karpen GH (1994) Position-effect variegation and the new biology of heterochromatin. Curr Opin Genet Dev 4(2):281–291. https://doi.org/10.1016/S0959-437X(05)80055-3
7. Sadelain M, Papapetrou EP, Bushman FD (2012) Safe harbours for the integration of new DNA in the human genome. Nat Rev Cancer 12(1):51–58
8. Zeng H, Horie K, Madisen L et al (2008) An inducible and reversible mouse genetic rescue system. PLoS Genet 4(5):e1000069. https://doi.org/10.1371/journal.pgen.1000069
9. Tasic B, Hippenmeyer S, Wang C et al (2011) Site-specific integrase-mediated transgenesis in mice via pronuclear injection. Proc Natl Acad Sci U S A 108(19):7902–7907. https://doi.org/10.1073/pnas.1019507108
10. Liu X, Wu H, Byrne M et al (1995) A targeted mutation at the known collagenase cleavage site in mouse type I collagen impairs tissue remodeling. J Cell Biol 130(1):227–237. https://doi.org/10.1083/jcb.130.1.227
11. Blake JA, Eppig JT, Kadin JA et al (2017) Mouse genome database (MGD)-2017: community knowledge resource for the laboratory mouse. Nucleic Acids Res 45(D1): D723–D729. https://doi.org/10.1093/nar/gkw1040
12. Zambrowicz BP, Imamoto A, Fiering S et al (1997) Disruption of overlapping transcripts in the ROSA βgeo 26 gene trap strain leads to widespread expression of β-galactosidase in mouse embryos and hematopoietic cells. Proc Natl Acad Sci U S A 94(8):3789–3794. https://doi.org/10.1073/pnas.94.8.3789
13. Liu C, Du Y, Xie W, Gui C (2013) Purification of plasmid and BAC transgenic DNA constructs. Methods Mol Biol 1027:203–215. https://doi.org/10.1007/978-1-60327-369-5_9
14. Yao X, Zhang M, Wang X et al (2018) Tild-CRISPR allows for efficient and precise gene knockin in mouse and human cells. Dev Cell 45(4):526–536. https://doi.org/10.1016/j.devcel.2018.04.021
15. Soriano P (1999) Generalized lacZ expression with the ROSA26 Cre reporter strain. Nat Genet 21(1):70–71
16. Chu VT, Weber T, Graf R et al (2016) Efficient generation of Rosa26 knock-in mice using CRISPR/Cas9 in C57BL/6 zygotes. BMC Biotechnol 16(1):1–5. https://doi.org/10.1186/s12896-016-0234-4
17. Gu B, Posfai E, Rossant J (2018) Efficient generation of targeted large insertions by microinjection into two-cell-stage mouse embryos. Nat Biotechnol 36(7):632–637. https://doi.org/10.1038/nbt.4166
18. Wong GT (2002) Speed congenics: applications for transgenic and knock-out mouse strains. Neuropeptides 36(2-3):230–236. https://doi.org/10.1054/npep.2002.0905
19. Fielder TJ, Barrios L, Montoliu L (2010) A survey to establish performance standards for the production of transgenic mice. Transgenic Res 19(4):675–681. https://doi.org/10.1007/s11248-009-9335-3
20. Takeo T, Nakagata N (2016) Immunotherapy using inhibin antiserum enhanced the efficacy of equine chorionic gonadotropin on superovulation in major inbred and outbred mice strains. Theriogenology 86(5):1341–1346. https://doi.org/10.1016/j.theriogenology.2016.04.076

21. Takeo T, Nakagata N (2015) Superovulation using the combined administration of inhibin antiserum and equine chorionic gonadotropin increases the number of ovulated oocytes in C57BL/6 female mice. PLoS One 10(5): e0128330. https://doi.org/10.1371/journal.pone.0128330
22. Abe T, Inoue KI, Furuta Y, Kiyonari H (2020) Pronuclear microinjection during S-phase increases the efficiency of CRISPR-Cas9-assisted Knockin of large DNA donors in mouse zygotes. Cell Rep 31:107653. https://doi.org/10.1016/j.celrep.2020.107653
23. Nakagata N (1992) Embryo transfer through the wall of the fallopian tube in mice. Jikken Dobutsu 41(3):387–388. https://doi.org/10.1538/expanim1978.41.3_387

Chapter 8

CRISPR-on for Endogenous Activation of *SMARCA4* Expression in Bovine Embryos

Virgilia Alberio, Virginia Savy, and Daniel F. Salamone

Abstract

The CRISPR-on system is a programmable, simple, and versatile gene activator that has proven to be efficient in cultured cells from several species and in bovine embryos. This technology allows for the precise and specific activation of single endogenous gene expression and also multiplexed gene expression in a simple fashion. Therefore, CRISPR-on has unique advantages over other activator systems and a wide adaptability for studies in basic and applied science, such as cell reprogramming and cell fate differentiation for regenerative medicine.

In this chapter, we describe the materials and methods of the CRISPR-on system for activation of the endogenous SMARCA4 expression in bovine embryos.

Key words CRISPR-dCasVP160, CRISPR-on, Gene expression, Bovine, Embryo

1 Introduction

In recent years, precise manipulation of mammalian genomes has been facilitated by engineered endonucleases like Zinc Finger Proteins (ZFPs) and Transcription Activator-Like Effector Nucleases (TALENs). More recently, a new endonuclease system called CRISPR-Cas (Clustered Regularly Interspaced Short Palindromic Repeats/CRISPR-associated) has been reported [1] as a more efficient, less expensive, and technically easier-to-obtain tool [2–4]. Since its development as a tool, CRISPR/Cas has been efficiently used to induce precise modifications in the genome of a wide spectrum of organisms [2–4].

The advantages of CRISPR/Cas over other engineered nucleases are in terms of simplicity and versatility. The CRISPR/Cas system can be programmed to edit different genes by simply changing the nucleotide sequence of a short RNA molecule that

Virgilia Alberio and Virginia Savy should be considered join first author.

Paul John Verma, Huseyin Sumer and Jun Liu (eds.), *Applications of Genome Modulation and Editing*,
Methods in Molecular Biology, vol. 2495, https://doi.org/10.1007/978-1-0716-2301-5_8,

guides the Cas9 nuclease to its target site through complementary base pairing, known as sgRNA [2–4].

CRISPR/Cas is not just a gene-editing tool. Through point mutations, a catalytically inactive nuclease version of Cas9 (dCas9, from "*dead–Cas9*") was obtained and then fused to different epigenetic effectors like DNA methylating domains, histone acetylases/deacetylases, or transcription activators/repressors *(reviewed by* [5]*)*. These effectors were originally tested in TALENs and ZFPs platforms [6–9] to turn on or off the expression of individual genes in a precise manner in cells of various species [9–14]. However, the laborious and expensive procedures to design them hindered their widespread use [15].

The dCas9-mediated artificial transcription factors, also known as CRISPR-on, have been improved over the years to achieve higher levels of induction. A dCas9 fused to 4 tandem repeats of the VP16 activator (a protein from Herpes Simplex virus involved in gene activation) resulted in the dCas9-VP64 artificial transcription factor that induces immediate gene transcription in cells [11–13, 16]. The so-called next-generation activators fused 10 tandem repeats to obtain the dCas9-VP160 version with improved activity [17]. These effectors induce histone acetylation indirectly, through the recruitment of transcription factors such as histone acetyltransferases [18]. Finally, by fusing dCas9 to three different activation domains the gene activation strategy was improved, and a robust gene induction was achieved in cells [19]. Moreover, the use of pools of sgRNAs to target a single gene results in a synergistic effect on the level of transcriptional activation [11–13, 19]. Finally, CRISPR-on can simultaneously activate multiple genes when different genes are targeted at the same time, similarly to what was observed when using the dCas9 nuclease in gene editing [17].

Regarding nonspecific effects, changes associated with gene transcription, epigenetic marks, and chromatin accessibility are often limited to the intended target in a very specific manner [2, 11, 13, 14, 20–24]. This is because pools of sgRNAS are necessary to induce a significant change and, therefore, off-target effects are reduced. In any case, as occurs for the CRISPR/Cas9 tool, there are algorithms to reduce the probability of such nonspecific unions to design precise sgRNAs [25–27].

Particularly in bovine embryos, the CRISPR-on system was recently used for the transient modulation of endogenous gene expression at the early stages of development. Thus, it could be an attractive approach to correct deficiencies in the gene expression of in vitro produced embryos from in vitro fertilization (IVF), intracytoplasmic sperm injection (ICSI), and somatic cell nuclear transfer (SCNT). In this chapter, we describe the materials and methods of the CRISPR-on system for the activation of the endogenous SMARCA4 (SM) expression in bovine embryos that was recently reported for the first time by our laboratory [28].

2 Materials

2.1 sgRNAs Design and Cloning into an Expression Vector

1. Restriction enzyme *Bsa I* and optimal buffer.
2. Ethidium Bromide (Working Solution 0.5 mg/mL).
3. 10× Loading Buffer.
4. Forward and reverse primers for sgRNA reconstruction.
5. T4 Ligase Buffer.
6. T4 DNA Ligase.
7. Competent bacteria (*e.g., E. coli* DH5α).
8. Kanamycin.
9. M13 Forward primer: 5′[GTAAAACGACGGCCAG]3′.
10. M13 Reverse primer: 5′[CAGGAAACAGCTATGAC]3′.
11. High-fidelity Taq Polymerase and optimal buffer.
12. RNase Inactivation Reagent.
13. Plasmid DNA Miniprep Kit.
14. DNA Gel Extraction Kit.
15. PCR Fragment Purification Kit.
16. 100 bp DNA molecular weight marker.
17. Plasmid: pUC57-sgRNA expression vector (Addgene, #51132).
18. Standard de-salted oligos: forward oligo (*Fwd*): 5′-TAGGG-N20–3′, reverse oligo (*Rv*): 3′-C-N20-CAAA-5′. The following table shows the specific sequences used for this work:

Oligos	**Sequence (5′–3′)**
SM1 Fwd	**TAGGG**CGGGGGCGCGGGCAGCGTG
SM2 Fwd	**TAGGG**AAGCGAGAGAGGGAGTTCG
SM3 Fwd	**TAGGG**CACGCGCGCTAGGAGCGGA
SM4 Fwd	**TAGGG**TTGTCTGGGAGAGGTGGGT
SM1 Rv	**AAAC**CACGCTGCCCGCGCCCCCGC
SM2 Rv	**AAAC**CGAACTCCCTCTCTCGCTTC
SM3 Rv	**AAAC**TCCGCTCCTAGCGCGCGTGC
SM4 Rv	**AAAC**ACCCACCTCTCCCAGACAAC

2.2 In Vitro Transcription (IVT)

1. Restriction enzyme Dra I and respective buffer.
2. BSA 100×.
3. RNase Inactivation Reagent.
4. RNase decontamination solution.

5. Ethidium Bromide.
6. Plasmid DNA Miniprep Kit.
7. PCR Purification Kit.
8. In vitro Transcription Kit for small sequences.
9. In vitro Transcription Kit for large sequences.
10. RNA Purification Kit for small sequences.
11. RNA Purification Kit for large sequences.
12. Plasmid: pAC154-dual-dCas9VP160-sgExpression (Addgene #48240).
13. High-fidelity DNA polymerase and buffer.
14. Universal Primer BGH Reverse.
15. Specific Forward primer containing the T3 promoter sequence (T3_Fwd): 5′- AATTAACCCTCACTAAAGGGAGAcaggttggaccggtgccaccA-3′. Uppercase letters indicate the T3 minimum promoter sequence.
16. DNAse.
17. *E. coli* Poly(A) Polimerase (EPAP).
18. 1 kb DNA molecular weight marker.
19. RNA molecular weight marker.
20. RNA Gel Loading Dye.

2.3 Microinjection of Bovine Zygotes

1. Denuded in vitro- or in vivo-derived zygotes.
2. 35 mm-, 60 mm-, 100 mm-Petri dish.
3. Mouth pipette or Drummond pipette.
4. Microinjection 9 mm pipette.
5. Holding pipette.
6. Tyrode's Lactate-Pyruvate-HEPES (TALP-H) medium: 114 mM NaCl, 3.2 mM KCl, 0.5 mM $MgCl_2 \cdot 6\ H_2O$, 2 mM $CaCl_2 \cdot 2H_2O$, 10 mM Sodium Lactate, 0.1 mM Sodium Pyruvate, 2 mM $NaHCO_3$, 5 mg/mL Phenol Red, 3 mg/mL BSA, 10 mM HEPES, and 1% Antimycotic Solution (100×, A5955 Sigma) in embryo tested water. Adjust pH to 7.3–7.4, Osmolarity must be 260–270 mOsm/kg. Store at 4 °C [29].
7. Synthetic Oviductal Fluid with amino acid (SOFaa) medium: 107.63 mM NaCl, 7.16 mM KCl, 1.19 mM KH_2PO_4, 1.51 mM $MgSO_4$, 1.78 mM $CaCl_2 \cdot 2H_2O$, 5.35 mM Sodium Lactate, 25 mM $NaHCO_3$, 7.27 mM Sodium Pyruvate, 45 mL/mL BME amino acids, 5 mL/mL MEM amino acids, 0.34 mM Tri-Sodium Citrate, 2.77 mM MyoInositol, 10 mg/mL Phenol Red, 7.27 mg/mL BSA, 1% Antimycotic Solution (100×, A5955 Sigma), and 2.5% FBS in embryo tested water. Store at 4 °C [30, 31].

8. Embryo culture-tested Mineral Oil.
9. Microinjection mixture (dCas9VP160 RNA + pool of sgRNAs).
10. Polyvinyl pyrrolidone (PVP).
11. RNA Stabilization and Storage Solution.
12. Gas mixture (5% CO_2, 5% O_2, 90% N_2).
13. Gas chamber for embryo culture.

2.4 RNA Isolation and RT-qPCR Analysis

1. RNA Extraction kit.
2. Random primers and reverse transcriptase for cDNA synthesis.
3. Specific primers for the analyzed genes (amplicon size 140 bp): Fwd: 5′-CTGCAGGAACGGGAATACAG-3′, and Rv: 5′-CTGGAAGTTCAGCAGTCTGAG-3′.
4. SYBR Green.

2.5 Immuno-cytochemistry (ICC)

1. Glass slides and coverslips.
2. 4-well and 96-well plates.
3. Transparent nail polish.
4. Petroleum jelly.
5. Antifade Mounting Medium.
6. PBS/PVP: 0.2% PVP in PBS. Store at 4 °C.
7. PBST/BSA: 0.1% Tween® 20 and 1% BSA in PBS. Store at 4 °C.
8. 4% paraformaldehyde (PFA): Prepare it when used.
9. Wash Solution: 0.1% Tween® 20, and 0.1% BSA in PBS. Store at 4 °C.
10. Permeabilization Solution: 0.25% Triton X-100 in PBS. Store at 4 °C.
11. Blocking Solution: 5% BSA in PBS. Store at 4 °C.
12. DAPI solution: 1:100 dilution of 100 mg/mL DAPI in PBS/PVP. Prepare it when used.
13. Primary Antibody: 1:300 dilution of Rabbit Anti-HA tag antibody (#9110, Abcam) in Blocking Solution.
14. Secondary Antibody: 1:1000 dilution of Goat Anti-rabbit IgG in PBST/BSA.

2.6 Buffers

1. Electrophoresis buffer TAE 50×: 242.2 g/L of Tris base, 52.1 mL/L of Acetic acid, 18.612 g/L of EDTA in ddH_2O [32].

2. Agarose Gel: addition of the desired amount of agarose to 1× TAE buffer to make 1.2% and 2% agarose gel [32].
3. LB medium: 5 g/L yeast extract, 10 g/L peptone tryptone, and 10 g/L NaCl in ddH_2O [33].
4. LB–agar medium: 5 g/L yeast extract, 10 g/L peptone tryptone, 10 g/L NaCl, 15 g/L agar ddH_2O [34].
5. 1% SDS: 1% (w/v) Sodium Dodecyl Sulfate (SDS) in distilled water.
6. Agarose gel with 1% (w/v) formaldehyde: Melt 1 g agarose in 72 mL water. Cool to 60 °C. In a chemical fume hood, add 10 mL of 10× MOPS buffer and 18 mL 37% formaldehyde. Assemble the gel.
7. 10× MOPS electrophoresis buffer: 0.2 MOPS (adjust to pH 7 with NaOH), 20 mM sodium acetate, 10 mL EDTA (pH 8). Sterilize the solution by filtration through a 0.45-mm Millipore filter, and store at room temperature protected from light.
8. Gel Loading Buffer: 95% formamide, 0.025% xylene cyanol, 0.025% bromophenol blue, 18 mM EDTA, 0.025% SDS in distilled water.

2.7 Equipment

1. Thermocycler for PCR and qPCR.
2. Laminar flow cabin.
3. Extraction hood.
4. Refrigerated ultracentrifuge.
5. Incubator for bacteria culture.
6. Shaker incubator for bacteria liquid culture.
7. UV Transilluminator.
8. −80 °C Freezer.
9. Horizontal gel electrophoresis apparatus.
10. Stereoscopic microscopy.
11. Micromanipulator and injector.
12. Inverted microscope.
13. Warm plate.
14. 5% CO_2 incubator for embryo culture.
15. Sanger sequencing platform and software for analysis.
16. Confocal Laser Scanning Microscope.

3 Methods

3.1 Template-Specific sgRNAs Design and Cloning into an Expression Vector

1. Determine the proximal promoter region of the target gene using available genome databases (*see* **Note 1**). The regulatory region ranging from −400 to −50 bp upstream of the TSS is considered as the optimal activation window [11, 14, 17]. Use either the template or non-template DNA strand (*see* **Note 2**).
2. Use any guide RNA design tool to design four non-overlapping sgRNAs targeting the proximal promoter region of the target gene (*see* **Note 3**).
3. Select the best four non-overlapping oligonucleotides from the output of the sgRNA design tool (*see* **Notes 2** and **4**).
4. Confirm that the selected sgRNAs do not contain restriction enzyme sites for *Dra I* (5′-TTTAAA-3′) or *Bsa I* (5′-GGTCTC-3′).
5. Dilute *Fwd* and *Rv* oligos to 100 mM in ddH_2O.
6. Prepare one annealing reaction for each sgRNA oligo pair:

Component	Amount
Oligo_*Fwd* (100 mM)	1 mL
Oligo_*Rv* (100 mM)	1 mL
T4 Ligation Buffer 10×	1 mL
ddH_2O	6.5 mL
T4 Ligase	0.5 mL
Final volume	10 mL

7. Place the samples in a thermocycler and set up the following conditions: 37 °C for 30 min, 95 °C for 5 min, and then ramp down to 25 °C at 5 °C/min.
8. For in vitro transcription, we use the pUC57-sgRNA expression vector (Fig. 1) that contains a T7 promoter and a kanamycin-selectable marker (*see* **Note 5**). Digest the expression vector at 37 °C for 2 h.

Components	Amount
pUC57-sgRNA	1–5 mg
B*saI*	2 mL
Buffer 10× (Cutsmart)	10 mL
BSA 100×	1 mL
ddH_2O	× mL
Final volume	100 mL

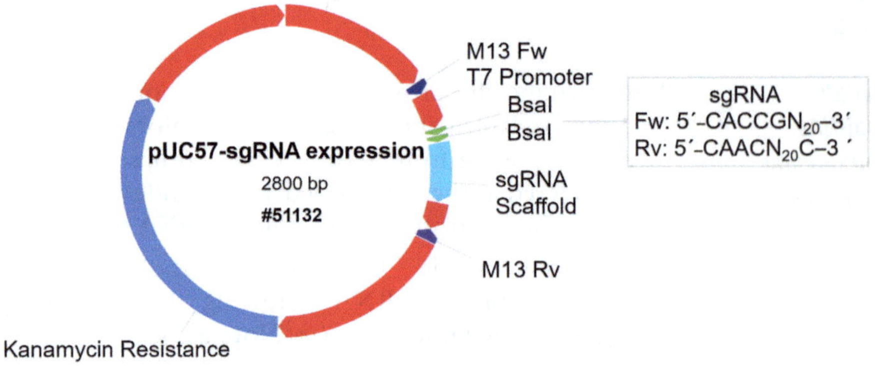

Fig. 1 Expression vector *pUC57–sgRNA expression* contains sequences for kanamycin resistance, M13 primers *(Fwd* and *Rv)* to confirm the presence of the guides, sgRNA scaffold, T7 promoter, and *BsaI* restriction sites, where each guide is inserted with the sequence shown on the right

9. Heat-inactivate the reaction at 65 °C for 20 min.
10. Use 1% (w/v) agarose gel electrophoresis to remove the 25 bp product of the *Bsa I* digestion and verify the fragment size (2760 bp) using a suitable DNA ladder.
11. Purify the linearized plasmid using the commercial gel extraction kit according to the manufacturer's directions. Elute the DNA in 30 mL of the kit's buffer or water.
12. Set up the ligation reaction as follows. Incubate the reaction at 16 °C for overnight.

Component	Amount
Bsa I-digested plasmid	50 ng
Annealed oligo duplex (1:200 dilution)	1 mL
10× T4 Ligase buffer	1 mL
ddH_2O	× mL
T4 DNA ligase	1 mL
Final volume	10 L

13. Include a digested plasmid only with no annealed oligo reaction as control.
14. Transform each ligation into a competent *E. coli* strain (*see* **Note 6**). Include a negative control of vector-only ligation.
15. Plate the transformation on LB agar plates supplemented with 50 μg/mL kanamycin and incubate them at 37 °C overnight.
16. Evaluate the plates for colony growth. Inoculate a single colony into a 3 mL culture of LB medium with 50 μg/mL kanamycin and shaking culture for overnight at 37 °C.

17. Check two or three colonies of each sgRNA for the correct insertion. Isolate the plasmids using a commercial miniprep kit following the manufacturer's instructions.
18. Confirm ligation of the annealed oligos by *Bsa I* digestion or colony-PCR (*see* **Note 7**).
19. Verify the selected plasmids by capillary Sanger sequencing as oligos may be mutated.
20. Inoculate the correct clones in 3–5 mL LB medium with kanamycin and shaking culture overnight at 37 °C.
21. Prepare bacteria glycerol stocks for each clone as follows and store them at −80 °C.

Component	Amount
Overnight culture	350 mL
100% (v/v) sterilized glycerol	150 mL

3.2 In vitro transcription (IVT) of sgRNA

1. Use the extra volume of overnight culture for stocks to isolate the plasmids using a commercial miniprep kit following the manufacturer's instructions.
2. Digest 1 mg of each sgRNA expression vector as follows:

Component	Amount
Confirmed pUC57-sgRNA vectors	1–5 μg
Dra I	1 mL
Buffer 10× (Cutsmart)	1 mL
ddH_2O	× mL
BSA (100×)	1 mL
Final volume	100 mL

3. Incubate the samples at 37 °C overnight.
4. Treat the digestion reactions with RNase Inactivation Reagents (*see* **Note 8**) and incubate at 60 °C for 10–20 min (*see* **Note 9**).
5. Clean the bench with a commercially available RNase decontamination Solution (*see* **Note 10**).
6. From this step forward, use gloves and maintain RNase-free technique during all RNA synthesis and manipulation.
7. Clean up the digestion reaction using a commercial PCR Purification Kit and elute in 30 mL of RNAse-free water.
8. Measure the DNA concentration using a NanoDrop spectrophotometer.
9. Set the IVT reaction using a commercial IVT kit (*see* **Note 11**).

10. Use DNAse treatment for 10 min to remove contaminating DNA.
11. Purify the synthesized sgRNAs using commercially available kits (*see* **Note 12**).
12. Elute in 30 mL of RNAse-free water.
13. Measure the RNA concentration using a NanoDrop spectrophotometer.
14. Check for single–band product in a 1% agarose gel electrophoresis (no denaturing).
15. Prepare appropriate volume aliquots and store at −80 °C (*see* **Note 13**).

3.3 IVT of dCas9–Act

We use pAC154-dual-dCas9VP160-sgExpression plasmid as a template for IVT of protein dCas9 fused to 10-tandems repeats of VP16 domain (dCas9VP160) (Fig. 2).

1. Design a specific *Fwd* primer containing the T3 sequence to PCR-amplify the coding region of dCas9 (T3-dCas9VP160-Fwd, *see* **Note 14**).
2. Set up the following PCR reaction (*see* **Note 15**):

Component	Amount
Buffer 5×	10 mL
dNTPs (10 mM)	2 mL
T3_dCas9VP160_Fw	1 mL
BGH *Rv* Primer	1 mL
High-fidelity DNA polymerase	0.5 mL
Plasmid (#48240)	~5 ng
ddH_2O	× mL
Final volume	50 mL

3. Use the following cycle parameters: 1 cycle at 95 °C for 5 min, followed by 95 °C for 30 s, 58 °C for 30 s, and 68 °C for 5 min, during 35 cycles, and a final extension at 68 °C for 5 min.
4. Use 1% (w/v) agarose gel electrophoresis to check for a single–band product of the appropriate length (~4700 bp).
5. Treat the PCR product with RNase Inactivation Reagents (*see* **Note 8**) and incubate at 60 °C for 10–20 min.
6. Purify the PCR product using a commercial PCR Purification Kit.
7. Clean the bench with any commercially available RNase decontamination Solution (*see* **Note 10**).

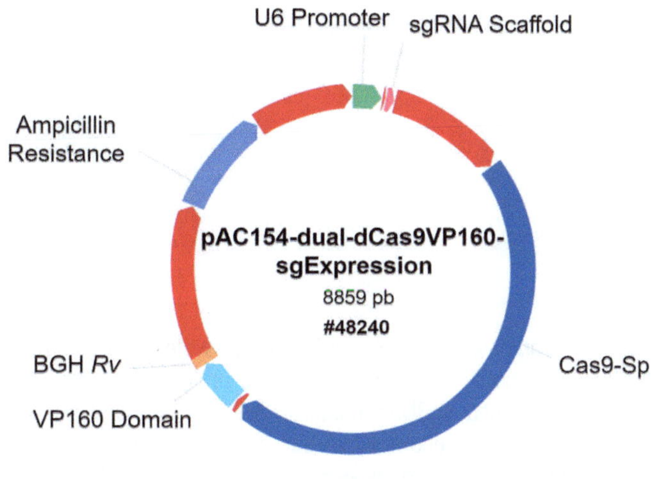

Fig. 2 Expression vector pAC154–dual–dCas9VP160–sgExpression contains sequences for dCas9 protein from *Streptococcus pyogenes* (Sp), the VP160 domain, ampicillin resistance as a selection antibiotic and BGH Rv primer

8. From this step forward, use gloves and maintain RNase-free technique during all RNA synthesis and manipulation.
9. Synthesize the capped mRNA using a commercial T3 Transcription Kit and follow the manufacturer's instructions (*see* **Note 16**).
10. Use 0.1–0.2 mg of the purified PCR product as templates for in vitro transcription and 1 mL of DMSO to prevent secondary structure of the DNA in a 20 mL final reaction. Incubate at 37 °C for 2 h.
11. Polyadenylate the 3′-termini of in vitro transcribed RNA using the Poly(A) Tailing Kit and following the manufacturer's instructions.
12. Purify the synthesized mRNA using commercially available column-based RNA clean up kits. Keep the tube in ice.
13. Measure the RNA concentration using a NanoDrop spectrophotometer.
14. Prepare a 1% (w/v) formaldehyde-containing agarose gel.
15. Prepare the RNA samples for denaturing agarose-formaldehyde gel electrophoresis: mix 2 volumes of Gel Loading Buffer and 1 volume of RNA sample. Heat at 70 °C for 2 min and chill immediately on ice to prevent secondary structure formation.
16. Load the samples in the gel (*see* **Note 17**). Include an RNA size marker (*see* **Note 18**).

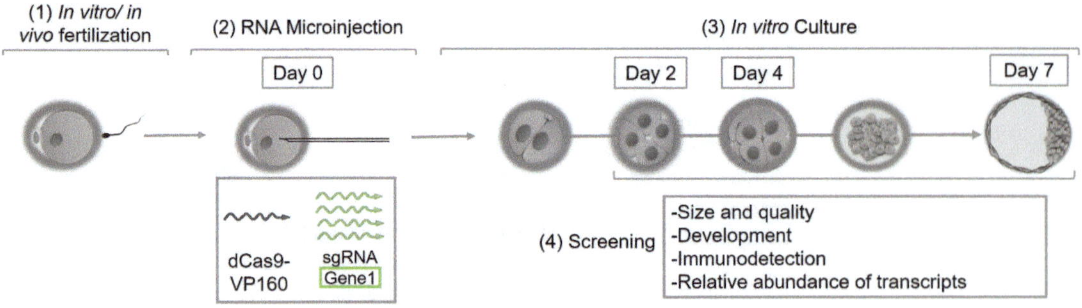

Fig. 3 General workflow scheme of CRISPR-on system for the RNA microinjection in bovine embryos. Presumed bovine zygotes are produced in vivo or in vitro (1) and then microinjected at day 0 (D0) with dCas9-VP160 mRNA and sgRNAs mixture. Embryos are cultured in vitro until D7 post-fertilization (3). Alternatively, samples are recovered at day 2, 4, and 7 to evaluate different parameters such as embryo size, development and quality, gene expression, and immunodetection (4)

3.4 Microinjection of Bovine Zygotes

The procedures described below aim to evaluate the modulation of targeted endogenous gene expression in preimplantation bovine embryos, using CRISPR-on system. For this, it is necessary to follow a series of steps that are briefly described in Fig. 3. Presumptive zygotes produced in vitro or in vivo are cytoplasmically injected with a mixture of dCas9VP160 mRNA and four non-overlapping sgRNAs. At days 2, 4, and 7 of in vitro culture, parameters of embryo quality such as size, morphology, development, and relative abundance of transcripts are evaluated. It is also recommended to assess and quantify the presence of downstream proteins related to the transcription activation by immunodetection.

1. Immediately after IVF, select those denuded zygotes with at least one polar body.
2. Place the selected embryos in a 65-mm Petri dish containing 50 μL drops of SOFaa medium covered with mineral oil. Keep the dish in an embryo CO_2 incubator until microinjection.
3. As an IVF control, separate at least 40 embryos and place them in groups of 20 in 50 μL drops of SOFaa medium covered with mineral oil.
4. Place the holding and microinjection pipettes in the micromanipulator arms, each one attached to a microinjector.
5. Set up the location parameters of the pipettes under the light of the microscope.
6. Place a 100-mm Petri dish with two 100 mL drops of TALP-H medium on the microscope's warm plate.
7. Aspirate medium with both pipettes until a suction equilibrium is reached.
8. Adjust the injection pressure.
9. Prepare CRISPR-on injection mixture as follows (*see* **Note 19**):

Injection mixture	Component	Final concentration
CRISPR-on mixture	dCas9VP160 mRNA Mix of 4 sgRNAs PVP	100 ng/mL 50 ng/mL 10% (v/v)
SHAM control mixture	dCas9VP160 mRNA Mix of 4 sgRNAs PVP	100 ng/mL – 10% (v/v)

10. Place four drops of 100 mL TALP-H medium in the center of a 100-mm Petri dish forming a square shape and two or three more drops around, in case of eventual washes.
11. Above the upper left drop, not far from it, place the 3 μL CRISPR-on mixture on the plate and, in the same way, above the upper right drop, place the 3 μL SHAM control mixture.
12. Add embryo culture-tested mineral oil to the plate until the drops are covered.
13. With a mouth pipette, transfer groups of 20 embryos to the top of the upper drop of the micromanipulation plate (*see* **Note 20**).
14. Load 10 pL of SHAM mixture into the microinjection pipette (*see* **Note 21**).
15. Take a zygote with the holding pipette and insert the microinjection pipette inside the embryo.
16. Gently, aspirate to break the plasma membrane and then apply positive pressure to deposit the content of the pipette into the cytoplasm.
17. Remove the microinjection pipette and place the embryo at the bottom of the drop to identify that it has already been microinjected.
18. When all 20 embryos have been microinjected, place them in a 50 μL drop of SOFaa medium, and incubate them inside the CO_2 incubator.
19. Proceed microinjecting the rest of the embryos.
20. Place groups of 20 microinjected embryos into 50 μL drops of SOFaa medium at 38.5 °C in humidified gas mixture (5% CO_2, 5% O_2, 90% N_2) inside a gas chamber.
21. Make sure the Petri dish is well labeled with the correct treatments.
22. Renew the culture medium on day 2 of in vitro culture.
23. Record cleavage, morulae, and blastocyst rates on days 2, 5, and 7, respectively.

24. Collect pools of 10 embryos at days 2 (D2) and 4 (D4) of in vitro culture.
25. Collect pools of five blastocysts at day 7 (D7) of in vitro culture.
26. Place each pool of embryos in a RNAse-free 1.5 mL tube with 8 μL of RNA stabilization and storage solution (*see* **Note 22**) for subsequent *RT-qPCR* analysis (*see* **Notes 23** and **24**). Store them at −80 °C until analysis.

3.5 Embryo Quality Assessment

1. The embryonic developmental rate is determined by the ratio of the total number of cleaved embryos, morulae, or blastocysts obtained and the total number of oocytes used.
2. Use the millimeter eyepiece or scale attached to a microscope to determine the size of the embryos in each stage, measured in μm.
3. Use the IETS Manual as a guide of the different embryo morphologies for its classification (*see* **Notes 24** and **25**).

3.6 Gene Transcription Analyses Using RT-qPCR

1. Design gene-specific primers for RT-qPCR analysis (*see* **Note 26**). Include specific primers for internal standards, such as Actin and Gapdh.
2. Isolate RNA from the pools of 10 D2 and D4 embryos and five D7 blastocysts using a commercial kit, according to manufacturer's instructions (*see* **Note 27**).
3. Use isolated RNA as templates for cDNA synthesis.
4. Prepare the mix for cDNA synthesis with random primers and a reverse transcriptase, with 10 μL of total RNA in a 20 μL of final reaction.
5. Perform the qPCR analysis of the samples as plicates (technical replicates) in a 50 μL reaction, containing Sybr green, 20 mmol/μL of each primer, cDNA template, and Milli-Q water.
6. Analyze the data using the ddCT method with the geometric mean of Actin and Gapdh as internal standards

3.7 Immunocytochemistry to Determine dCas9VP160 Protein

This procedure is specific to determine the dCas9VP160 mRNA decay and translation at days 2, 4, and 7 of bovine embryo development [28]. However, it is a general protocol that can be adapted to other mRNAs and proteins by using specific antibodies (and their dilutions). As an example, it is shown the effect of the modulation of endogenous gene expression on embryo quality in Fig. 4. The immunocytochemistry was performed to assess the expression of a downstream-protein of the target gene endogenously activated by the CRISPR-on system.

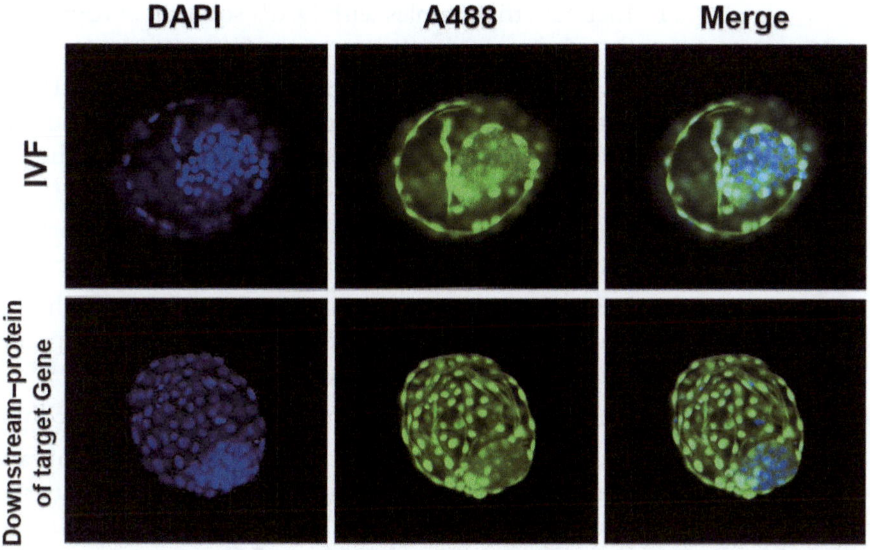

Fig. 4 Effect of the modulation of endogenous gene expression on embryo quality. Top row: IVF non-microinjected control. Bottom row: ICC of the downstream-protein of the target gene endogenously activated by the CRISPR-on system. Images represent a single Z plane of individual D7 embryos

1. Remove the zona pellucida of the embryos with 15 mg/mL of protease (*see* **Note 28**) in TALP-H, until total degradation is observed.
2. Fix the preimplantation embryos in 100 mL drops of fresh 4% PFA at room temperature for 20 min.
3. Wash the embryos in a 300 mL drops of PBS/PVP. Repeat two times. Samples can be stored up to 1 week at 4 °C.
4. Incubate the samples in 300 mL drops of permeabilization solution for 15 min.
5. Wash the embryos in a 300 mL drop of Wash Solution for 15 min. Repeat 2 times.
6. Incubate the embryos in Blocking Solution for 1 h to block nonspecific immunoreactions.
7. In a 96-well plate, incubate the samples in at least 50 mL of diluted primary antibody overnight at 4 °C (*see* **Notes 29** and **30**).
8. Wash the embryos in a 300 mL drop of Wash Buffer for 15 min. Repeat two times.
9. Place the embryos into a 1.5 mL tube and incubate them in 100 mL of the diluted secondary antibody in the dark at room temperature, for 1 h (*see* **Notes 29** and **31**).
10. Wash the embryos in a 300 mL drop of Wash Buffer for 15 min. Repeat two times.

11. Incubate the samples with DAPI solution at room temperature for 5 min.
12. Using a mouth pipette, put five embryos in a 10 mL drop of antifade mounting medium (*see* **Note 32**) within a Petri dish, taking as little volume as possible to minimize the presence of other media (*see* **Notes 33** and **34**).
13. Place two very fine lines of petroleum jelly on both sides of the glass slide at a distance equal to the width of the coverslip.
14. Between the two lines of petroleum jelly on the glass slide, place drops (maximum five drops per coverslip) of antifade mounting medium in a vertical way on the glass slide.
15. Place each embryo in one antifade mounting medium drop.
16. Using the mouth pipette, remove the excess of antifade mounting medium until the embryo is fixed to the glass slide and place the coverslip over them with gentle pressure to prevent them from collapse (*see* **Notes 35**).
17. Using a 10-uL pipette, fill in the space between the glass slide and the cover slide with antifade mounting medium by capillarity.
18. Seal the cover slide with transparent nail polish.
19. Make sure the glass slides are correctly labeled.
20. Samples can be stored no more than 1 week at 4 °C in a humidified chamber, prevented from light.
21. Observe the samples in a confocal microscope.

4 Notes

1. A good example of Genome database for most species but particularly for bovines is Ensembl Genome Browser (https://www.ensembl.org/index.html, [35]).
2. Please note that it is possible to use different versions of dCas9 with activator domains (i.e., p300, [36]) that require the selection of different ranges of regulatory regions and the use of different amounts of sgRNAs.
3. CRISPR online tools for sgRNA design:

Online tool	Reference
E-CRISPR design tool	[37]
Breaking-Cas	[38]
CRISPR design	[39][a]

[a] No longer available, although used in [28]

The sgRNA design tools identify and ranks the most suitable 20 bp oligonucleotides.

4. Verify that the PAM sequence (5′-NGG-3′) is not included in the oligo sequence. The PAM sequence mentioned is specific for the Cas9VP160 protein from *Streptococcus pyogenes*, described in this protocol. Always check the PAM sequence corresponding to the Cas nuclease to be used, as this varies depending on the species from where the protein was obtained.
5. For any other expression vector, verify and adapt the restriction enzymes needed.
6. *DH5a* strain.
7. *Bsa I* sites are eliminated by digestion/ligation, so no-digestion indicates possible insertion of the annealed oligo duplex. IMPORTANT: include a positive control by digesting pUC57–sgRNA empty vector. In case of performing the colony-PCR, use M13 Rv universal primer and the oligo Fwd used for the annealing.
8. 4 mL of RNAsecure 25×, AM7005, Ambion.
9. This step also heat-inactivates the digestion reaction.
10. RNaseZap®, R2020, Sigma.
11. We recommend using the MEGAshortscript Kit (AM1354, Ambion). Use at least 0.5 mg of *Dra I*-digested plasmids as template. Incubate for 4 h.
12. We recommend using the MEGAclear Kit (AM1908, Ambion).
13. Minimize the number of freeze-thaw cycles.
14. *Fwd* primer containing the T3 sequence:

    T3_dCas9VP160_F: 5′-AATTAACCCTCACTAAAGGGAGAcaggttggaccggtgccaccA-3′.

    Uppercase letters indicate the T3 minimum promoter sequence.
15. Use a high-fidelity DNA polymerase to minimize mutations (i.e., Q5 High fidelity, M0491, New England Biolabs).
16. We recommend using mMESSAGE mMACHINE™ T3 Transcription Kit (Ambion, AM1348).
17. cRNA should appear as a single band of ~4700 bp, whereas smearing indicates RNA degradation.
18. Millennium RNA Marker, AM7150, Ambion.
19. If you are willing to activate simultaneously more than one gen, you have to add the appropriates sgRNAs to activate each gene (i.e., to activate two genes simultaneously, add the eight sgRNAs to the mixture to a 50 ng/mL final concentration) always in at least 3 mL of final mixture.

20. Start always microinjecting with the SHAM mixture to avoid contamination in the microinjection pipette and then proceed working with the *CRISPR-on mixture*. Wash the pipette in PVP three times, in between groups.
21. Avoid placing the holding pipette into the *RNA mixture* drop, as the mix may be diluted.
22. We recommend using *RNAlater* (AM 7020, Ambion Co., Austin, TX, USA).
23. Avoid diluting the *RNA later* with the embryo culture.
24. It is important that the whole process previously described is performed at in least three biological replicates.
25. IETS, International Society of embryo transfer. https://www.iets.org/.
26. Design the primers to target different exons of the gene to be studied, to avoid amplification from genomic DNA contamination. Add negative controls, using water as a template.
27. At this point, it is very important to always keep the RNA tubes on ice.
28. We recommend using Protease from Streptomyces griseus (Sigma, P-8811).
29. Make sure the plate is completely hermetic to avoid evaporations.
30. Negative controls skip this step as they are only incubated with the second antibody.
31. From this point on, try to keep the lights as dim as possible.
32. We recommend using Vectashield® Mounting Medium (H-1000, Vector Laboratories).
33. If necessary, make two or three other drops to avoid diluting the antifade mounting medium.
34. As soon as the embryos are placed in the antifade mounting medium, keep an eye on them as they become transparent.
35. Do not apply pressure with your fingers. Take a p100 tip and apply gentle pressure on the left and right sides of the coverslip, where the petroleum jelly is located. Once the embryos have touched the coverslip, stop the pressure to prevent them from collapsing. If embryos collapse, the images of the ICC will not be representative of an embryo shape.

References

1. Garneau JE et al (2010) The CRISPR/Cas bacterial immune system cleaves bacteriophage and plasmid DNA. Nature 468(7320):67–71
2. Mali P et al (2013) RNA-guided human genome engineering via Cas9. Science 339(6121):823–826
3. Wang H et al (2013) One-step generation of mice carrying mutations in multiple genes by CRISPR/Cas-mediated genome engineering. Cell 153(4):910–918
4. Joung JK, Sander JD (2013) TALENs: a widely applicable technology for targeted genome editing. Nat Rev Mol Cell Biol 14(1):49–55
5. Thakore PI et al (2016) Editing the epigenome: technologies for programmable transcription and epigenetic modulation. Nat Methods 13(2):127–137
6. Beerli RR et al (1998) Toward controlling gene expression at will: specific regulation of the erbB-2/HER-2 promoter by using polydactyl zinc finger proteins constructed from modular building blocks. Proc Natl Acad Sci U S A 95(25):14628–14633
7. Beerli RR, Dreier B, Barbas CF 3rd (2000) Positive and negative regulation of endogenous genes by designed transcription factors. Proc Natl Acad Sci U S A 97(4):1495–1500
8. Zhang F et al (2011) Efficient construction of sequence-specific TAL effectors for modulating mammalian transcription. Nat Biotechnol 29(2):149–153
9. Miller JC et al (2011) A TALE nuclease architecture for efficient genome editing. Nat Biotechnol 29(2):143–148
10. Larson MH et al (2013) CRISPR interference (CRISPRi) for sequence-specific control of gene expression. Nat Protoc 8(11):2180–2196
11. Gilbert LA et al (2013) CRISPR-mediated modular RNA-guided regulation of transcription in eukaryotes. Cell 154(2):442–451
12. Maeder ML et al (2013) CRISPR RNA-guided activation of endogenous human genes. Nat Methods 10(10):977–979
13. Perez-Pinera P et al (2013) RNA-guided gene activation by CRISPR-Cas9-based transcription factors. Nat Methods 10(10):973–976
14. Konermann S et al (2015) Genome-scale transcriptional activation by an engineered CRISPR-Cas9 complex. Nature 517(7536): 583–588
15. van der Oost J (2013) Molecular biology. New tool for genome surgery. Science 339(6121): 768–770
16. Chakraborty S et al (2014) A CRISPR/Cas9-based system for reprogramming cell lineage specification. Stem Cell Rep 3(6):940–947
17. Cheng AW et al (2013) Multiplexed activation of endogenous genes by CRISPR-on, an RNA-guided transcriptional activator system. Cell Res 23(10):1163–1171
18. Hall DB, Struhl K (2002) The VP16 activation domain interacts with multiple transcriptional components as determined by protein-protein cross-linking in vivo. J Biol Chem 277(48): 46043–46050
19. Chavez A et al (2015) Highly efficient Cas9-mediated transcriptional programming. Nat Methods 12(4):326–328
20. Mendenhall EM et al (2013) Locus-specific editing of histone modifications at endogenous enhancers. Nat Biotechnol 31(12):1133–1136
21. Polstein LR et al (2015) Genome-wide specificity of DNA binding, gene regulation, and chromatin remodeling by TALE- and CRISPR/Cas9-based transcriptional activators. Genome Res 25(8):1158–1169
22. Morita S et al (2016) Targeted DNA demethylation in vivo using dCas9-peptide repeat and scFv-TET1 catalytic domain fusions. Nat Biotechnol 34(10):1060–1065
23. Liu XS et al (2016) Editing DNA methylation in the mammalian genome. Cell 167(1): 233–247.e17
24. Amabile A et al (2016) Inheritable silencing of endogenous genes by hit-and-run targeted epigenetic editing. Cell 167(1):219–232.e14
25. Hsu PD et al (2013) DNA targeting specificity of RNA-guided Cas9 nucleases. Nat Biotechnol 31(9):827–832
26. Doench JG et al (2014) Rational design of highly active sgRNAs for CRISPR-Cas9-mediated gene inactivation. Nat Biotechnol 32(12):1262–1267
27. Doench JG et al (2016) Optimized sgRNA design to maximize activity and minimize off-target effects of CRISPR-Cas9. Nat Biotechnol 34(2):184–191
28. Savy V et al (2020) CRISPR-on for activation of endogenous SMARCA4 and TFAP2C expression in bovine embryos. Reproduction 159(6):767–778
29. Bavister BD, Yanagimachi R (1977) The effects of sperm extracts and energy sources on the motility and acrosome reaction of hamster spermatozoa in vitro. Biol Reprod 16(2): 228–237

30. Tervit HR, Whittingham DG, Rowson LE (1972) Successful culture in vitro of sheep and cattle ova. J Reprod Fertil 30(3):493–497
31. Holm P et al (1999) High bovine blastocyst development in a static in vitro production system using SOFaa medium supplemented with sodium citrate and myo-inositol with or without serum-proteins. Theriogenology 52(4):683–700
32. Voytas D (2001) Agarose gel electrophoresis. Curr Protoc Immunol Chapter 10:Unit 10.4
33. Miller, J.H., Experiments in molecular genetics. 1972
34. Sambrook J et al. (2006) The condensed protocols from molecular cloning: a laboratory manual. Cold Spring Harbor, N.Y. Cold Spring Harbor Laboratory Press
35. Yates AD et al (2020) Ensembl 2020. Nucleic Acids Res 48(D1):D682–d688
36. Hilton IB et al (2015) Epigenome editing by a CRISPR-Cas9-based acetyltransferase activates genes from promoters and enhancers. Nat Biotechnol 33(5):510–517
37. Heigwer F, Kerr G, Boutros M (2014) E-CRISP: fast CRISPR target site identification. Nat Methods 11(2):122–123
38. Oliveros JC et al (2016) Breaking-Cas-interactive design of guide RNAs for CRISPR-Cas experiments for ENSEMBL genomes. Nucleic Acids Res 44(W1):W267–W271
39. Ran FA et al (2013) Genome engineering using the CRISPR-Cas9 system. Nat Protoc 8(11):2281–2308

Part III

Applications of Genome Manipulation

Chapter 9

CRISPR/Cas9 Mutagenesis to Generate Novel Traits in *Bactrocera tryoni* for Sterile Insect Technique

Amanda Choo, Elisabeth Fung, Thu N. M. Nguyen, Anzu Okada, and Peter Crisp

Abstract

Sterile Insect Technique (SIT) is a biocontrol strategy that has been widely utilized to suppress or eradicate outbreak populations of insect pests such as tephritid fruit flies. As SIT is highly favored due to it being species-specific and environmentally friendly, there are constant efforts to improve the efficiency and efficacy of this method in particular at low pest densities; one of which is the use of genetically enhanced strains. Development of these desirable strains has been facilitated by the emergence of the CRISPR/Cas genome-editing technology that enables the rapid and precise genomic modification of non-model organisms. Here, we describe the manual microinjection of CRISPR/Cas9 reagents into tephritid pest *Bactrocera tryoni* (Queensland fruit fly) embryos to introduce ideal traits as well as the molecular methods used to detect successful mutagenesis.

Key words Embryo microinjections, CRISPR/Cas mutagenesis, *Bactrocera tryoni*, Queensland fruit fly, Tephritids, Genetic sexing strains, Sterile insect technique

1 Introduction

Recent advancement in genome engineering tools such as the Clustered Regularly Interspaced Short Palindromic Repeats (CRISPR)/CRISPR-associated (Cas) technology has allowed for rapid and targeted genetic modification in non-model organisms, including horticultural pests, using a more efficient and less laborious method. The CRISPR/Cas9 system is a ribonucleoprotein (RNP) complex consisting of the bacterial endonuclease Cas9 and a guide RNA complex made up of two RNA structures—the CRISPR RNA (crRNA) and the trans-activating CRISPR RNA (tracrRNA) [1]. The crRNA structure encodes a unique CRISPR guide RNA (gRNA) sequence that is complementary to the target

Amanda Choo and Elisabeth Fung contributed equally with all other contributors.

Paul John Verma, Huseyin Sumer and Jun Liu (eds.), *Applications of Genome Modulation and Editing*, Methods in Molecular Biology, vol. 2495, https://doi.org/10.1007/978-1-0716-2301-5_9,

sequence in the genome, leading to the binding of the whole complex to the loci of interest while the Cas9 endonuclease will cleave the DNA 3–4 bases upstream of a specific protospacer adjacent motif (PAM) resulting in double-stranded breaks at the target site [1]. Repair of the DNA double-stranded breaks will then occur, and this often results in nonspecific indels (insertions or deletions) introduced through the non-homologous end-joining (NHEJ) repair mechanism [2]. The introduction of indels may cause a frameshift in the genetic code resulting in either the absence of protein or the production of a non-functional protein ("knock-out" mutation). Alternatively, in the presence of a donor template, the homology-directed repair (HDR) mechanism could be utilized to introduce a specific desired sequence ("knock-in" mutation) [2].

Tephritid fruit flies are one of the most economically devastating horticultural pests worldwide, with significant impacts on management costs, crop production, and market access [3]. *Bactrocera tryoni* (*B. tryoni*), commonly known as Queensland fruit fly, is the major tephritid pest species in Australia affecting more than a hundred different host plants [4, 5] and has the ability to tolerate a wide climatic range [6]. Organophosphates such as fenthion and dimethoate, which were previously used to control *B. tryoni* populations, are now restricted [7], resulting in a need for an environmentally friendly, alternative method such as Sterile Insect Technique (SIT) to manage this destructive polyphagous pest. SIT involves the release of millions of a sterilized pest species into a target area, where the sterile males can mate with wild females which will then produce non-viable embryos resulting in suppression or complete eradication of the pest population [8]. The current *B. tryoni* SIT program in Australia involves the release of both sterile males and females, and a detectable fluorescent dye is used to label the released flies [9]. However, replacement of the fluorescent dye with visual markers [10] and use of a male-only release strain would significantly increase the efficiency and cost-effectiveness of the SIT program [11]. While efforts have previously been made to develop a male-only strain as well as strains carrying a genetic fluorescent marker using random mutagenesis via gamma irradiation and genetic transformation [12, 13], suitable non-transgenic strains are yet to be produced. The availability, efficiency, and precision of the CRISPR/Cas gene-editing technology now provides the opportunity to feasibly develop such strains to be incorporated into the SIT program.

We have demonstrated successful application of the CRISPR/Cas9 technology in *B. tryoni* and generated knock-out strains containing phenotypic markers, such as white eyes and yellow body color [14–16]. We have also introduced a precise single base substitution into an endogenous gene using the HDR mechanism and established a user method for knocking-in specific desirable mutations in the first step towards generating a *B. tryoni* male-only strain

[17]. Here, we discuss in detail the methods we have used to generate these knock-out and knock-in mutations in the tephritid pest species *B. tryoni,* including our manual embryo microinjection technique which can be easily set up and the non-lethal genotyping assays (Polymerase Chain Reaction, T7 Endonuclease I Assay, restriction enzyme digests) suitable for the rapid screening of a large number of insects to identify desired mutants, all of which can be applied to other non-model organisms.

2 Materials

2.1 CRISPR/Cas9 (See Note 1)

1. Alt-R CRISPR/Cas9 crRNA.
2. Alt-R CRISPR/Cas9 tracrRNA.
3. Alt-R S.p. Cas9 Nuclease.
4. Single-stranded DNA oligo as the HDR single-stranded donor (ssODN) template.
5. Nuclease-free Duplex Buffer.
6. Nuclease-free water.
7. 1× Phosphate-Buffered Saline (PBS): Dissolve 8 g of NaCl, 0.2 g of KCl, 1.44 g of Na_2HPO_4, 0.24 g of KH_2PO_4 in 800 mL of MilliQ water. Adjust solution to pH 7.4 with HCl. Add water to a total volume of 1 L. Sterilize the solution by autoclaving for 20 min at 121 °C. Store at room temperature.

2.2 Bactrocera tryoni *Embryo Microinjection (See Note 2)*

1. Embryo collection devices: 30 mL perforated plastic portion cup with lid containing a piece of absorbent cloth soaked in diluted apple juice (1:1) to attract female flies and stimulate oviposition. The absorbent cloth has to be kept moist to prevent desiccation of the eggs.
2. Sieve for rinsing embryos.
3. Fine paintbrushes for collecting and lining embryos.
4. 1% bleach: Add 25 mL of 4% bleach into 75 mL of MilliQ water to make 100 mL of solution.
5. MilliQ water.
6. Paper towels.
7. Agarose plates: To make 1% agarose, add 1 g of agarose powder in 100 mL of MilliQ water in a large bottle or flask. Dissolve the agarose completely using a microwave and allow it to cool on the bench. Pour the agarose solution into 90 mm plastic petri dishes and allow to solidify. Store at 4 °C.
8. Timer.
9. Microscope slides and coverslip.

10. Rubber cement/double-sided tape (*see* **Note 3**).
11. Oil (*see* **Note 4**).
12. Borosilicate capillaries with filament (*see* **Note 5**).
13. Microloader pipettes for loading pulled capillaries.
14. Scalpel blades.
15. 10× Injection Buffer: 1 mM Sodium Phosphate buffer pH 6.8 and 50 mM KCI. Mix 0.5 mL of 0.1 M Sodium Phosphate buffer pH 6.8 (*see* **Note 6**), 2.5 mL of 1 M KCI, and 47 mL of Nuclease-free water.
16. Microinjection setup: micromanipulator, inverted microscope, and syringe.

2.3 Bactrocera tryoni Rearing

1. Insectary facility with control environment rooms (*see* **Note 7**).
2. Mesh insect rearing cages or pint-sized insect pots.
3. Donut lids for pint-sized insect pots.
4. Fine mesh nylon netting stockinet sleeve for donut lid (40 × 15.5 cm).
5. Larval gel diet [18]: Combine in one large beaker 30.6 g Brewer's yeast, 18.2 g sugar, 0.3 g Methyl ρ-Hydroxy Benzoate (Methyl Paraben), 3.5 g Citric acid, and 0.3 mL Wheat Germ / vegetable oil. In another beaker, mix 0.3 g Sodium Benzoate into 75 mL of hot water, add it to the dry ingredients. Combine 1.5 g Bacteriological agar powder and 75 mL of boiling water in another beaker and bring it to boil in a microwave to dissolve the agar completely. Mix the agar into the dry ingredients and blend it for 1–2 min with stick mixer. Pour into suitable container or plate and leave it to solidify at room temperature and until cool. Store at 4 °C.
6. Water devices: Fill a small plastic container with water, create a small slit in the lid of the container, and fit a piece of absorbent sponge through the slit. Keep the absorbent sponge touching the base of the water container to guarantee constant water access for the flies.

2.4 Polymerase Chain Reaction (PCR)

1. Phire Animal Tissue Direct PCR Kit from ThermoFisher Scientific: Dilution Buffer, DNA Release Additive, Phire Hot Start II DNA Polymerase, and Phire Animal Tissue PCR Buffer (*see* **Note 8**).
2. PCR tubes.
3. DNase-free water.
4. Target-specific primers.
5. Thermal cycler.

2.5 Gel Electrophoresis

1. Agarose powder.
2. 1× TAE Buffer: 1 in 50 dilution of 50× TAE Buffer (242 g Tris base, 57.1 mL Acetic acid, and 100 mL of 0.5 M EDTA pH 8.0 made up to 1 L with MilliQ water).
3. Nucleic acid staining solution.
4. Molecular weight markers.
5. Gel cast, combs, and electrophoresis tanks.
6. Gel imaging system (*e.g.*, ChemiDoc from Bio-Rad).

2.6 Restriction Enzyme Digest and T7 Endonuclease I (T7EI) Assay

1. Target-specific restriction enzyme and corresponding buffer.
2. T7 Endonuclease I and buffer.
3. DNase-free water.
4. PCR tubes.
5. Thermal cycler.
6. DNA quantification instrument (*e.g.*, NanoDrop Spectrophotometer or Qubit Fluorometer).

3 Methods

3.1 Designing CRISPR Guide RNA and HDR Donor Template Sequences

1. Identify target region and nature of mutation (knock-out or knock-in) to be induced.
2. Extract DNA from 6 to 8 individual flies of the laboratory culture and sequence the region of interest (*see* Subheadings 3.5 and 3.6 for protocol) to identify single nucleotide polymorphisms (SNPs) that are present in the laboratory population (*see* **Note 9**).
3. Design the 20 nt guide RNA sequence(s) (gRNA) by firstly identifying available PAM sequences (NGG) at the region of interest and the corresponding 20 nt of DNA sequence upstream of each PAM (*see* **Note 10**). Select the most ideal gRNA based on the following criteria:
   (a) No SNPs present within the gRNA sequence.
   (b) Presence of PAM cut site near the target loci where the mutation is to be introduced.
   (c) Little or no off-targets predicted (off-targets can be identified by running the 20 nt gRNA sequence in a blastn search against the *B. tryoni* genome).
4. For knock-in mutations, determine if there will be any restriction enzyme sites that are abolished or introduced with the desired mutation—this will enable a restriction digest assay to be used as a genotyping method to identify mutants (*see* Subheading 3.8).

5. Design the HDR donor template sequence for knock-in mutations based on the following criteria:
   (a) Homology arms of approximately 150 nt overall (75 nt on each homology arm flanking the desired mutation).
   (b) Introduction of the desired mutation.
   (c) Introduction of a silent mutation in the PAM site or proximal to the PAM site within the gRNA sequence (if possible) (*see* **Note 11**).
   (d) Strand orientation (evaluate the target mutation and its location relative to the cut site).

3.2 Preparation of Alt-R CRISPR/Cas9 Reagents and Microinjection Mix (See Note 1)

1. Mix the Alt-R S.p. Cas9 enzyme stock solution (10 μg/μL) thoroughly by inverting the tube several times and briefly centrifuging the tube.
2. Dilute the stock solution to a working concentration of 1 μg/μ L with 1× PBS as follows:

Reagents	Volume	Final concentration
Alt-R S.p. Cas9 (10 μg/μL)	1 μL	1 μg/μL
1× PBS	9 μL	–
Final volume	**10 μL**	–

3. Spin down the Alt-R CRISPR-Cas9 RNA and DNA oligo pellets.
4. Resuspend each RNA oligo (for the Alt-R CRISPR-Cas9 crRNA and tracrRNA) in Nuclease-free Duplex Buffer to obtain 100 μM stock solutions accordingly:

Reagents	Normalized amount (nmol)	Volume of resuspension buffer (μL)
Alt-R CRISPR-Cas9 crRNA	2 nmol 10 nmol 50 nmol 100 nmol	20 μL 100 μL 500 μL 1000 μL
Alt-R CRISPR-Cas9 tracrRNA	5 nmol 20 nmol 100 nmol	50 μL 200 μL 1000 μL

5. Resuspend the Ultramer® DNA Oligo with Nuclease-free water to obtain a 1 μg/μL stock solution.
6. Store all resuspended oligos at −20 °C.
7. Prepare the crRNA-tracrRNA duplex with a final duplex concentration of 40 μM:

Reagents	Volume	Final concentration
Alt-R CRISPR-Cas9 crRNA (100 μM)	1 μL	~472 ng/μL
Alt-R CRISPR-Cas9 tracrRNA (100 μM)	1 μL	~887 ng/μL
Nuclease-free Duplex Buffer	0.5 μL	–
Final volume	**2.5 μL**	**40 μM (duplex RNA)**

8. Heat the duplex RNA at 95 °C for 5 min in the thermal cycler and then allow to cool to room temperature (20–25 °C).
9. Prepare the injection mix by assembling the ribonucleoprotein (RNP) complex in the injection buffer:

For knock-out experiments (with no HDR donor template)		
Reagents	**Volume**	**Final concentration**
Complexed crRNA:tracrRNA oligos (40 μM)	2.5 μL	10 μM
Diluted Alt-R S.p. Cas9 enzyme (1 μg/μL)	3 μL	300 ng/μL
10× Injection Buffer	1 μL	1×
Nuclease-free water	3.5 μL	–
Final volume	**10 μL**	–

For knock-in experiments (with HDR donor template)		
Reagents	**Volume**	**Final concentration**
Complexed crRNA:tracrRNA oligos (40 μM)	2.5 μL	10 μM
Diluted Alt-R S.p. Cas9 enzyme (1 μg/μL)	3 μL	300 ng/μL
ssODN (1 μg/μL)	2 μL	200 ng/μL
10× Injection Buffer	1 μL	1×
Nuclease-free water	1.5 μL	–
Final volume	**10 μL**	–

10. Incubate at room temperature for 5 min to assemble the RNP complexes prior to use for embryo microinjections (*see* **Note 12**).

3.3 Bactrocera tryoni *Embryo Microinjection*

1. Set up egging devices in cages containing flies of optimal egging laying age to collect 0–1 h embryos (*see* **Note 13**).

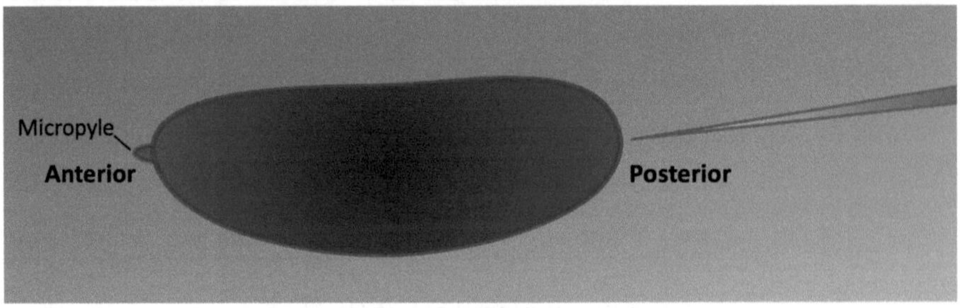

Fig. 1 Simplified diagram of *B. tryoni* embryo. The anterior of the embryo is easily recognizable by the presence of the micropyle. Microinjections are performed by penetrating the embryo at the posterior end using a needle

2. Transfer all embryos onto a sieve by washing them with water.
3. Rinse embryos with MilliQ water several times.
4. Pour a small quantity of 1% bleach into the small cup/container and using a paintbrush, transfer all or a large number of embryos into the bleach solution and leave it for 3 min to dechorionate the embryos. Give it a swirl occasionally to aid in the dechorionating process (*see* **Note 14**).
5. Rinse the embryos onto a new sieve with MilliQ water several times to get rid of the bleach and let the embryos sit in fresh MilliQ water for 15 min.
6. Blot the sieve on some paper towel to remove the remaining liquid and then transfer the embryos using a slightly wetted paintbrush to a piece of agarose gel (*see* **Note 15**).
7. Using a fine paintbrush, line the embryos in a straight line on the piece of agarose gel in the same orientation with the micropyle (on the anterior side) all facing the same direction (microinjections to be performed through the posterior side (Fig. 1).
8. Line a coverslip with a strip of rubber cement or double-sided tape and transfer the lined embryos from the piece of agarose gel onto the coverslip by pressing the surface with the glue/tape firmly onto the embryos.
9. Leave embryos on coverslip for a few minutes at room temperature to let embryos desiccate slightly. Cover the embryos with oil (Paraffin oil for rubber cement, Halocarbon oil if double-sided tape is used).
10. To prepare the needle (capillary) for injections (*see* **Note 16**): using a microloader pipette, carefully load injection mix into the needle.
11. Place needle into the microcapillary holder attached to the syringe. Ensure the syringe has been pulled all the way up beforehand so that there is pressure on the needle.

12. Add oil onto a microscope slide and place it onto the microscope stage. Adjust the needle and lower it until the closed end is under the oil. The open end should be bevelled (*see* **Note 5**) in order to inject into the embryos. Keep the needle under oil when not in use.
13. Place the coverslip on a microscope slide with the embryos facing upward (*see* **Note 17**) and then place the slide onto the microscope stage in the orientation where the posterior of the embryos is facing the injection needle.
14. To perform microinjections: bring the embryos into view and focus.
15. Lower the needle until the tip is under the oil and bring it close to the posterior end of the embryos.
16. Position it at the first embryo, adjust the angle until it is parallel and in the same focal plane as the embryo (might require several attempts to get the angle right). Adjust angle by using the different knobs on the micromanipulator to lower the needle or to bring the needle forward.
17. Once ready, gently move the needle forward to penetrate the embryo. Move the needle so that it is 1/3 into the embryo. Ideally, the needle should be able to enter the embryo fairly easily—the tip of the embryo should be pressed back slightly but bounce back once the needle is in.
18. Press gently on the syringe to release pressure and the injection mix should flow into the embryo (which should be visible; *see* **Note 18**).
19. Pull the needle out quickly but gently. Ideally not too much cytoplasm will be displaced (it will leak out at the injection site).
20. Move onto the next embryo quickly and repeat until the line of embryos have all been injected. The angle of the needle may need to be adjusted for different embryos.
21. Remove or eliminate any uninjected embryos in order to prevent false positives. Count the number of successfully injected embryos. Discount any that look unlikely to survive (too much leakage of cytoplasm).
22. Remove as much Halocarbon oil 700 covering the embryos as possible using a Kimwipe (*see* **Note 4**). Place the coverslip containing the injected embryos onto an agarose gel plate and seal it with parafilm.
23. Place the agarose gel plate containing injected embryos at the following conditions: humidity of 65–75% and temperature of 25 °C for 3 days or at 21 °C for 4–5 days.

24. Rescue any free or developing first instar larvae with a paintbrush and gently transfer them to larval diet for larval development.

3.4 Mating Crosses to Obtain G_1 Progeny for Molecular Genotyping

1. Designate each surviving adult from the microinjections as G_0 (*i.e.*, G_0#1, G_0#2. . . .) and separate males and females within 3 days after emergence to prevent any mating.
2. Set up a mating cross for each individual virgin G_0 fly with 4–6 virgin wild type flies of the opposite gender.
3. Collect eggs (G_1 progeny) from each individual mating cross when the flies are at the optimal egg laying age (usually starting from 12 days after emergence; *see* **Note 19**). Designate the G_1 progeny according to the mating cross they came from (*i.e.*, for mating crosses of G_0#1 and G_0#2, progeny will be labeled G_1#1.1, G_1#1.2. . . and G_1#2.1, G_1#2.2. . ., respectively).
4. Rear the G_1 progeny to adulthood and separate males and females upon emergence to prevent any mating. Treat each G_1 fly as an individual for molecular genotyping to identify the G_1 flies that carry the desired mutation (*see* **Note 20**, Fig. 2).

3.5 Molecular Genotyping: Non-Lethal Genomic DNA Extraction from Single Fly Leg

1. For each sample, prepare the following DNA extraction mix in a PCR tube:

Reagents	Volume
Dilution buffer	20 μL
DNA release additive	0.5 μL
Final volume	**20.5 μL**

2. Anesthetize the flies and for each individual fly, remove one of the mid legs using forceps (*see* **Note 21**). Immerse each leg in the DNA extraction mix in an individual PCR tube. Gently transfer fly to an individual vial containing gel diet and make a record to correspond each fly-containing vial to its DNA extract/PCR tube.
3. Spin samples down briefly to ensure that the legs remain immersed in the liquid and incubate at room temperature for 2–5 min.
4. Incubate at 98 °C for 2 min in a thermal cycler.
5. Allow sample to cool to room temperature before using for molecular analyses. Store samples at −20 °C.

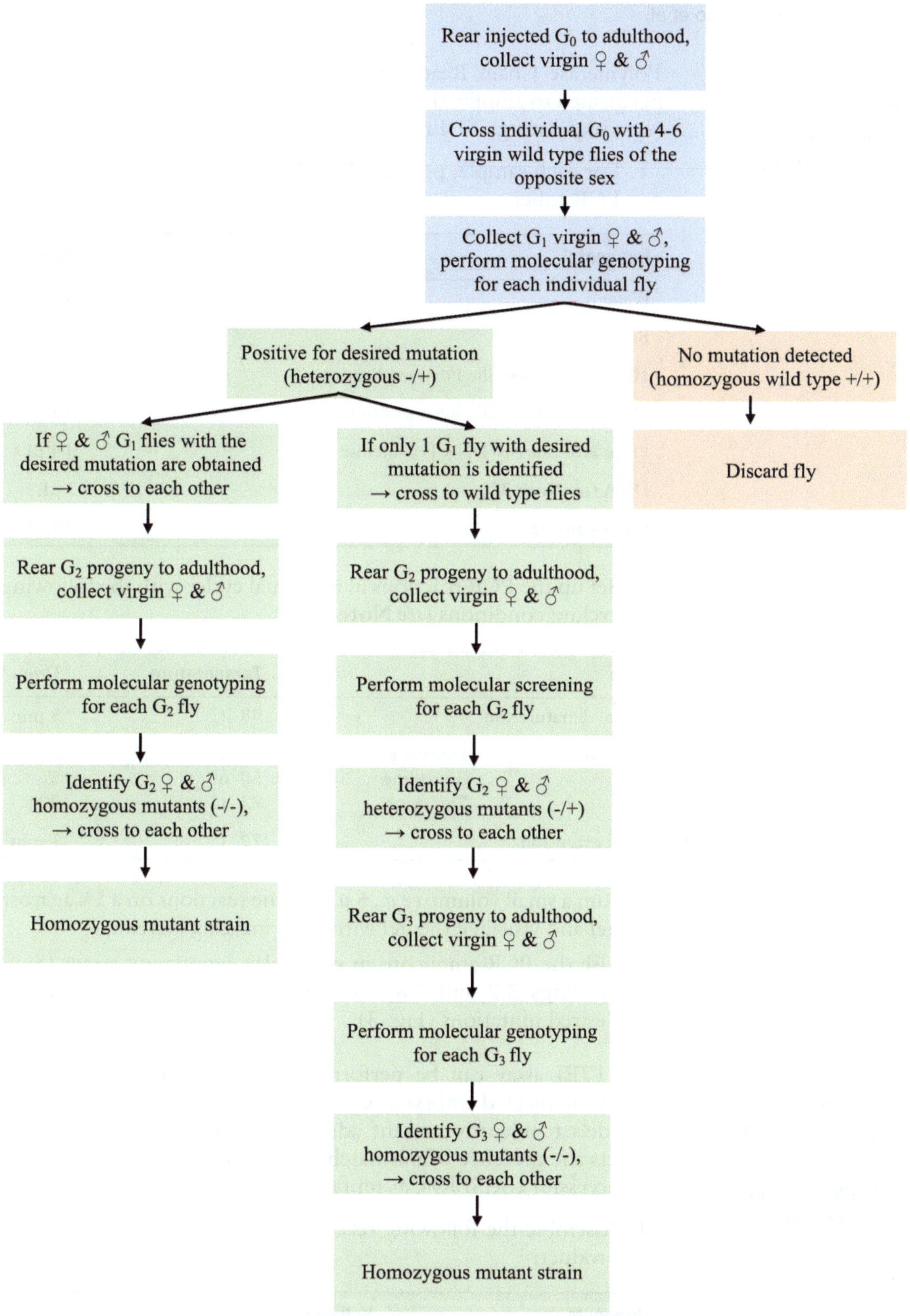

Fig. 2 Schematic of the post-microinjection workflow performed to generate a stable homozygous mutant strain. The purple boxes show the mating scheme of G_0 individuals required to obtain G_1 progeny for identification of flies carrying the desirable mutation(s). The green boxes show the mating scheme performed over the following generation(s) in order to produce a homozygous mutant strain (either using the G_2 or G_3 progeny depending on the number of G_1 mutants that were obtained). ♀ and ♂ represent female and males respectively

3.6 Molecular Genotyping: Polymerase Chain Reaction and Gel Electrophoresis

Polymerase Chain Reaction using target-specific primers will be performed to amplify the region of target DNA for further analyses (*e.g.*, sequencing, T7 Endonuclease I assay, restriction digest).

1. For each sample, prepare the following PCR reaction mix in a PCR tube:

Reagents	Volume
Water	6.6 μL
Buffer	10 μL
10 μM target-specific Forward Primer	0.5 μL
10 μM target-specific Reverse Primer	0.5 μL
Phire Hot Start II DNA Polymerase	0.4 μL
DNA (*see* **Note 22**)	2 μL
Final volume	**20 μL**

2. Set up the PCR reactions in a thermal cycler with the following cycling conditions (*see* **Note 23**):

Step		Temperature	Time
Initial denaturation		98 °C	5 min
x 35 cycles	Denaturation Annealing Extension	98 °C 50–60 °C 72 °C	5 s 5 s 20 s
Final extension		72 °C	1 min

3. Run a small volume (*e.g.*, 5 μL) of the reactions on a 1% agarose gel and visualize the gel with a gel imaging system.
4. Use the PCR amplicons in molecular genotyping assays (Subheadings 3.7 and 3.8) to identify samples that contain the desired mutations (Fig. 3).

3.7 Molecular Genotyping: T7 Endonuclease I (T7EI) Assay for Assessing CRISPR Targeting Success and Identifying Presence of indels

The T7EI assay can be performed to assess CRISPR targeting success in injected embryos (*see* **Note 24**) or to detect the presence of indels to identify mutant adult flies. The T7 Endonuclease I detects the presence of mismatches in the DNA amplicons caused by successful CRISPR/Cas mutagenesis at the target loci.

1. Assemble the following reaction in a PCR tube for each PCR product:

Reagents	Volume
PCR product	200 ng
NEB Buffer 2	2 μL

(continued)

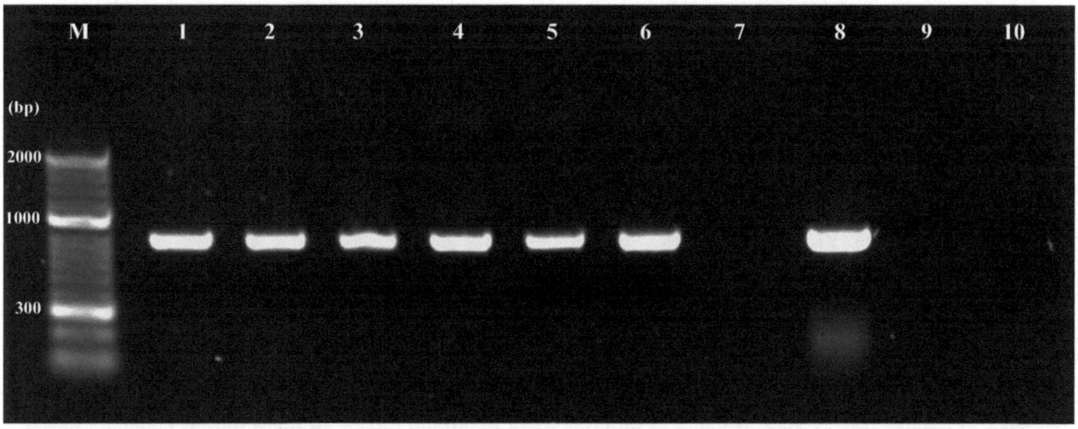

Fig. 3 An example of a gel image showing the result from a PCR analysis performed using DNA extracts of single fly leg from different individuals (Lanes 1–6). *M* = molecular weight marker, Lane 8 = positive control (DNA extracted with Qiagen DNeasy® Blood & Tissue Kit Qiagen kit), Lane 10 = negative control (no DNA), Lane 7 and 9 = blank/no samples

Reagents	Volume
Water	Add to make final overall volume of 19 μL
Final volume	**19 μL**

2. Anneal the PCR products in a thermocycler using the following conditions:

Step	Temperature	Time
Initial denaturation	95 °C	5 min
Annealing	Ramp down to 85 °C Ramp down to 25 °C	−2 °C/s −0.1 °C/s

3. Add 1 μL of T7 Endonuclease I to each annealed PCR product and incubate at 37 °C for 15 min in a thermal cycler.
4. Run the reactions immediately on a 2% agarose gel and visualize gel using a gel imaging system.
5. Send PCR products of samples that have been identified to contain the desired mutations or samples with non-conclusive results to a sequencing facility to confirm the presence/absence and nature of the mutations (Fig. 4).

3.8 Molecular Genotyping: Restriction Enzyme Digest to Identify Specific Mutations

The restriction enzyme digest assay is performed to identify desired mutants through the detection of the removal or addition of a restriction enzyme cut site at the genomic target loci.

1. Assemble the following components in a PCR tube:

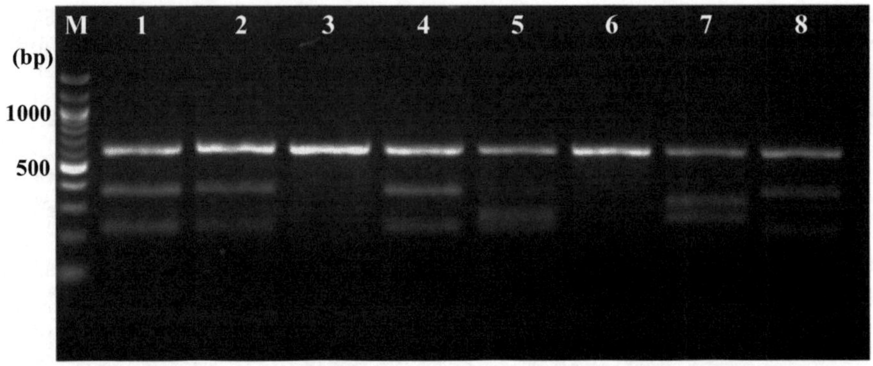

Fig. 4 An example of a gel image showing the result from a T7EI assay used to identify mutant flies by detection of indels. Three different types of mutants were identified based on the different cleavage patterns (Mutant #1—Lanes 1, 2, 4, and 8; Mutant #2—Lane 5; Mutant #3—Lane 7). The genotypes were all subsequently confirmed by Sanger sequencing. *M* = molecular weight marker

Reagents	Volume
Water	Add to make total reaction volume 10 μL
10× Buffer	1 μL
Restriction enzyme	0.2 μL
PCR product (*see* **Note 22**)	~70 ng
Final volume	**10 μL**

2. Incubate reaction at the recommended temperature for chosen restriction enzyme (generally at 37 °C) in thermocycler for 5 min–1 h (*see* **Note 25**).
3. Stop digestion, if necessary, by heat inactivation (at the recommended temperature and duration for chosen restriction enzyme) when possible or by adding 10 mM final concentration of EDTA.
4. Run the reactions on a 1% agarose gel and visualize the gel with a gel imaging system.
5. Send PCR products of samples that have been identified to contain the desired mutations or samples with non-conclusive results to a sequencing facility to confirm the presence/absence and nature of the mutations (Fig. 5).

3.9 Generation of Stable Homozygous Mutant Strains

1. Set up mating crosses between heterozygous (−/+) G_1 mutants that had been identified from the molecular genotyping assays (*see* Subheadings 3.6, 3.7, and 3.8) (*see* **Note 26**, Fig. 2).
2. Collect eggs (G_2 progeny) from the mating cross and rear G_2 progeny to adulthood.

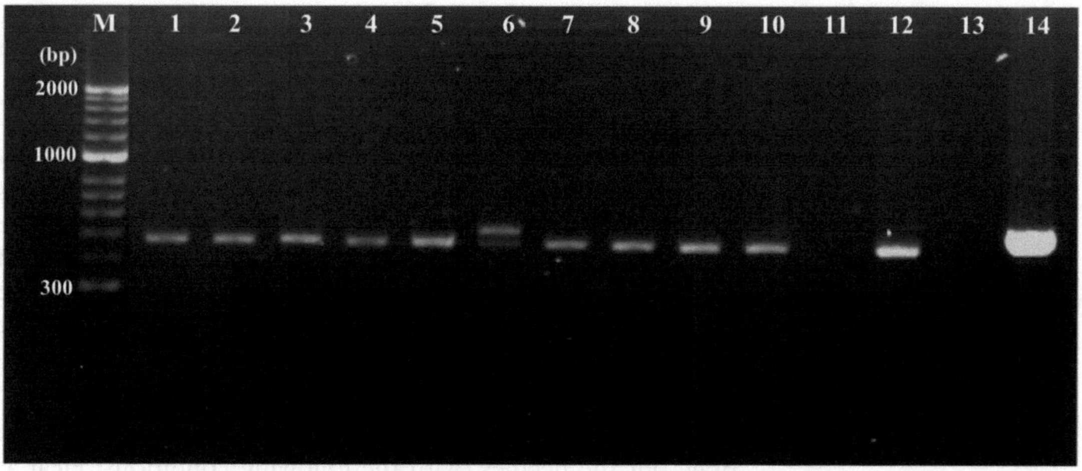

Fig. 5 An example of a gel image showing the result from a restriction enzyme digest assay used to identify mutant flies. Complete digestion with BaeGI restriction enzyme is indicative of homozygous wild type genotype (Lanes 1–5, 7–10), as observed with the positive control wild type sample (Lane 12) due to the presence of the BaeGI recognition sequence in the wild type amplicons. A heterozygous mutant was identified (Lane 6) through the undigested PCR amplicon of its mutant allele which abolishes the BaeGI recognition sequence. The genotype of the identified heterozygous mutant was subsequently confirmed by Sanger sequencing. *M* = molecular weight marker, Lane 14 = undigested sample, Lanes 11 and 13 = blank/no samples

3. Perform the same genotyping assays (*see* Subheadings 3.6, 3.7, and 3.8) on the G_2 adults to identify homozygous (−/−) mutants (*see* **Note 27**).
4. Set up mating crosses between the identified homozygous (−/−) G_2 mutants to establish a homozygous (−/−) mutant strain.

4 Notes

1. The CRISPR reagents can be generated in the laboratory or purchased from various companies; we are using the CRISPR/Cas9 system manufactured by Integrated DNA Technologies (IDT) and the protocol presented has been optimized for this system.
2. Most laboratories use electronic micromanipulation workstations such as the Eppendorf InjectMan® and FemtoJet®; here, we describe the method for manual microinjections using a micromanipulator, pulled capillaries, and a syringe which can be easily set up in the absence of any electronic microinjectors and/or micromanipulators.
3. We recommend the Maruni Earth Rubber Cement (Japan), Scotch Permanent Double-sided Tape or Whiteliner 3 M #1522 (Clear Tape). We have tested various rubber cements

and double-sided tapes and have found that these work the best in regard to sticking embryos down for microinjections and with the least toxicity to embryos.

4. We have found that the Faulding Liquid Paraffin oil works the best in conjunction with the Maruni rubber cement. The Halocarbon oil 700 is ideal for using with the double-sided tape as it is a heavier oil compared to the paraffin oil and reduces the movement of the embryos during microinjection. Prolonged exposure to Halocarbon oil 700 can be toxic to embryos, hence removal of the oil straight after microinjection is necessary for embryo survival.
5. Borosilicate capillary needles have to be drawn out/"pulled." We used a Sutter P-87 flaming/brown micropipette puller to draw out the needles (with the following conditions: Heat at 600, Velocity at 150, Pull at 150, and Time at 250) and manually created an opening in the needle with a scalpel blade. A beveller machine can also be used to create an ideal needle for microinjections. The needles should be drawn out in advance and kept under sterile conditions.
6. To prepare of 0.1 M Sodium Phosphate buffer pH 6.8, mix 51 mL of 0.2 M NaH_2PO_4 + 49 mL of 0.2 M Na_2HPO_4 + 100 mL of Nuclease-free water and verify or adjust to pH 6.8.
7. Controlled environment conditions that are required are as such 25 ± 2 °C, 65 ± 10% relative humidity (RH) and under a 14:10 light/dark cycle. Our rearing is done in a biological control insectary under strict quarantine regulations.
8. Specialized polymerase to be used with the corresponding extraction reagents; buffer includes dNTPs and $MgCl_2$.
9. It is important to identify SNPs present in the laboratory culture that is being used for the embryo microinjections and CRISPR/Cas9 targeting in order to prevent mismatches in the design of the CRISPR guides. Presence of mismatches may reduce the frequency and efficiency of the CRISPR/Cas9 complex binding to the genomic region of interest.
10. The CRISPR guide RNA can be designed manually as described or using available programs such as CRISPOR (www.crispor.tefor.net) [19] for existing genomes. It should be noted that the PAM sequence is not included in the 20 nt guide RNA sequence. The efficiency of the guide RNAs can be assessed by performing T7EI assays on injected embryos (*see* **Note 24**).
11. Silent mutations are mutations that result in changes to the nucleotide sequence but not the amino acid sequence. Introduction of silent mutations in the PAM site or proximal to the PAM site within the gRNA sequence could abolish or reduce

frequency of the CRISPR/Cas9 complex recognizing and binding to the genomic DNA, hence preventing recleaving of the DNA after the desired mutation has been introduced using the HDR donor template. We try to avoid introducing silent mutations that result in non-canonical PAMs such as -NGA as it has been reported to also be recognized by Cas9 in human cells [20].

12. We are usually able to inject approximately 600–700 embryos with 10 μL of the microinjection mix during successful microinjection sessions. The 10 μL microinjection mix can be divided into smaller aliquots to be used across different days/microinjection sessions.
13. Embryo quality can be affected by age of the parental flies—if flies are too old, the quality of embryos produced is poor and not ideal for microinjections. The optimal time for embryo collection (to obtain maximum number of embryos) has been found to be during the middle of the day whereas less embryos are obtained if collections are done too early in the morning or too late in the afternoon.
14. The bleach dechorionation step is fairly toxic to embryos. Factors such as the brand of bleach and duration of bleaching have to be considered and trialed beforehand to determine the conditions that will result in maximum survival of embryos. We find that cleaning/household bleach is less harsh for the embryos compared to laboratory grade bleach. We proposed an approximate duration of 3 min duration for the bleaching—embryos may be killed if left too long in the bleaching solution but if the bleaching step is too short, the embryos may not be fully dechorionated. However, we have also noticed that the duration may have to be adjusted slightly from time to time-newly diluted bleach solution tends to be stronger and may require shorter time to dechorionated embryos. Other factors that may affect the required time of bleaching are egg quality and temperature/humidity of room. While manual dechorionation is possible with some other species, it is difficult and time-consuming to manually dechorionate *B. tryoni* embryos.
15. Texture of agarose gel may affect the lining process/success of transferring embryos to coverslip, for example, the embryos will not transfer properly if the surface is too moist.
16. The needle is very fragile and can break easily hence it has to be handled carefully. When breaking the tip of the needle or bevelling the needle, the size of the opening/thickness of the tip is very important—if the tip is too blunt (too much of the needle has been broken off), it will create a big hole in the embryos, resulting in cytoplasmic leakage and poor survival. In such a case, it will be better to discard that needle and use a new needle. If the tip is too fine (not enough of the needle has been

broken off or not sufficiently bevelled), it would be difficult to penetrate the embryos and the needle will bend when pressed against the embryos.

17. Ensure that the coverslip is firmly stuck onto the slide so that it does not move during the injection process by adding a drop of water on the slide and placing the coverslip on the drop of water and/or sticking down the coverslip with tape if necessary.
18. When injecting into the embryo, a fine control of the syringe is required—do not inject too much otherwise too much of cytoplasm will be displaced or the embryo will explode. Ensure that there is no flowback of cytoplasm into the needle. If you see cytoplasm/liquid moving up the needle, remove the liquid from the needle by injecting it into the oil.
19. Collection of eggs (G_1 progeny) from each mating cross should be performed until the females stop laying eggs or >100 viable eggs are collected in order to obtain sufficient G_1 progeny for screening and to be able to identify the maximum number of G_1 mutants to generate a stable strain. It is also possible that in some cases no viable eggs will be produced from the mating crosses as the microinjection process and/or CRISPR/Cas mutagenesis can cause sterility in some of the G_0 injected flies.
20. If a mutation is detected within a G_1 population, the G_0 parental fly (the progenitor of the G_1 population) is considered a germline mutant. It should be noted that for each germline mutation, not every G_1 progeny may carry the mutation hence a minimum of 50–100 G_1 flies should be genotyped. Any G_1 fly that is detected to carry a mutation would be heterozygous (−/+) for the mutation as the G_0 parental fly had been out-crossed to a wild type fly.
21. Flies can be anesthetized by exposure to cold or CO_2 (if such setup is available). It should be noted that prolonged exposure to those conditions may result in death of the fly; hence, a suitable duration of exposure needs to be predetermined. We have found that exposure to −20 °C for 1 min 15 s is sufficient to temporarily knock the flies out without killing them. We then keep the flies chilled (by placing them on a cold petri dish) to prevent them from waking up while we remove the leg. During removal of the legs, caution must be taken to not puncture or injure the fly.
22. As this has been set up to be a quick-and-easy method for screening a large number of flies (for presence/absence of a mutant allele), we use a fixed volume of the DNA extract (2 μL) rather than a set concentration of DNA in our PCR reaction. We have found that with our non-lethal genomic DNA extraction method from a single fly leg using the Phire Animal Tissue Direct PCR Kit, 2 μL of the extracted genomic DNA is consistently sufficient and optimal for all of our PCR reactions. We

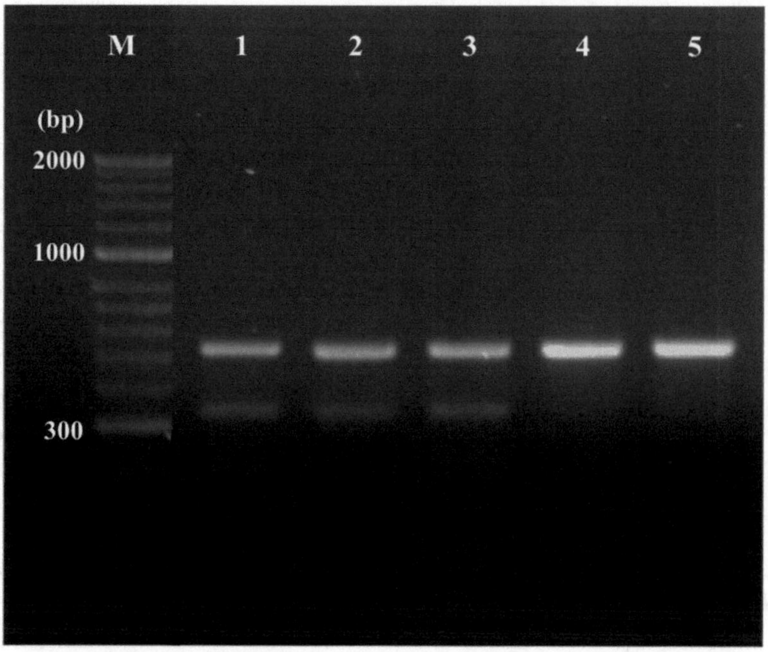

Fig. 6 An example of a gel image showing the result from a T7EI assay used to confirm that successful CRISPR/Cas mutagenesis can be achieved with the CRISPR guide. DNA cleavage is observed for the three samples containing injected embryos (Lanes 1–3) but not in the control non-injected embryo samples (Lanes 4 and 5). 30–40 embryos were pooled for each sample. The size of the lower two bands in the injected samples corresponds to the products produced by cleavage of DNA at the mutation loci (when the DNA mismatch would occur). *M* = molecular weight marker

have previously quantified a number of single fly leg Phire DNA extracts and found a yield range of 0.3 ng/μL–4.6 ng/μL. With the PCR products, we usually use 1–4 μL of the unpurified PCR product for the restriction digest and 7 μL for the T7 endonuclease I assays.

23. Adjustments will have to be made to the annealing temperature (based on length and composition of primers used) and extension time (recommended 20 s for amplicons ≤1 kb and 20 s/kb for amplicons >1 kb).

24. The T7EI assay can be performed on sacrificed 24 h embryos post-microinjection to confirm successful targeting using the designed CRISPR guides. DNA extraction can be performed on the embryos using the Phire Animal Tissue Direct PCR Kit, with 30–40 embryos extracted in the 20.5 μL DNA extract solution per sample. Control samples containing non-injected embryos should be included. This assay should be carried out for each CRISPR guide to confirm the efficiency of the guide prior to the actual CRISPR/Cas mutagenesis experiment (Fig. 6).

25. An initial trial digest should be run to determine the optimal incubation time for the particular restriction enzyme and PCR product—while incubation time is generally 1 h, certain restriction enzymes have been shown to be able to digest the product in 5–15 min.
26. In the case that only one G_1 mutant fly was identified, the G_1 fly should be mated to wild type flies (*see* Fig. 2) to produce more heterozygous mutant flies in the following generation (G_2). The genotyped G_2 heterozygous mutants can then be mated to each other to produce homozygous mutant progeny (G_3) and subsequently a homozygous mutant strain.
27. According to Mendel's law of segregation, the genotypes of the G_2 progeny should follow the ratio 1:2:1 for homozygous mutants (−/−): heterozygous mutants (−/+): wild type (+/+).

Acknowledgments

MT13059—This project has been funded by Hort Innovation, using the Apple & Pear, Strawberry, Citrus, Cherry, Summerfruit, Table Grape, and Vegetable research and development levies and contributions from South Australian Research and Development Institute (SARDI), Primary Industries and Regions South Australia (PIRSA), and the Australian Government. Hort Innovation is the grower-owned, not-for-profit research and development corporation for Australian horticulture.

References

1. Jinek M, Chylinski K, Fonfara I et al (2012) A programmable dual-RNA-guided DNA endonuclease in adaptive bacterial immunity. Science 337(6096):816–821. https://doi.org/10.1126/science.1225829
2. Doudna JA, Charpentier E (2014) The new frontier of genome engineering with CRISPR-Cas9. Science 346(6213): 1077–1086. https://doi.org/10.1126/science.1258096
3. Qin YJ, Paini DR, Wang C et al (2015) Global establishment risk of economically important fruit fly species (Tephritidae). PLoS One 10(1). https://doi.org/10.1371/journal.pone.0116424
4. Hancock D, Hamacek E, Lloyd A et al (2000) The distribution and host plants of fruit flies (Diptera: Tephritidae) in Australia. Queensland Department of Primary Industries, Australia
5. Clarke AR, Powell KS, Weldon CW et al (2011) The ecology of *Bactrocera triyoni* (Diptera: Tephritidae): what do we know to assist pest management? Ann Appl Biol 158(1):26–54. https://doi.org/10.1111/j.1744-7348.2010.00448.x
6. Yonow T, Sutherst RW (1998) The geographical distribution of the Queensland fruit fly, *Bactrocera* (Dacus) *tryoni*, in relation to climate. Aust J Agric Res 49(6):935–953. https://doi.org/10.1071/a97152
7. Dominiak BC, Ekman JH (2013) The rise and demise of control options for fruit fly in Australia. Crop Protect 51:57–67. https://doi.org/10.1016/j.cropro.2013.04.006
8. Knipling EF (1955) Possibilities of insect control or eradication through the use of sexually sterile males. J Econ Entomol 48(4):459–462. https://doi.org/10.1093/jee/48.4.459
9. Weldon CW (2005) Marking Queensland fruit fly, *Bactrocera tryoni* (Froggatt) (Diptera: Tephritidae) with fluorescent pigments: pupal emergence, adult mortality, and visibility and

persistence of marks. Gen Appl Ent J Entomol Soc New South Wales 34:7–13

10. Dominiak BC, Sundaralingam S, Jiang L et al (2010) Impact of marker dye on adult eclosion and flight ability of mass produced Queensland fruit fly *Bactrocera tryoni* (Froggatt) (Diptera: Tephritidae). Aust J Entomol 49:166–169. https://doi.org/10.1111/j.1440-6055.2010.00745.x
11. Rendón P, McInnis D, Lance D et al (2004) Medfly (Diptera : Tephritidae) genetic sexing: large-scale field comparison of males-only and bisexual sterile fly releases in Guatemala. J Econ Entomol 97(5):1547–1553. https://doi.org/10.1603/0022-0493-97.5.1547
12. Meats A, Maheswaran P, Frommer M et al (2002) Towards a male-only release system for SIT with the Queensland fruit fly, *Bactrocera tryoni*, using a genetic sexing strain with a temperature-sensitive lethal mutation. Genetica 116(1):97–106. https://doi.org/10.1023/a:1020915826633
13. Raphael KA, Shearman DCA, Gilchrist AS et al (2014) Australian endemic pest tephritids: genetic, molecular and microbial tools for improved Sterile Insect Technique. BMC Genet 15(2):1–3. https://doi.org/10.1186/1471-2156-15-s2-s9
14. Choo A, Crisp P, Saint R et al (2018) CRISPR/Cas9-mediated mutagenesis of the white gene in the tephritid pest *Bactrocera tryoni*. J Appl Entomol 142(1–2):52–58. https://doi.org/10.1111/jen.12411
15. Nguyen TNM, Mendez V, Ward C et al (2021) Disruption of duplicated yellow genes in *Bactrocera tryoni* modifies pigmentation colouration and impacts behaviour. J Pest Sci 94(3): 917–932. https://doi.org/10.1007/s10340-020-01304-9
16. Ward CM, Aumann RA, Whitehead M et al (2021) White pupae phenotype of tephritids is caused by parallel mutations of a MFS transporter. Nat Commun 12(1):1–2
17. Choo A, Fung E, Chen IY et al (2020) Precise single base substitution in the shibire gene by CRISPR/Cas9-mediated homology directed repair in *Bactrocera tryoni*. BMC Genet 21(2): 127. https://doi.org/10.1186/s12863-020-00934-3
18. Moadeli T, Taylor PW, Ponton F (2017) High productivity gel diets for rearing of Queensland fruit fly, *Bactrocera tryoni*. J Pest Sci 90(2): 507–520. https://doi.org/10.1007/s10340-016-0813-0
19. Concordet JP, Haeussler M (2018) CRISPOR: intuitive guide selection for CRISPR/Cas9 genome editing experiments and screens. Nucleic Acids Res 46(W1):W242–W245. https://doi.org/10.1093/nar/gky354
20. Zhang YL, Ge XL, Yang FY et al (2014) Comparison of non-canonical PAMs for CRISPR/Cas9-mediated DNA cleavage in human cells. Sci Rep 4. https://doi.org/10.1038/srep05405

Chapter 10

CRISPR/Cas9 Genome Editing in the New World Screwworm and Australian Sheep Blowfly

Daniel F. Paulo, Megan E. Williamson, and Maxwell J. Scott

Abstract

Blowflies are of interest for medical applications (maggot therapy), forensic investigations, and for evolutionary developmental studies such as the evolution of parasitism. It is because of the latter that some blowflies such as the New World screwworm and the Australian sheep blowfly are considered major economic pests of livestock. Due to their importance, annotated assembled genomes for several species are now available. Here, we present a detailed guide for using the *Streptococcus pyogenes* Cas9 RNA-guided nuclease to efficiently generate both knockout and knock-in mutations in screwworm and sheep blowfly. These methods should accelerate genetic investigations in these and other closely related species and lead to a better understanding of the roles of selected genes in blowfly development and behavior.

Key words CRISPR/Cas9, Genome editing, Gene disruption, Knockout, Homologous donor repair, Knock-in, Reverse genetics, T7 RNA transcription, Non-lethal DNA isolation, T7E1, Cas9 in vitro cleavage assay

1 Introduction

Blowflies (Diptera: Calliphoridae) are a diverse group of flies distributed worldwide that are of ecological, forensic, medical, and economic importance. Regarding the latter, myiasis-causing blowflies, such as the New World screwworm, *Cochliomyia hominivorax* (Coquerel, 1858), and the Australian sheep blowfly, *Lucilia cuprina* (Wiedemann, 1830), are obligate or facultative ectoparasites of warm-blooded vertebrates [1]. Females of these species lay their eggs on dried margins of wounds and bodily orifices of their hosts. After hatching, their larvae feed on the live tissues of animals to complete development [2]. Odors released from the colonized wounds act as chemical cues, attracting more females to oviposit, which can lead to severe infestations and host death if untreated

Daniel F. Paulo and Megan E. Williamson contributed equally to this work.

Paul John Verma, Huseyin Sumer and Jun Liu (eds.), *Applications of Genome Modulation and Editing*,
Methods in Molecular Biology, vol. 2495, https://doi.org/10.1007/978-1-0716-2301-5_10,

[3]. These infestations have destructive consequences on domestic welfare, accounting for significant economic losses in the livestock industry, while impacts on wildlife and humans are largely unknown [4]. Because of their economic importance, the genomes of *C. hominivorax* and *L. cuprina* have been assembled and annotated [5, 6], allowing the selection and functional evaluation of genes of interest. While RNAi methods have been used to investigate gene function in blowflies [7, 8], CRISPR/Cas9 (Clustered Regularly Interspaced Palindromic Repeats, and the CRISPR-associated endonuclease 9)-based gene editing offers the potential for precise gene knockouts for reverse genetics [9]. At present, efficient systems for making transgenic *L. cuprina* and *C. hominivorax* have been developed using *piggyBac* transposon vectors and fluorescent protein marker genes [10–13]. Likewise, the CRISPR/Cas9 system can be used to "*knock-in*" a gene construct into a precise location in the genome. In contrast to Cas9-induced modifications, *piggyBac*-mediated transgenesis is relatively random, which often results in significant differences in the levels of gene expression at different locations, and therefore several independent lines are typically maintained for each gene construct.

In its simplest form, the CRISPR/Cas9 system relies on two main components; a single guide RNA (sgRNA) complementary to a target sequence on the genome, and the *Streptococcus pyogenes*-derived endonuclease Cas9 which by the guidance of the sgRNAs will promote DNA double-stranded brakes (DSBs) at a specific site in the genome [9]. These DSBs are then recognized and repaired by cells using either the nonhomologous end-joining (NHEJ) or the homology-directed repair (HDR) pathways. The error-prone NHEJ pathway often results in a variety of insertions and/or deletions (collectively called indels) at the targeted position of the genome, which can be used to disrupt gene function by shifting its open reading frame (ORF). Strains developed through this method are called knockouts (KO), and they are extremely valuable to reverse genetics through the observation of functional consequences of gene disruption in vivo. On the other hand, the HDR pathway relies on homologous templates to repair DSBs. The natural template is the opposing chromosome, but synthetic donor DNA sequences can also be provided. When a synthetic template is used, such as double- or single-stranded DNA (dsDNA and ssDNA, respectively) the HDR results in the exchange of the original genomic content by a new one at the specific genome-targeted site, and therefore strains developed through this method are called knock-ins (KI).

In this chapter, we describe a comprehensive guide for genome editing in blowflies using the CRISPR/Cas9 system (Fig. 1). The methods described here were successfully used to disrupt and analyze gene function in vivo, as well to develop stable germline mutant strains of *C. hominivorax* and *L. cuprina* species [14–17].

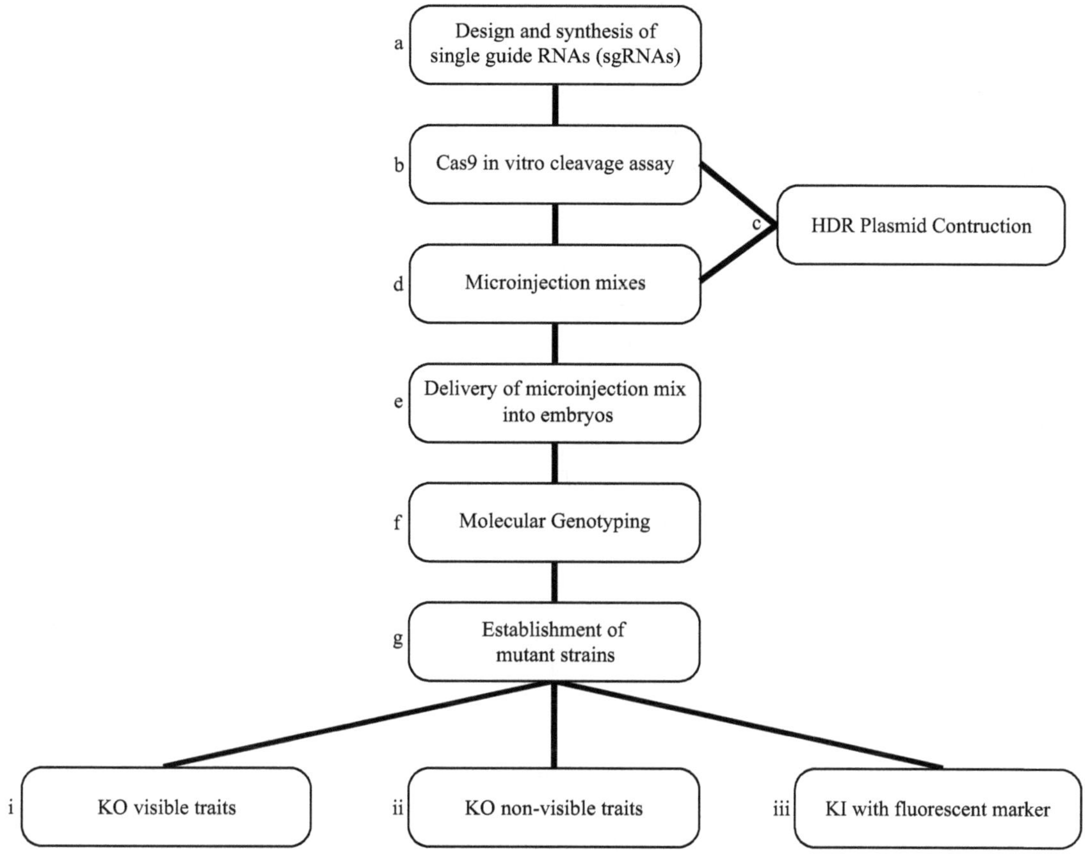

Fig. 1 Flowchart summarizing the steps to induce genome edits in the screwworm and sheep blowfly using CRISPR/Cas9. (**a**) Experiments begin with the design of single guide RNAs (sgRNAs) complementary to a target region in the gene of interest (GOI). Bioinformatic tools are used to assist sgRNAs design and off-target predictions. Selected candidate sgRNAs are then synthesized using an in vitro T7 transcription method. (**b**) A Cas9 in vitro assay can be used to evaluate candidate sgRNAs activity. Active sgRNAs can be directly used in knockouts (KO) experiments. (**c**) If aiming for a knock-in (KI), a donor plasmid construction should be designed and synthesized. The plasmid will be offered as a template for the homology-directed repair (HDR) pathway. (**d**) Microinjection experiments begin with the pre-assembly of Cas9-sgRNAs ribonucleoprotein complexes (RNPs) in an injection buffer. The mix can be supplemented with fluorescent markers to enhance microinjection experiments and a donor plasmid for KIs. (**e**) Custom-made needles are loaded with the microinjection mix and used in microinjection procedures to deliver RNPs into blowfly embryos. (**f**) Surviving flies can be screened for genome modifications through a number of molecular genotyping strategies, including the T7 endonuclease 1 assay (T7E1) and direct Sanger sequencing followed by peak scan. (**g**) Different crossing schemes (i, ii, iii) are then used to establish germline mutant strains depending on the type of modification induced in the genome and its consequences on species' traits

These site-directed mutagenesis methods provide not only an efficient, flexible, and inexpensive way to study gene function in non-traditional model insects [18], but also to develop highly efficient Cas9-based genetic methods for control of insect pests, such as the blowflies that are major pests of livestock [19, 20] (*see* also Chapter 20 on gene drive).

2 Materials

2.1 Synthesis of Single Guide RNAs (sgRNAs)

1. Q5® High-Fidelity 2× Master Mix (New England Biolabs) or similar reagents for high-fidelity polymerase chain reaction (PCR), such as Phusion® High-Fidelity PCR Kit (New England Biolabs).
2. sgRNA universal reverse primer (PAGE purified): 5′-AAA AGC ACC GAC TCG GTG CCA CTT TTT CAA GTT GAT AAC GGA CTA GCC TTA TTT TAA CTT GCT ATT TCT AGC TCT AAA AC-3′.
3. sgRNA custom forward primer (PAGE purified, *see* Subheading 3.1.1 on how to design this primer): 5′-GAA ATT AAT ACG ACT CAC TAT **AGG** [(N)$_{17–20}$ *Cas9-target sequence excluding PAM*] GTT TTA GAG CTA GAA ATA GC-3′.
4. PCR Purification Kit such as from QIAGEN or Zymo Research.
5. 10× TBE (or TAE): Dilute to 1× by mixing 100 mL of 10× TBE with 900 mL dH_2O. Important: For gels run with RNA samples, create a fresh 1× buffer in RNase-free containers and use RNase-free or DEPC-treated water.
6. 1 kb Plus DNA Ladder (New England Biolabs) or similar ladder that can distinguish band sizes down to 100 bp and up to the size of your largest HDR fragment. 1 kb Plus DNA ladder ranges from 100 bp to 10 kb.
7. Benchmark Scientific SmartGlow™ Loading Dye with Safe Green Stain (Accuris Instruments): Add 0.5 μL to every 10 μL of sample. Add 1 μL to every 10 μL of ladder. Alternative loading dye and visualization stains can be used. For RNA gel loading, use the same volume of Gel Loading Buffer II (Invitrogen).
8. For quantifications using the Qubit™ platform: (DNA) Qubit™ dsDNA HS Assay Kit (Invitrogen); (RNA) Qubit™ RNA HS Assay Kit (Invitrogen). Both should be used with Qubit™ Assay Tubes (Invitrogen).
9. MEGAshortscript™ T7 Transcription Kit (Invitrogen).
10. Phenol: Chloroform: Isoamyl Alcohol (25:24: 1, v/v).
11. Chloroform.
12. 5PRIME Phase Lock Gel™- Heavy (Quanta bio).
13. Isopropanol or 100% Ethanol.
14. 70% Ethanol.

2.2 Cas9 In Vitro Cleavage Assay

1. Amplifications (PCR), gel electrophoresis, and nucleic acid quantifications: Refer to Subheading 2.1 (**items 1**, and **5–8**).

2. EnGen® Spy Cas9 NLS (New England Biolabs) or Guide-it™ Genotype Confirmation Kit (Takara Bio).
3. Template DNA: Design forward and reverse genotyping primers to amplify a 600 bp to 1 kb region with the Cas9-targeted cut sites towards the center of the fragment (*see* **Note 1** for genotyping primer design tips).
4. Synthesized target-specific sgRNA(s) (obtained in Subheading 3.1.2).

2.3 Homology-Directed Repair (HDR) Plasmid Construction

1. Amplifications (PCR), gel electrophoresis, and nucleic acid quantifications: Refer to Subheading 2.1 (**items 1**, and **5–8**).
2. Fragment-specific forward and reverse primers with 15–40 bp overlap with adjacent fragments. These overhangs are critical for proper assembly of fragments. Use the NEBuilder assembly tool (available at https://nebuilder.neb.com/#!/) to generate primers with proper overhangs (*see* **Note 2** for HDR primer design tips).
3. NEB® High-Fidelity Restriction Endonucleases (when available) or any comparable restriction endonucleases.
4. NEBuilder® HiFi DNA Assembly Cloning Kit (New England Biolabs).
5. DNA Clean & Concentrator™-5 Kit (Zymo Research).
6. Zymoclean™ Gel DNA Recovery Kit (Zymo Research).
7. Competent cells such as NEB® 10-beta Competent *E. coli* (High Efficiency).
8. LB Growth Medium: Dissolve 25 mg LB Broth in 1000 mL milliQ water in a 1 L glass bottle. Prepare four clean glass bottles and add 100 mL LB to each bottle (save remaining LB Growth to make selection plates). Label bottles "*LB growth*". Loosely cover bottles with lid and autoclave for 30 min. Store LB Growth at room temperature for up to 6 months. Do not use the medium if it becomes cloudy; it is a sign of contamination.
9. Selection plates: Add 9 g of agar to the remaining 600 mL LB (above) and mix well. Loosely cover bottles with lid and autoclave for 30 min. Incubate agar LB at 55 °C for 20 min or until you can pick the flask up with gloved hands. Add 900 μL of 100 mg/mL Ampicillin and mix well by inversion. Carefully pour out into sterile plastic petri dishes. Dry for at least 45 min at room temperature before storing at 4 °C for up to a month.
10. Zymo ZR Plasmid Miniprep™—Classic (Zymo Research).
11. ZymoPURE™ II Plasmid Midiprep (Zymo Research).

2.4 Microinjection Mixes

1. Cas9 protein from *Streptococcus pyogenes* with NLS (PNA Bio or New England Biolabs). If supplied as dried protein, reconstitute to 1 μg/μL in nuclease-free water. Make aliquots of 3 μL in 0.5 mL centrifuge tubes on ice. Store at −80 °C for long term or −20 °C for use on a shorter term.
2. M Sodium phosphate buffer, pH 6.8: Dissolve 1.31 g of disodium phosphate (Na_2HPO_4) and 0.704 g of monosodium phosphate (NaH_2PO_4) in 80 mL of dH_2O by stirring. Adjust solution pH to 6.8 using hydrochloric acid (HCl) or sodium hydroxide (NaOH). Add dH_2O to a final volume of 100 mL. Make aliquots of 50 mL in conical centrifuge tubes and store at 4 °C for up to 6 months.
3. M Potassium chloride (KCl): Dissolve 14.91 g of KCl in 100 mL of dH_2O by stirring. Store at room temperature indefinitely.
4. 10× Injection buffer: Mix 0.5 mL of 0.1 M Sodium phosphate buffer and 7.5 mL of 2 M KCl in 42 mL of nuclease-free water for a final volume of 50 mL. Filter the entire volume through 0.22 μm Millipore filters (Fisher Scientific), make aliquots of 500 μL, and store at −20 °C indefinitely. Final concentrations should be 300 mM of KCl and 1 mM sodium phosphate buffer (*see* **Note 3**).
5. Phenol red solution (Sigma-Aldrich). Store at room temperature.
6. Microloader™ Pipette Tips (Eppendorf).

2.5 Delivery of Microinjection Mix into Embryos

1. Customized needles: Quartz glass capillary with filament (Sutter Instrument, Cat# QF100-70-10; Outside ∅ 1.00 mm, inside ∅ 0.70 mm, length 10 cm). Wrap capillaries in foil and bake at 200 °C for 4 h to remove any RNase. Pull needles with a Sutter P-2000 micropipette puller. The ideal needle should have a short-to-medium taper (2–5 mm long) with a thin tip to penetrate the chorion of blowfly embryos. For *Cochliomyia* embryos a standard setting is: heat = 650–690; fill = 04; velocity = 50–60; delay = 150–165; and pull = 155–180. Needle settings for *Lucilia cuprina*: heat = 690; fill = 04; velocity = 60; delay = 155; pull = 170.
2. For screwworm injections only, 4% (w/v) potassium hydroxide (KOH): Dissolve 40 g of KOH in one liter of dH_2O. Store at room temperature indefinitely.
3. Precleaned Premiere® micro slides No. 7104. Double Concavity 3″ × 1″ (1.2 mm).
4. Desiccator and dry silica gel beads.
5. Halocarbon oil 27 (Sigma-Aldrich).

6. Microinjection Station such as XenoWorks™ Micromanipulator and Digital Microinjector (Sutter Instrument) with either an inverted compound microscope or stereo microscope containing sufficient magnification to readily visualize injection needles and embryos.
7. Modular incubator chamber (Billups-Rothenberg).

2.6 Genotyping by T7E1

1. Proteinase K: Dilute 100 mg of Proteinase K (Invitrogen) in 5 mL of ddH_2O for a final concentration of 20 mg/mL. Make aliquots of 200 μL in 1.7 mL centrifuge tubes and store at −20 °C.
2. Proteinase K Solution (PKS): *per* sample; Mix 1.25 μL of 20 mg/mL Proteinase K, 5 μL of 100 mM Tris–HCL, 1 μL of 50 mM EDTA, and 12.5 μL of 100 mM NaCl. Raise the volume to 50 μL with dH_2O. Final concentrations should be: 0.5 mg/mL PK, 10 mM Tris-HCL, 1 mM EDTA, and 25 mM NaCl. Always use fresh prepared PKS.
3. T7E1 buffer: *per* sample; dilute 2 μL of 10× NEBuffer™ 2 (New England Biolabs) in 6 μL of dH_2O. Store at −20 °C indefinitely.
4. T7E1 enzyme: ten-fold dilute the T7 Endonuclease I (New England Biolabs) in dH_2O for a final concentration of 1 U/μL. Store at −20 °C indefinitely.
5. Stopping buffer: In a 1.7 mL tube combine 0.8 mL of 6× Gel Loading Dye, Purple, no SDS (New England Biolabs) with 0.4 mL of 0.5 M EDTA (pH 8). Store at 4 °C.

2.7 Genotyping by Sequencing

1. Injected, unhatched embryos and non-injected, unhatched embryos of the same age. Alternatively, tissue from a fly that develops from an injected embryo and a control DNA from a non-edited fly. If not to be processed immediately after injection, the embryos are quickly frozen in liquid nitrogen and stored at −80 °C.
2. Taq DNA Polymerase, recombinant (Invitrogen), or similar polymerase that leaves a 3′-end A-overhang, allowing for direct TA-cloning. If directly sequencing a PCR product, 3′-end A-overhang is unnecessary and any preferred polymerase can be used.
3. pGEM®-T Easy Vector System I (Promega) or pGEM®-T Easy Vector System II (Promega) with competent cells.
4. QIAprep® Spin Miniprep Kit (Qiagen) or Zymo Quick-DNA™ Miniprep Kit (Zymo Research).
5. KCM transformation buffer: 1 mL of 10× KCM (1 M KCl, 0.3 M $CaCl_2{\cdot}2H_2O$, 1 M $MgCl_2$) mixed with 1.5 mL 10% PEG in 10 mL of dH_2O.

6. Bacterial culture supplies: Refer to Subheading 2.3 (**items 8** and **9**).
7. X-Gal Solution.
8. Standard sequencing primers M13-F (5′-TGT AAA ACG ACG GCC AGT-3′) and M13-R (5′-CAG GAA ACA GCT ATG ACC-3′).

3 Methods

3.1 Design and Synthesis of Single Guide RNAs (sgRNAs)

3.1.1 Design of sgRNAs

Estimated time required: 1 d

1. Search for PAM (protospacer adjacent motif) sequences within your gene of interest (GOI). For the *S. pyogenes* Cas9 nuclease, the PAM sequence is 5′-NGG -3′, where "N" is any nucleotide.
2. Select 17–24 bp sequences directly upstream of the PAM site to be used as the protospacer for your GOI-specific sgRNA. Following a list of considerations for target sequence selection:
   (a) Select protospacer sequences with 40–60% GC content. Protospacer sequences with a "GG" before the PAM are also expected to enhance sgRNA activity [21].
   (b) Check the sgRNA for off-target effects against the genome of your model species. Several programs are available and recommended to assist design sgRNAs, including CHOPCHOP [22] and CRISPOR [23]. However, keep in mind that the accuracy of these programs will greatly be influenced by the reference sequence database used (e.g., quality of the genome assembly for your species, proxy genomes available, or completeness of ESTs databases).
   (c) Discard any sgRNAs complementary to nonspecific genomic regions that have a downstream PAM sequence. Avoid sequences with few mismatches (1–3 nts) between the distal 5′-end of sgRNAs and off-targets, as Cas9 can tolerate them. Mismatches within the first 10 nucleotides upstream to PAM (the so-called seed region [24]) will compromise DNA binding and cleavage, therefore they can be overlooked.
3. Assembly of sgRNA custom forward primers: Forward primers must include a T7 promoter region at the 5′-end and a complementary sequence to universal reverse primer at the 3′-end. Specifically, sgRNAs forward primers are composed by:
   (a) T7 promoter sequence: 5′-GAA ATT AAT ACG ACT CAC TAT A[GG]-3′.
   (b) Protospacer: N_{17-24} (protospacer changes depending on your target and does not include PAM site).

(c) Sequence complementary to reverse primer: 5′-GTT TTA GAG CTA GAA ATA GC-3′.

4. Following is a list of considerations for sgRNA custom forward primer design:
   (a) There should be a GG between the T7 sequence and protospacer. This can come from the protospacer sequence or added in to facilitate in vitro transcription using T7 kits.
   (b) If PAM is on the + strand, select the same sequence as the + strand for protospacer.
   (c) If PAM is on the − strand, select the reverse complement of the + strand.
   (d) PAM (while not included in the forward primer) should always fall between the protospacer and sequence complementary to reverse primer (i.e., PAM should always be towards 3′-end of protospacer in primer sequence).
   (e) Primers should be ordered with 100 nM concentration and PAGE purified.

Your final custom sgRNA forward primer should look like this (*see* Subheading 2.1 for the universal reverse primer sequence): 5′-[GAA ATT AAT ACG ACT CAC TAT A] [GG] [GOI-specific sequence, 17–24 bp excluding PAM] [GTT TTA GAG CTA GAA ATA GC]-3′.

3.1.2 Synthesis of sgRNAs

Estimated time required: 2 d

1. Prepare a PCR reaction containing 20 μL of 5× Phusion® HF buffer, 2 μL of 10 mM dNTP mix, 1 μL of 2 U/μL Phusion DNA polymerase (or similar high-fidelity polymerase, *see* Subheading 2.1), 5 μL of 10 μM sgRNA universal reverse primer, 5 μL of 10 μM sgRNA custom forward primer and nuclease-free water to a final volume of 100 μL.
2. Synthesize templates for sgRNAs T7-transcription through a step-up PCR in a preheated thermocycler under the following conditions: 98 °C for 2 min, 10 cycles of [98 °C for 10 s, 60 °C for 30 s, and 72 °C for 15 s], followed by 25 cycles of [98 °C for 10 s, 65 °C for 30 s, and 72 °C for 15 s], and a final extension step at 72 °C for 10 min. We have also used cycling conditions of 98 °C for 2 min, 35 cycles of 98 °C for 10 s, 60 °C for 30 s, and 72 °C for 30 s, followed by a final extension step at 72 °C for 5 min. Allow reactions to cool at 10 °C before further handling. Amplified templates can be kept overnight at 4 °C before purification or stored at −20 °C indefinitely.

3. Resolve 5 μL of the amplification product in a 2% agarose gel in 1× TAE Buffer. Amplifications of sgRNA templates are expected to be about 125 bp.
4. Purify the remaining PCR volume using the QIAquick® PCR Purification Kit or Zymo DNA Clean & Concentrator™ for PCR products, according to the manufacturer's instructions. Usually, 35 μL of nuclease-free water is enough to elute the purified templates, but this volume might be adjusted based on the PCR yield.
5. Measure DNA concentration with NanoDrop™ Spectrophotometer or Qubit™ fluorometer. Adjust the concentration of sgRNA templates to nearly 100 ng/μL in nuclease-free water. Templates can be kept overnight at 4 °C before proceeding with T7 transcription or stored at −20 °C indefinitely.
6. Prepare a MEGAshortscript™ in vitro RNA transcription reaction as follows: 2 μL of 10× Reaction Buffer, 8 μL of the dNTPs mix (prepared with dUTP instead of dTTP), 2 μL of T7 Enzyme Mix, 300–600 ng of template DNA (up to 8 μL) and nuclease-free water for a final volume of 20 μL. Transcriptions are performed at 37 °C for at least 4 h in a preheated thermocycler. Transcriptions can also be done overnight.
7. Reserve 1.5 μL of the transcription product, mix with 3 μL of Gel Loading Buffer II, and store at 4 °C for control purposes.
8. Add 1 μL of TURBO DNase to the remaining transcription product, mix gently by pipetting, and incubate at 37 °C for 15 min in a thermocycler. Proceed immediately with the purifying step.
9. Increase the volume of the treated transcription product to ~200 μL by adding 160 μL of nuclease-free water, and 20 μL of ammonium acetate and mix. Transfer to a 2 mL phase gel lock (PGL) tube that has been prespun or a standard 1.7 mL tube.
10. In a chemical fume hood, add 200 μL of Phenol: Chloroform: Isoamyl Alcohol, and mix well by inversion. The tube can be mixed by vortex if not using a PGL tube.
11. Centrifuge at 16,000 × *g* for 5 min at 4 °C to separate the organic bottom phase and the aqueous top phase. Carefully transfer the upper aqueous fraction to a clean PGL or 1.7 mL tube using a pipette.
12. Add 200 μL of Chloroform: Isoamyl Alcohol, mix and centrifuge as before. Carefully transfer the aqueous fraction to a new 1.7 mL tube using a pipette.
13. Add 400 μL of cold Isopropanol or two volumes of 100% ethanol to the aqueous fraction and mix by inversion.

Precipitate the transcribed RNA at −20 °C for at least 1 h. Alternatively, precipitations can be done overnight.

14. Centrifuge the samples at 16,000 × *g* for 20–30 min at 4 °C. The precipitated RNA should appear as a translucent pellet. Carefully discard the supernatant by inversion and then wash the pellet with 0.5–1 mL of 70% cold ethanol.
15. Centrifuge at 16,000 × *g* for 10 min at 4 °C, carefully remove the supernatant as before, and allow the pellet to air-dry for 10–20 min at room temperature. Resuspend the pellet in 20 μL of nuclease-free water.
16. Quantify the purified samples using NanoDrop™ Spectrophotometer or Qubit™ fluorometer (*see* Subheading 2.1, **item 8**). Aliquot and store in 0.5 mL tubes at −80 °C.
17. Verify the sgRNA transcription quality by electrophoresis in a 2% agarose gel in 1× TAE Buffer. Heat selected sgRNA and control samples (**steps 7** and **17**) at 75 °C for 5 min in a preheated thermocycler and let them rest on ice for 2 min before loading into the gel.

3.2 Cas9 In Vitro Cleavage Assay

Estimated time required: 1 d

1. Prepare a PCR reaction containing 25 μL of 2× Terra™ PCR direct buffer, 1.5 μL of each 10 μM forward and reverse primer, 1 μL of 1.25 U/μL Terra™ PCR Direct Polymerase Mix (or alternatively Q5® High-Fidelity 2× Master Mix), 1–1000 ng of genomic DNA and nuclease-free water to a final volume of 50 μL. Reactions should be performed on ice. Use 1.7 mL tubes to prepare the mix and 0.2 mL tubes or 96-well plates for reactions.
2. Perform amplifications in a preheated thermocycler under the following conditions: 98 °C for 2 min, 35 cycles of [98 °C for 10 s, 50–72 °C for 15 s, and 68–72 °C for 1 min], followed by a final extension step at 68–72 °C for 2 min. For control purposes, reserve 5 μL of each reaction mixed with gel loading dye. Keep control samples at 4 °C.
3. Prepare a reaction to complex Cas9 and sgRNAs for a final volume of 27 μL containing 20 μL nuclease-free water, 3 μL of 1× NEBuffer™ 3. 1, 600 ng of sgRNA, and 1 μL of 20 μM EnGen Cas9. Gently mix by pipetting and do a quick spin in a microcentrifuge. An alternative low volume protocol we have used is to mix 1 μL of 100 ng/μL target-specific sgRNA and 0.5 μL of 500 ng/μL Guide-it Recombinant Cas9 Nuclease for a final volume of 1.5 μL.
4. Incubate 10 min at room temperature followed by 10 min at 37 °C. This step is critical for loading the sgRNA onto Cas9 protein.

5. Add 50–100 ng of template DNA for a final volume of 30 μL. Gently mix by pipetting and do a quick spin in a microcentrifuge. For the low volume protocol, mix the entire pre-assembled RNPs with 1 μL of 15× Cas9 Reaction Buffer, 1 μL of 15× BSA, and 2.5 μL of the unpurified PCR product for a final volume of 15 μL in nuclease-free water.
6. Perform the in vitro cleavage assays at 37 °C for 1 h in a preheated thermocycler. Inactivate Cas9 by the addition of 2 μL of 10 mg/mL proteinase K followed by heating at 80 °C for 5 min. Allow tubes to cool at 10 °C before further handling. The inactivation step can be skipped for the low volume protocol.
7. Mix the entire volume with 3 μL of gel loading buffer. Resolve the entire reaction volume and the control samples (reserved in **step 2**), in a 2% agarose gel in 1× TBE Buffer. Alternatively, reactions can be stored at 4 °C for up to a month (*see* **Note 4** for assay details and expected results).

3.3 HDR Plasmid Construction

Estimated time required: 7 d

1. Select the sgRNA to be used along with the HDR construct and determine the Cas9-cut site (3 bp upstream of PAM).
2. Select ~1000 bp both upstream and downstream of the Cas9-cut site to be used as the left and right homology arms of the construct. Ideal homology arms will start as close as possible or within 10 bp of the cut site.
3. Design primers for restriction digests to amplify all necessary fragments of HDR construct using NEBuilder Assembly tool (available at https://nebuilder.neb.com/#!/). This includes the Left Homology Arm, Right Homology Arm, cassette to be inserted into the genome (ideally including a fluorescent marker for easy tracking) and a vector backbone with proper 15–40 bp overhangs to ligate the fragments together. (*see* **Note 2** for HDR primer design tips).
4. Perform PCR amplification of products as follows (**steps 4–7**): Prepare a PCR for a final volume of 50 μL containing 12.5 μL of 2× Q5® High-Fidelity Master Mix, 1.25 μL each 10 μM target-specific forward and reverse primer, 1–1000 ng of genomic or 1 pg–10 ng of plasmid containing the desired fragment.
5. Perform amplification in preheated thermocycler under the following conditions: 98 °C for 30 s, 35 cycles of 98 °C for 10 s, 50–72 °C for 10 s, and 72 °C for 1 min/kb, followed by a final extension step at 72 °C for 2 min. Allow the reactions to cool down at 10 °C before proceeding. Reactions can be stored at 4 °C for up to a month, and at −20 °C indefinitely.

6. Run 5 μL of PCR product on 1% agarose gel in 1× TBE Buffer to confirm single product of correct size.
7. Purify the remaining 45 μL of the PCR using Zymo DNA Clean & Concentrator™ following manufacturer's instructions for PCR products (ratio 5: 1 for DNA binding buffer and sample, respectively). Elute sample with 10 μL Elution Buffer.
8. Perform restriction enzyme digestions as follows (**steps 8–12**): Prepare a 50 μL digest reaction with 5 μL digest buffer, 5 μL (total) restriction enzyme, 1–2 μg of plasmid with desired fragment, up to 50 μL of water. Important: If using more than 1 enzyme, do not exceed 5 μL total in reaction or one-tenth of reaction volume. Incubate at the recommended temperature of enzyme for 1 h.
9. Run all 50 μL of reactions on 1% agarose gel in 1× TBE buffer. Excise the proper band using a clean and sharp scalpel. Place the gel fragment in a 1.5 mL centrifuge tube and weigh the gel slice in grams.
10. Gel purify excised band using Zymoclean™ Gel DNA Recovery Kit following manufacturer's instructions. Elute sample with 10 μL Elution Buffer.
11. Quantify the samples using Qubit™ fluorometer (*see* Subheading 2.1, **item 8**).
12. Assemble HiFi DNA assembly reaction as follows: For 2–3 fragment assembly (Use a 1:2 ratio of vector:insert): 0.03–0.2 pmols total amount of fragments in reaction, 10 μL of 2× NEBuilder HiFi DNA Assembly Master Mix, and nuclease-free water up to 20 μL; For 4–6 fragment assembly (Use a 1: 1 ratio of vector:insert): 0.2–0.5 pmols total amount of fragments in reaction, 10 μL of 2× NEBuilder HiFi DNA Assembly Master Mix, and nuclease-free water up to 20 μL.
13. Incubate reactions in a preheated thermocycler at 50 °C for 15 min (2–3 fragment assembly) or 60 min (4–6 fragment assembly).
14. Samples can either be used immediately for transformation into *E. coli* or stored at −20 °C until ready to use.
15. Transform NEB® 10-beta competent cells with 2 μL of assembly product and follow the transformation protocol as written.
16. Select several colonies from transformation, grow each colony overnight (12–16 h) in 4 mL of LB Growth supplemented with 4 μL of 100 mg/mL Ampicillin. On the next day, perform minipreps using Zymo ZR Plasmid Miniprep™-Classic according to the manufacturer's instructions. Important: Make sure to use a proper antibiotic for your plasmid backbone.

17. Confirm plasmids carrying correct inserts with restriction enzyme screening. Choose enzymes that will give you unique bands with the properly assembled product. Assemble restriction enzyme screen as follows (**items 19–21**).
18. Mix 1 μL restriction enzymes total (adjust based on the number of enzymes you are screening without going over 1 ul) with 1 μL of 10× digest buffer, and 100–200 ng of plasmid for a volume up to 10 μL in nuclease-free water.
19. Incubate the reaction for 1 h at proper temperature for enzymes used. Resolve reactions on 1% agarose gel in 1× TBE and select plasmid with proper banding patterns.
20. Confirm plasmid with at least three different enzyme sets. Additionally, plasmid can be confirmed via Sanger sequencing by sequencing across junctions.
21. Transform NEB® 10-beta competent cells with a small amount of the correct plasmid (10–100 ng) and follow the transformation protocol per the manufacturer's instructions.
22. Select 1–2 colonies for miniprep by growing overnight in 50 mL of LB Growth with 50 μL of 100 mg/mL Ampicillin or proper antibiotic for plasmid backbone. All colonies should be correct since they were transformed off of pure plasmid, but it is always good practice to confirm with a restriction digests (*see* **steps 19–21**).
23. Midiprep the overnight cultures using ZymoPURE™II Plasmid Midiprep Kit following the manufacturer's instructions.
24. Quantify the purified samples using a NanoDrop™ Spectrophotometer or using a Qubit™ fluorometer (*see* Subheading 2.1, **item 8**).

3.4 Microinjection Mixes

Estimated time required: 1 h.

1. For Cas9-mediated mutagenesis (KO): In a 0.2 mL tube, add 1.5–5 μL of Cas9 protein (final concentration 480–500 ng/μL), 1 μL Injection Buffer, 0.75 or 1.5 μL 2 M KCl, 0.5–1 μL of 2 μg/μL sgRNA (final concentration 100–200 ng/μL) and nuclease-free water to a final volume of 10 μL (*see* **Note 3** for details on microinjection mix). The lower concentration KCl is used for *Lucilia* and the higher concentration for *Cochliomyia*. Optional: Include 0.5 μL phenol red dye to the mix to aid in visualizing the solution when injecting embryos. Alternatively, include 0.5 μL of a marker plasmid with a constitutive promoter driving a fluorescent protein marker gene to a final concentration of 200–500 ng/μL (include the plasmid after Cas9-sgRNA incubation). The transient expression of the marker the day after injections can be used to select embryos for further analysis. Mix gently by pipetting and incubate in a

preheated thermocycler at 37 °C for 10 min, with the lid opened. Alternatively, perform the incubation step at 37 °C for 30 min in a water bath. Prepare one reaction *per* sgRNA to be used.

2. For HDR (KI): In a 0.2 mL tube, add 1.5–5 μL of Cas9 protein (final concentration 480–500 ng/μL), 1 μL Injection Buffer, 1.5 μL KCl, 1 μL of 2 μg/μL sgRNA (final concentration 100–200 ng/μL) and nuclease-free water to a final volume of 8–9 μL. Optional: Include 0.5 μL phenol red dye to the mix. Incubate reaction at 37 °C for 30 min in a water bath. After incubation add 1–2 μL HDR plasmid DNA (total of 5 μg) bringing final volume up to 10 μL.
3. Transfer total volume to Millipore ultrafree-MC centrifugal filter 0.45 μM pore size and centrifuge at 12,000 × *g* for 4 min at 4 °C. Alternatively, centrifuge the RNPs solution at 16,000 × *g* for 10 min at room temperature. Transfer most of the upper portion of the mix (about 8.5 μL) to a new 0.2 mL tube.
4. Keep the injection mix on ice until ready to use. Alternatively, injection mix can be stored at −80 °C. However, the efficiency appears to decrease after freeze thaws.

3.5 Loading and Opening the Needles

1. Load needles with ~3 μL of prepared microinjection mix using a pipette with an Eppendorf femtotips capillary pipette tip or similar capillary pipette tip.
2. Carefully insert the pipette tip into the opening of the needle (opposite the pulled point). Slowly pipette the injection mix into the needle while slowly pulling out of the needle avoiding loading air bubbles into the needle.
3. Hold the end of the needle and flick the needle downwards to move all the injection mix towards the tip of the needle and remove any air bubbles that have formed. Optional: Pre-load a few needles before starting microinjections. This will save time in case you need to replace the needle during experiments. Keep the pre-loaded needles on ice.
4. Tighten a loaded needle into the holder of the micromanipulator when ready to start microinjection experiments.
5. Two methods can be used to open needles: (i) The tip of the needles is preferably opened by using a Sutter BV-10 beveler. Needles are beveled at a 25 ° angle for 10 s using a fine Diamond Abrasive Plate (Sutter Instrument #104D); (ii) If a beveler is not available, the needles can be opened by touching the needle on the embryo chorion. While repeatedly pressing the "*clear*" button, gently move the embryo up and down using the microscope stage controls to open the needle tip. You will notice when it is open whenever droplets of liquid start to be released from the tip.

3.6 Delivery of Microinjection Mix into Embryos

3.6.1 Cochliomyia Embryo Preparation

1. Stimulate egg laying by offering to females an oviposition device consisting of warm raw ground meat mixed with spent larval media [25]. Place the oviposition device inside a 6-day--old colony cage containing nearly 200 individuals (males and females, ratio about 1: 1). Transfer the cage into a BOD incubator at 37 ± 1 °C, relative humidity 80 ± 5% (RH), in darkness. Check the cage for laid eggs every 15 min. Collect new batches of eggs every 15 min after the first oviposition. Ideally, you should inject the eggs within 30–45 min after placing the oviposition device into the cage.
2. Carefully harvest the embryo clumps from the oviposition device with the help of fine forceps and transfer them to a clean glass vial (about 25 mL; 15 mm ø × 60 mm height). Fill two-third of the vial with 4% (w/v) KOH, lid it tight and shake vigorously for 2 min in order to separate the embryos. Transfer the embryos to a clean mesh basket and wash them with a strong and abundant amount of dH_2O using a squirt bottle. Keep the mesh basket in a petri dish filled with a bit of dH_2O.
3. Fix a small stripe of thin double-sided tape across the middle of each depression in microscope slides. Align the embryos in the slides by placing 20–25 eggs onto each side with the help of a fine moist painting brush. Align the eggs so their posterior poles face towards the nearest slide edge.
4. Dehydrate the embryos in a desiccator for 6 min. This is a key step and should be optimized during microinjections. If eggs are too "*moist,*" they might detach from slides or confer more resistance to the needles. On the other hand, if the eggs are too "*dry*" they will bend over preventing needle penetration. In both cases, embryo survival will be negatively affected.
5. After dehydration, cover the aligned eggs with one or two drops of Halocarbon oil 27. Wait about 5 min before injecting. The oil will clear the embryos making it possible to visualize injections and confirm embryos are pre-blastoderm, ideally stage 1 or 2. Do not inject embryos that have developed beyond the cellular blastoderm stage.

3.6.2 Lucilia Embryo Preparation

1. Fill 35 mm petri dishes with raw ground beef and press meat so it is compact. Place a finger indent in the middle of the meat as an area for the flies to lay eggs. Place 1 meat dish per cage containing 8- to 12-day-old mated female flies for 10 min at a time until flies begin to lay eggs. Remove meat dish after 10 min and use eggs for lining up slides. Ideally, you should inject the eggs within 30–45 min after placing the meat into the cage.

2. Place a small square (10 mm × 10 mm) of kimwipe on a flat slide. Place 1 drop of dH_2O on the kimwipe to keep the eggs moist as you are separating them. Pick up a small clutch of eggs using fine forceps and put the clutch on the wet kimwipe. Carefully use the forceps to separate the eggs from one another by rolling the eggs around. Be gentle to not damage the eggs.
3. On a new depression microscope slide, place a small piece of double-sided tape across the middle of the depression with the smooth side facing the center of the depression.
4. Pick up the wet kimwipe with forceps and dab it on a dry kimwipe until the kimwipe with the eggs is left damp. Transfer kimwipe with eggs to the prepared slide with double-sided tape, placing it adjacent to the double-sided tape for easy transfer of eggs.
5. Using a forceps or paintbrush, carefully start transferring eggs from the kimwipe to the double-sided tape in such a way that the slightly wider posterior end of the egg is hanging off the tape into the depression well (side to be injected) and all eggs are positioned on the same plane and depth (approximately).
6. Desiccate eggs for 6 min in a desiccation chamber (time will vary depending on humidity in room). Remove the slide from the desiccation chamber and cover eggs with halocarbon oil 27 ensuring oil stays within the depression well. This usually requires 3–4 gravity drops of the oil from a transfer pipette. Wait several minutes until the oil permeates the chorion and clarifies the embryo before injecting.

3.6.3 Microinjections

1. Place the egg lined slide on the microscope stage and stabilize it with the stage clip. The posterior end of the eggs should face the injection needle. Turn on the microscope light beam and orient the focus so that the eggs are in complete focus and their posterior ends are aligned with the center of the microscope eyepiece.
2. Using the micromanipulator, orient the needle towards the first embryo in the row, as close as possible but without touching the oil surface. While monitoring through the eyepiece, slowly hover the needle back and forth over the embryos until you detect the shadow of the needle. Align the shadow with the first egg. Set the "*work*" button so that the micromanipulator will automatically adjust the needle to the correct position for subsequent slides. Important: When setting the "*work*" button, set it so that it is completely out of the oil to avoid potential damage to eggs if the slides are slightly different.
3. Slowly bring down the needle until its tip enters the oil. Adjust the focus so the needle tip is nearly in the same focal plane as

the embryos and fine-tune the needle position so its tip is centered with the first embryo in the row.

4. With the needle inside the oil, press the "*clear*" button to get the injection mix flowing. If nothing is coming out, it is possible that the tip of your needle is not open enough or there are air bubbles present in the needle: (i) If the tip of your needle is not open enough, you can gently try rubbing the tip of the needle across the edge of the tape. This can sometimes lead to an opening too big which will decrease the success of your injections. Alternatively, you can remove the injection mix from the needle with a microloader pipette tip and load a new needle. (ii) If there are air bubbles in the needle, remove the needle from the holder and give it several hard flicks towards the ground (holding the end opposite the point). Do this until no more air bubbles are present.
5. Set the controls to "*continuous*" mode. Set the transfer pressure until you begin to see a steady stream of slow droplets exiting the needle. Set the pressure so that enough injection mix is coming out during each injection. These settings will be different depending on your equipment and species. Both transfer pressure and injection pressure will likely need to be adjusted during microinjections, mainly if the needle gets clogged or the opening becomes too big.
6. With all setup, microinjections can now begin. Using the stage controls of the microscope or using the micromanipulator, repeat the following procedures (**items 7–10**).
7. Carefully insert the tip of the needle into the posterior end of the embryo. Ideally, the needle should be just inside the embryo. Going in too deep will result in possibly missing where the pole cells will form and damaging the embryo.
8. Inject a small volume of the injection mix by gently pressing the "*inject*" button. You will notice a small cloud lightly distinct from the embryo turbidity due to displacement of cells within the egg. If phenol red was included in the injection mix (*see* Subheading 3.4), you should also see a small amount of red colored fluid injected at the posterior end of the embryo. You might also notice the embryo inflating. Repeat this step if you see none of these.
9. Carefully remove the needle from the embryo. A small amount of liquid may leak out of the embryo, but not much.
10. Press the "*inject*" button in between each embryo injection to help ensure the needle does not get clogged with debris (*see* **Note 5** for microinjection tips and basic troubleshooting). Center the next embryo in front of the needle and repeat the process (from **item** 7) until all embryos are injected.

11. After performing microinjections, place the slides in the modular incubator chamber lined with humidified paper towels. Close the chamber and fill with oxygen. Do not exceed the specifications of your chamber (usually around 2 psi). Leave the chamber at room temperature or desired temperature for your species.
12. Removing oil covering screwworm embryos (not Lucilia): After concluding the microinjections, remove the excess of oil by pouring dH_2O directly into the embryos with a squirt bottle. Drain the excess of dH_2O using a piece of clean tissue paper applied on the edge of the embryos. Transfer the slide into a petri dish lined with a humidified paper towel. Surround the embryos in the slide with a small amount of moist larval diet. Place the plates in the modular incubator chamber (as in **step** 7) and incubate overnight at 37 ± 1 °C with $80 \pm 5\%$ RH.
13. On the next day, screen the slides for the presence of first instar larvae using a stereo microscope under a white light. Gently pick the larvae with fine forceps and transfer them into a small plastic container supplied with larvae diet. Keep note of the number of larvae recovered to access the hatching rate (*see* **Note 6**). Alternatively, simply transfer the entire content of the slide into a container supplied with larvae diet.
14. If the injection mix contained a fluorescent marker, first transfer all the hatching larvae into a petri dish filled with a bit of dH_2O. Screen for fluorescent larvae using the appropriate filter under a stereo fluorescence microscope. Keep a note of the number of marked larvae recovered to access the microinjection success rate (*see* **Note 6**). Transfer only the marked larvae into a small plastic container supplied with larvae diet. Proceed with your routine rearing protocols until flies reach adulthood. Alternatively, you can screen for fluorescent larvae while they are still on the slide and transfer once hatched to a container with larval diet.

3.7 Genotyping of Flies Carrying Cas9-Mediated Mutations

3.7.1 Non-lethal DNA Isolation

Non-lethal DNA extractions are made necessary when KO experiments are not expected to generate any visible trait. Therefore, hundreds of flies will need to be genotyped in every generation, while maintaining them alive and individualized. The following protocol is an inexpensive and less time-consuming option to commercially available kits.

1. Upon adult emergence, individually capture each fly to be genotyped with the help of a culture glass tube. Gently hold the captured fly and dissect a single midleg using a fine forceps. Transfer the dissected fly to an identified individual cage, provided with water and food. Place the sampled leg in an

identified 1.7 mL centrifuge tube. Maintain the collect tubes in ice during the sampling process.

2. Add 50 μL of freshly prepared PKS in each sample and homogenize using sterilized grinding pestles. Perform this step at room temperature. Optional: Better results are achieved when samples are frozen at −80 °C for at least 20 min prior to the extractions.
3. Transfer the homogenized solutions, including remaining tissue debris, to 0.2 mL centrifuge tubes or 96-well plates using a pipette. Incubate the tubes in a preheated thermocycler at 37 °C for 60 min. Inactivate PK enzyme at 95 °C for 10 min, and let the reactions cool at 10 °C before further handling. Samples can be stored at −20 °C indefinitely or directly used for PCR-based genotyping strategies. Sufficient PCR amplification products should be obtained by using 2 μL of PKS-isolated DNA as template.

3.7.2 DNA Isolation from Injected Embryos

1. Inject about 100 embryos with the injection mix containing desired sgRNAs.
2. Collect late stage developing embryos onto a small mesh filter and remove oil by rinsing several times with ddH_2O. Blot water off of the eggs by wiping the bottom of the mesh basket with a kimwipe.
3. Carefully transfer the eggs to a 1.5 mL tube using a paintbrush or fine forceps, and flash freeze the eggs in liquid nitrogen or freeze on dry ice. Prepare DNA from flash frozen samples using Zymo DNA miniprep kit (or similar DNA extraction method) following the manufacturer's instructions. Samples can be stored at −20 °C indefinitely or directly used for PCR-based genotyping strategies.
4. For *Cochliomyia* embryos, we have also used the PKS protocol (*see* Subheading 3.7.1) and the DNAreleasy Advance Direct Lysis Kit (Bulldog Bio, Cat #LS06). For the latter, homogenize the collected embryos in a 1.5 mL centrifuge tube with 30 μL of DNAreleasy buffer using sterilized grinding pestles. Transfer the solutions, including remaining tissue debris, to 0.2 mL centrifuge tubes and incubate in a preheated thermocycler using the protocol "*difficult to lyse samples*" according to the manufacturer's instructions. Samples can be stored at −20 °C indefinitely or directly used for PCR-based genotyping strategies.
5. Optional: Quantify the samples using Qubit™ fluorometer (*see* Subheading 2.1, **item 8**). If you continually get high enough concentrations for PCR with your DNA prep method or if following Bulldog Bio protocol in **step 4**, quantification can be skipped.

3.7.3 Genotyping by T7 Endonuclease 1 (T7E1) Cleavage Assay

Estimated time required: 1 d

1. Prepare a PCR for a final volume of 12.5 μL in nuclease-free water containing 2.5 μL of 5× Phusion HF buffer, 0.5 μL of 2 mM dNTP mix, 0.2 μL of 2 U/μL Phusion DNA polymerase, 0.25 μL of each 10 μM target-specific forward and reverse primers (*see* **Note 1** for genotyping primer design tips), and 2 μL of template DNA. Additionally, prepare a no-template control reaction (NTC) and a positive control reaction using a DNA sample extracted from a wild-type individual to be used as a reference. Reactions should be performed entirely on ice. Use 1.7 mL centrifuge tubes to prepare the mix and 0.2 mL tubes or 96-well plates for reactions.
2. Perform amplifications in a preheated thermocycler under the following conditions: 98 °C for 2 min, followed by 35 cycles of [98 °C for 10 s, 60 °C for 30 s, and 72 °C for 30 s], and a final extension step at 72 °C for 5 min. Allow the reactions to cool at 10 °C before proceeding. Reactions can be stored at 4 °C for up to a month, and at −20 °C indefinitely.
3. For control purposes, reserve 2.5 μL of each reaction mixed with gel loading dye. Store control sample at 4 °C.
4. Mix the remaining 10 μL of the unpurified PCR product with 8 μL of T7E1 buffer. Incubate the mix at 95 °C for 10 min in a preheated thermocycler. Carefully remove the samples from the thermocycler and place them at room temperature (around 25 °C) for at least 90 min to allow complete heteroduplex formation.
5. Add 2 μL of 1 U/μL T7E1 enzyme in each sample, followed by incubation at 37 °C for 20 min in a preheated thermocycler. Stop the reactions by adding 5 μL of Stopping buffer. Resolve the entire T7E1 reaction volume, and the reserved PCR samples (in **step 3**), in a 2% agarose gel in cold 1× TBE Buffer. Alternatively, reactions can be stored at 4 °C (*see* **Note 7** for expected results).

3.7.4 Genotyping by Sequencing

Estimated time required: 3 to 7 d.

1. Prepare a PCR for a final volume of 50 μL in nuclease-free water containing 5 μL of 10× PCR buffer, 5 μL of 5 mg/mL BSA, 3 μL of 25 mM $MgCl_2$, 1.5 μL of 2 mM dNTP mix, 1 μL of 10 μM each target-specific forward and reverse primers (*see* **Note** 1 for genotyping primer design tips), 0.25 μL of 5 U/μL *Taq* DNA Polymerase, and either 2 μL of template DNA from leg or 1 ng–1 μg from injected embryos. Additionally, prepare a no-template control reaction (NTC) and a positive control reaction using a DNA sample extracted from a wild-type individual to be used as a reference in further analysis.

2. Perform amplification in preheated thermocycler under following conditions: 98 °C for 30 s, 35 cycles of 98 °C for 10 s, 50–72 °C for 10 s, and 72 °C for 1 min/kb, followed by a final extension step at 72 °C for 2 min. Allow the reactions to cool down at 10 °C. Reactions can be stored at 4 °C for up to a month, and at −20 °C indefinitely.
3. Mix 5 μL of each PCR product with gel loading dye and resolve in a 1.5% agarose gel in 1× TAE Buffer.
4. Purify the remaining PCR volume using the QIAquick® PCR Purification Kit or Zymo DNA Clean & Concentrator™ for PCR products, according to the manufacturer's instructions. Usually, 35 μL of nuclease-free water is enough to elute the purified templates, but this volume might be adjusted based on the PCR yield.
5. From this point, two strategies can be used to access Cas9-induced allele modifications: Direct Sanger sequencing followed by the analysis of overlapping peaks, and subcloning and sequencing of PCR products. Both strategies are briefly described below:
   (a) *Direct Sanger sequencing*: Submit 10–40 ng of a 500–1000 bp purified PCR product with 6.4 pmol of either forward or reverse primer for DNA sequencing. Confirm concentrations required with your sequencing facility. Input the resulting "*.ab1*" files (experiment and wild-type reference file) into Synthego ICE (Inference of CRISPR edits) free online tool (available at https://ice.synthego.com/#/) along with the protospacer sequence of each sgRNA used in KO or KI experiments and the donor DNA sequence used (KI only). The analysis will fail if the "*.ab1*" files are too noisy or if the cut site is too close or too far from the beginning of sequencing. We also have achieved good results using the CRISP-ID online tool [26] (available at https://crispid.gbiomed.kuleuven.be/).
   (b) *Subcloning and sequencing*: Prepare ligation reactions in 0.5 mL tubes for a final volume of 10 μL in nuclease-free water containing 5 μL of 2× Rapid Ligation Buffer, 3 μL of the purified PCR product, 1 μL of 50 ng/μL pGEM-T Easy Vector, and 1 μL of T4 DNA Ligase. Gently mix the ligation reactions by pipetting and incubate at 4 °C overnight in a refrigerated water bath. Transform *E. coli* strain JM109 (Promega), plate on ampicillin/IPTG/X-gal plates and select white colonies. Inoculate the selected colonies in 3 mL of LB liquid medium (with ampicillin). Incubate cultures overnight at 37 °C under constant agitation. Perform a plasmid DNA isolation using the QIAprep Spin Miniprep Kit and confirm recombinant clones

by restriction enzyme digestion and agarose gel electrophoresis. Submit recombinant plasmids for Sanger DNA sequencing with standard M13 primers. Retrieve sequences using your favorite chromatogram viewer software and compare the results to the reference sequence using a multiple sequence alignment tool.

3.8 Establishment of Mutant Strains

Estimated time required: 3–4 months.

3.8.1 Crossing Scheme for KO of Visible (Phenotypic) Traits

1. To facilitate virgin collection, separate late-stage G_0 pupae that developed from injected embryos into individual small round plastic containers with small holes on the lid for air circulation. Alternatively, standard culture glass tubes with cotton wool plugs can be used. For *Cochliomyia*, cover the bottom of the containers with sawdust. It is also possible to simply collect young virgin flies from the container holding G_0 pupae if the flies are screened and collected at regular intervals.
2. Screen the G_0 mosaic adults upon emergence. Keep note of the number of wild-type and mosaic individuals obtained to calculate a "*mutagenesis rate*" (*see* **Note 6**). Transfer all obtained mosaic flies to a cage supplied with food and water. Let this colony inbreed freely. Discard wild-type flies.
3. Screen for G_1 biallelic mutant individuals upon adult emergence and discard all obtained wild-type flies. Keep note of the number of wild-type and mutant individuals to access the "*putative mutational inheritance rate*" (*see* **Note 6**). Randomly select biallelic mutant males to individual cages (one adult per cage) supplied with food and water. Keep the remainder of G_1 biallelic mutant adults in a separate cage, and let this colony inbreed freely (this is a backup colony).
4. To obtain lines homozygous for a specific mutation, set individual crosses of mutant G_1 males to several wild-type virgin females. Separate late-stage G_2 pupae as before (**item 1**).
5. Collect G_2 monoallelic mutant males and set cages each with a single male crossed with several wild-type virgin females as before. For each line, let G_3 inbreed freely (one cage per obtained line, these are the founder's colony).
6. For each founder cage, screen the G_4 for any homozygous biallelic mutant individuals upon the adult emergency. Transfer them to a new cage supplied with food and water. Let the colony inbreed freely to establish mutant homozygous strains.

3.8.2 Crossing Scheme for KO of Non-visible (Phenotypic) Traits

1. Crosses can simply be set with all viable G_0 flies. Alternatively, flies carrying mutations can be identified using the leg PCR assay. For the latter, upon adult emergence, collect G_0 males,

and proceed with the non-lethal DNA isolation (*see* Subheading 3.7.1) and T7E1 assay (*see* Subheading 3.7.3) protocols to identify mosaic mutant individuals (*see* **Notes** 7 and **8**). Transfer each dissected male into an individual cage supplied with food and water.

2. Cross 1 or 2 G_0 males with several wild-type virgin females. Similarly, cross up to 3 virgin G_0 females with wild-type males. If screening for mosaics (genotyping), individual crosses can be set while awaiting the results as following: 1 G_0 male × several wild-type virgin females, and 1 G_0 female × 2 wild-type virgin males. Crosses set with non-mosaic G_0 are discarded after obtaining the T7E1 genotyping results.
3. From each G_0 cross, randomly select G_1 putative monoallelic mutant males and perform molecular genotyping as before (**item 1**). Once again, set individual G_1 crossings with wild-type flies while awaiting the results of the genotyping assay, and then simply discard any crosses set with non-mutant G_1 males. Perform a genotyping by sequencing (*see* Subheading 3.7.4) to access genomic modifications hosted by the selected G_1 flies. Keep note of the number of wild-type and mutant individuals to access the "*putative mutational inheritance rate*" (*see* **Note 6**).
4. After obtaining G_1 genotyping results, select G_2 offspring carrying the same genomic modification and let inbreed freely (one cage per line). Upon adult emergence, collect at least 96 G_3 flies (sex ratio about 2:1, female and male), and proceed with non-lethal DNA isolation followed by Cas9 in vitro cleavage assay (*see* Subheading 3.2) to identify biallelic homozygous mutant siblings. Let them inbreed freely to establish a mutant strain homozygous for a specific mutation in the targeted gene.

3.8.3 Crossing Scheme for KI Mutation with Fluorescent Protein Marker

1. Set G_0 newly emerged adult crosses in bottles with food and water: Set 2 G_0 males with 8 wild-type virgin females per bottle and 3 G_0 virgin females with 3 wild-type males per cage.
2. Screen the embryos and first instar offspring for the presence of the fluorescent protein marker gene using a fluorescence stereo microscope such as the Leica M205.
3. Rear all first instars from any cage that had positive (fluorescent) G_1 larvae. Separate the positive G_1 third instar and discard all non-fluorescent larvae.
4. Set new crosses with one G_1 male backcrossed to 5 or 6 wild-type virgin females. Set several cages if possible. If none of the G_1 adults are males, the set crosses with G_1 females.
5. Screen G_2 third instar larvae under fluorescent stereomicroscope. About 50% of the larvae should appear fluorescent. Separate fluorescent larvae and discard non-fluorescent larvae.

6. Select several of the G_2 adults that developed from positive larvae and prepare genomic DNA using the Zymo DNA miniprep Kit following manufacturer's instructions. Quantify the samples using Qubit™ fluorometer (*see* Subheading 2.1, **item 5**).
7. Prepare a PCR reaction using the DNA from the G_2 adults to confirm KI location (*see* **Note 2** for KI confirmation primer design). Prepare a 25 μL reaction as follows: 12.5 μL Q5 High-Fidelity Master Mix (2×), 1.25 μL each target-specific forward and reverse primer (10 μM), 1 ng–1 μg of genomic containing site of interest, volume up to 25 μL with milliQ water.
8. Perform amplification in preheated thermocycler under following conditions: 98 °C for 30s, 35 cycles of 98 °C for 10 s, 50–72 °C for 10 s, and 72 °C for 1 min/kb, followed by a final extension step at 72 °C for 2 min. Allow the reactions to cool down at 10 °C for at least 10 min before proceeding. Reactions can be stored at 4 °C for up to a month, and at −20 °C indefinitely.
9. Run 10 μL of PCR reaction on a 1% agarose gel in 1× TBE to confirm that the G_2 flies have the correct insertion. If no band is produced or band of the incorrect size, the HDR construct was inserted into another region of the genome and does not contain your desired edit. Discard any lines giving nonspecific PCR products.
10. Proceed with backcrosses of KI males to wild-type virgin females for five generations before proceeding. These backcrosses aim to breed the KI to homozygosity, which helps to alleviate any edits made from off-target cutting of your sgRNAs.

4 Notes

1. Genotyping primers can be designed with the assistance of online tools, such as Primer-BLAST (available at https://www.ncbi.nlm.nih.gov/tools/primer-blast/). Melting temperature should be close to 60 °C to improve specificity. Using low primer concentrations (0.1–0.2 μM) also improves the amplification specificity and electrophoresis resolution. For T7EN1 and Cas9 in vitro cleavage assays, PCR amplification should ideally exclude intronic and polymorphic regions of the targeted genome sequence. That is because the T7E1 is prone to detect and cleave heterozygous single nucleotide polymorphisms (SNPs) although less efficiently. Optimal amplicon size should be around 600–800 bp, and up to 1 kb, with the Cas9-targeted region located asymmetrically within the PCR

product. Better resolutions are achieved when expected cleavage fragments are at least 200 bp in size, with ≥100 bp difference between them. The larger cleaved fragment should also be distinguishable from the original amplicon size.

2. Classic restriction digest cloning strategies can be used in place of NEBuilder HiFi Cloning. However, HiFi Cloning cuts down on the time it takes to build the construct. The NEBuilder Assembly tool (available at https://nebuilder.neb.com/#!/) helps ensure that you have the correct primer overhangs and correct final orientation of fragments. Another set of primers must also be designed to confirm the proper integration of KI. These can be designed by hand or with aid of a primer design program such as Primer3 [27] (available at https://bioinfo.ut.ee/primer3-0.4.0/). One primer should be designed within the KI construct but not within either homology arm. The second primer should be designed within the GOI but again not within the homology arm. This primer is essential to determine whether your KI has been inserted into the proper location. When KI is properly integrated, these primer sets will amplify. However, if KI went into an off-target location, no product should amplify. The minimum size that your product can be when designing these primers sets is the length of homology arm that the primers are spanning.

3. Using a KCl concentration near to 300 mM is prone to enhance Cas9 RNPs solubility [28] and helps to prevent needles from clogging so often. We found that survival of screwworm embryos is unaffected by this KCl concentration [15]. We have successfully induced somatic and germline mutations in *L. cuprina* and *C. hominivorax* species by using a wide-range concentration of the purified Cas9 protein, from 300 to 500 ng/μL (final concentration in the microinjection mix). However, higher rates of somatic mosaicism are expected when embryos are injected with high concentrations of Cas9 [15]. The sgRNAs concentrations should be set accordingly to Cas9. For the sheep blowfly, successful KOs have been achieved using sgRNA concentrations as low as 40 ng/ul (final concentration in the microinjection mix) when injecting multiple sgRNAs at once.

4. The Cas9 in vitro cleavage assay relies on the detection and cleavage of the unmodified (wild-type) allele by the specific sgRNA, while Cas9-modified alleles will not be cleaved, as the sgRNA targeted site no longer exists. As a result, for unmodified samples the PCR product will be cleaved into two fragments. Therefore, the assay can be used to test for sgRNAs efficiency in vitro. If used as a molecular genotyping tool (*see* Subheading 3.8.2, **item 4**) monoallelic mutants (e.g., individuals hosting one modified allele) should display three

fragments, one uncut and two smaller fragments, while biallelic mutants (e.g., individuals hosting indels in both alleles) should display only the large fragment (assuming that genotyping primers were designed as described in **Note 1**).

5. Following, a basic list of tips and troubleshooting for blowfly embryo injections:
   (a) Pole cell formation at the posterior end occurs prior to cellular blastoderm formation. Inject embryos at stage 1 or 2, which are well before pole cell formation.
   (b) If there is backflow into the needle from the embryo, increase the transfer pressure.
   (c) If there is significant cytoplasmic leakage after injection, this could indicate that the dehydration time should be increased. Conversely, if embryos show no resistance to injection this could indicate the embryos are overdessicated. Overdessication is probably the single biggest cause of very poor injection survival. For these reasons when first learning embryo microinjection, each injection session should include control slides of embryos that were uninjected and injected with the buffer (with phenol red dye) alone.
   (d) The needles can get clogged for two main reasons: Suspended particles in the microinjection mix and particles from injected embryos, such as from the chorion. It may be possible to unclog the needle by pressing the "clear" high pressure setting a few times. Alternatively, the opening of the needle can be enlarged by touching the embryo or tape on the slide surface. However, if the needle opening is so large that there is significant cytoplasmic leakage after injection, then the needle should be changed.
6. Keeping notes on microinjection experiments might help further troubleshooting and assess Cas9-induced genome modification efficiency. The following list contains simple statistics that can be retrieved from collected data:
   (a) **Hatching rate:**

       # hatching G0 larvae / # of injected embryos.
   (b) **Microinjection success rate:**

       # fluorescent marked G0 larvae/ # hatching G0 larvae.
   (c) **Mutagenesis rate:**

       # mosaic G0 adults / # surviving G0 adults.
   (d) **Putative mutational inheritance rate:**

       (*global*) # of G1 mutant offspring / # G0 crossings.

       (*per* crosses) # G1 mutant individuals / # all obtained G1 individuals.

7. The T7E1 genotyping assay relies on the detection and digestion of mismatched heteroduplex double-stranded DNA (dsDNA), which are generated by the hybridization between unmodified (wild-type) and mutated (edited) sequences. As a result, the cleavage usually produces two small fragments, in addition to a band with the expected amplification size (since wild-type alleles will be over-represented), revealing individuals hosting indels at the Cas9-targeted site (*see* supporting Fig. S5 in [15]). Molecular genotyping should be performed within the first 4 days after the adult emergence, as mating success tends to drop after this period. Additionally, T7E1 genotyping method can also be used to evaluate the efficiency and off-target effects of sgRNAs in vivo.
8. This first round of molecular genotyping can be skipped if using a fluorescent plasmid marker, such as the pB[Lchsp83-ZsGreen] [13], along with the microinjection mix. If embryos were injected with a validated high active sgRNA, the usage of a fluorescent marker will increase the likelihood of selecting mosaic flies at G_0.

Acknowledgments

Funding is gratefully acknowledged from specific cooperative agreements between the USDA-ARS and NCSU and the Panama-United States Commission for the Eradication and Prevention of Screwworm (COPEG) to MJS. This project was also supported by a grant from the São Paulo Research Foundation (FAPESP: 2017/05432-7 to DFP) and by an STRI Short-Term Fellowship (project award #4168 to DFP).

References

1. Stevens JR, Wallman JF (2006) The evolution of myiasis in humans and other animals in the old and new worlds (part I): phylogenetic analyses. Trends Parasitol 22:129–136
2. Hall M, Wall R (1995) Myiasis of humans and domestic animals. In: Advances in parasitology. Elsevier, pp 257–334
3. Hall MJ (1995) Trapping the flies that cause myiasis: their responses to host-stimuli. Ann Trop Med Parasitol 89:333–357
4. Hall MJ, Wall RL, Stevens JR (2016) Traumatic myiasis: a neglected disease in a changing world. Annu Rev Entomol 61:159–176
5. Anstead CA, Korhonen PK, Young ND et al (2015) Lucilia cuprina genome unlocks parasitic fly biology to underpin future interventions. Nat Commun 6:7344
6. Scott MJ, Benoit JB, Davis RJ et al (2020) Genomic analyses of a livestock pest, the New World screwworm, find potential targets for genetic control programs. Commun Biol 3:1–14
7. Concha C, Scott MJ (2009) Sexual development in Lucilia cuprina (Diptera, Calliphoridae) is controlled by the transformer gene. Genetics 182:785–798
8. Li F, Vensko SP, Belikoff EJ et al (2013) Conservation and sex-specific splicing of the transformer gene in the calliphorids Cochliomyia hominivorax, Cochliomyia macellaria and Lucilia sericata. PLoS One 8:e56303
9. Doudna JA, Charpentier E (2014) The new frontier of genome engineering with CRISPR-Cas9. Science 346:1258096

10. Concha C, Palavesam A, Guerrero FD et al (2016) A transgenic male-only strain of the New World screwworm for an improved control program using the sterile insect technique. BMC Biol 14:72
11. Concha C, Yan Y, Arp A et al (2020) An early female lethal system of the New World screwworm, Cochliomyia hominivorax, for biotechnology-enhanced SIT. BMC Genet 21:143
12. Yan Y, Scott MJ (2020) Building a transgenic sexing strain for genetic control of the Australian sheep blow fly Lucilia cuprina using two lethal effectors. BMC Genet 21:141
13. Concha C, Belikoff EJ, Carey B et al (2011) Efficient germ-line transformation of the economically important pest species Lucilia cuprina and Lucilia sericata (Diptera, Calliphoridae). Insect Biochem Mol Biol 41:70–75
14. Davis RJ, Belikoff EJ, Scholl EH et al (2018) No blokes is essential for male viability and X chromosome gene expression in the Australian Sheep Blowfly. Curr Biol 28:1987–1992. e3
15. Paulo DF, Williamson ME, Arp AP et al (2019) Specific gene disruption in the major livestock pests Cochliomyia hominivorax and Lucilia cuprina using CRISPR/Cas9. G3 Genes Genomes Genet 9:3045–3055
16. Williamson ME, Yan Y, Scott MJ (2021) Conditional knockdown of transformer in sheep blow fly suggests a role in repression of dosage compensation and potential for population suppression. PLoS Genet 17(10):e1009792
17. Paulo DF, Junqueira ACM, Arp AP, Vieira AS, Ceballos J, Skoda SR, Pérez-de-León AA, Sagel A, McMillan WO, Scott MJ et al (2021) Disruption of the odorant coreceptor Orco impairs foraging and host finding behaviors in the New World screwworm fly. Sci Rep 11 (1):11379
18. Dickinson MH, Vosshall LB, Dow JAT (2020) Genome editing in non-model organisms opens new horizons for comparative physiology. J Exp Biol 223:jeb221119
19. Champer J, Buchman A, Akbari OS (2016) Cheating evolution: engineering gene drives to manipulate the fate of wild populations. Nat Rev Genet 17:146–159
20. Scott MJ, Gould F, Lorenzen M et al (2018) Agricultural production: assessment of the potential use of Cas9-mediated gene drive systems for agricultural pest control. J Responsible Innov 5:S98–S120
21. Farboud B, Meyer BJ (2015) Dramatic enhancement of genome editing by CRISPR/Cas9 through improved guide RNA design. Genetics 199:959–971
22. Labun K, Montague TG, Krause M et al (2019) CHOPCHOP v3: expanding the CRISPR web toolbox beyond genome editing. Nucleic Acids Res 47:W171–W174
23. Concordet J-P, Haeussler M (2018) CRISPOR: intuitive guide selection for CRISPR/Cas9 genome editing experiments and screens. Nucleic Acids Res 46:W242–W245
24. Jiang F, Doudna JA (2017) CRISPR–Cas9 structures and mechanisms. Annu Rev Biophys 46:505–529
25. Chaudhury MF, Zhu JJ, Sagel A et al (2014) Volatiles from waste larval rearing media attract Gravid Screwworm flies (Diptera: Calliphoridae) to Oviposit. J Med Entomol 51:591–595
26. Dehairs J, Talebi A, Cherifi Y et al (2016) CRISP-ID: decoding CRISPR mediated indels by Sanger sequencing. Sci Rep 6:28973
27. Untergasser A, Cutcutache I, Koressaar T et al (2012) Primer3--new capabilities and interfaces. Nucleic Acids Res 40:e115
28. Burger A, Lindsay H, Felker A et al (2016) Maximizing mutagenesis with solubilized CRISPR-Cas9 ribonucleoprotein complexes. Development 143:2025–2037

Chapter 11

Generation of Gene Drive Mice for Invasive Pest Population Suppression

Mark D. Bunting, Chandran Pfitzner, Luke Gierus, Melissa White, Sandra Piltz, and Paul Q. Thomas

Abstract

Gene drives are genetic elements that are transmitted to greater than 50% of offspring and have potential for population modification or suppression. While gene drives are known to occur naturally, the recent emergence of CRISPR-Cas9 genome-editing technology has enabled generation of synthetic gene drives in a range of organisms including mosquitos, flies, and yeast. For example, studies in *Anopheles* mosquitos have demonstrated >95% transmission of CRISPR-engineered gene drive constructs, providing a possible strategy for malaria control. Recently published studies have also indicated that it may be possible to develop gene drive technology in invasive rodents such as mice. Here, we discuss the prospects for gene drive development in mice, including synthetic "homing drive" and X-shredder strategies as well as modifications of the naturally occurring *t* haplotype. We also provide detailed protocols for generation of gene drive mice through incorporation of plasmid-based transgenes in a targeted and non-targeted manner. Importantly, these protocols can be used for generating transgenic mice for any project that requires insertion of kilobase-scale transgenes such as knock-in of fluorescent reporters, gene swaps, overexpression/ectopic expression studies, and conditional "floxed" alleles.

Key words Gene drive, CRISPR, Genome editing, Mouse models, Genetic biocontrol

1 Introduction

Invasive species are a scourge on the planet. Damage to the environment, agriculture, and the cost of controlling or removing them has been estimated at approximately $120 billion USD per year [1]. The economic burden to the agricultural industry is enormous, costing tens of millions of dollars to individual countries each year [2]. For example, in Australia invasive pests are responsible for the loss of an estimated $1 billion per year through agricultural loss of revenue and environmental damage [3].

Invasive rodents, including mice, also pose a significant threat to biodiversity, particularly on islands, and are the likely cause of hundreds of species extinctions [4–7]. Previous attempts at invasive

Paul John Verma, Huseyin Sumer and Jun Liu (eds.), *Applications of Genome Modulation and Editing*, Methods in Molecular Biology, vol. 2495, https://doi.org/10.1007/978-1-0716-2301-5_11,

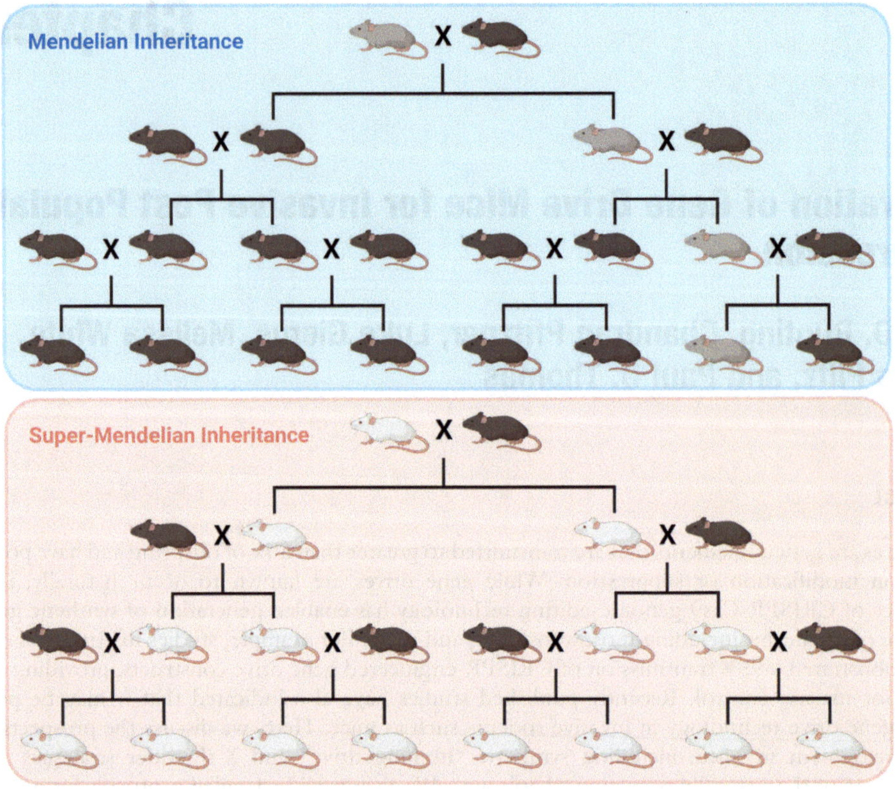

Fig. 1 Schematic comparing different inheritance modes. Top shows a hemizygous mouse carrying a typical (non-gene drive) transgene (gray). When released into the wild, it will breed with wild mice (black) and following the principles of Mendelian inheritance, it will not actively spread through the population. Bottom shows a mouse with a gene drive (white). In contrast to the above, when released the gene drive will rapidly spread through the population via Super-Mendelian inheritance. Image created with BioRender

vertebrate pest eradication using traditional methods such as poisoning, trapping, and hunting have met with some success [8]. Despite this, there are still many challenges. For example, non-selective toxins are often used which can impact non-target native species [8]. There are ethical concerns over the suffering caused by toxin ingestion [8]. The cost of failed eradication is also important, as this may inhibit further attempts to control invasive populations due to the perceived difficulty [8].

Gene drives are emerging as a new genetic biocontrol strategy with considerable potential for suppression or modification of wild populations. Gene drives are genetic elements that have "Super-Mendelian" (i.e., greater than 50%) transmission to offspring, enabling rapid spread through a target population [9] (Fig. 1). Remarkably, spread can occur even if the gene drive imparts a negative fitness cost to the animal [9]. Synthetic gene drive systems were originally proposed by Burt in 2003, based on the ability of site-specific selfish genes to copy themselves into specific DNA sequences [10]. Burt theorized that re-engineering these naturally occurring selfish genes to target alternative DNA sequences, such as essential host genes, could be a powerful tool for population-level manipulation. However, it wasn't until the discovery and development of the CRISPR-Cas9 system [11, 12] that comprehensive and systematic investigation of various strategies became possible.

There are many potential applications for gene drives that are described in detail elsewhere. Eradication of diseases such as malaria carried by wild animal hosts is a promising example [9]. This could be achieved by providing the host population with immunity to the disease pathogen, or even eradicating the host animal entirely [9]. The latter may be possible and environmentally feasible for certain species of mosquitoes although the benefits of eradication need to be weighed against the importance of their role in the ecosystem [9]. If a species is eradicated, it could also potentially be re-introduced once the pathogen is no longer present [9]. Applications also exist in the agriculture realm, taking the form of a gene drive removing pesticide resistance in crop pests, or introducing new alleles making them sensitive to different pesticides [9]. Given the ecological impact of invasive rodents on islands, there is also significant interest in developing gene drive strategies for mice and rats, although research in this area generally lags behind the insect field (see below).

1.1 CRISPR Gene Drives

The recent development of CRISPR genome-editing technology has enabled a host of new and innovative approaches in biomedical and agricultural research, including the development of genetic biocontrol strategies for wild populations. The CRISPR system functions as a pair of molecular scissors that can generate a double-stranded break at almost any genomic site. The system comprises the Cas9 endonuclease and a 20 nucleotide (or similar) single guide RNA (gRNA) that provides the positional address for Cas9-mediated DNA cleavage through complementary RNA:DNA base pairing with the target genomic sequence. Target site selection is limited by the availability of a Protospacer Adjacent Motif (PAM), which is located immediately adjacent to the gRNA-binding sequence and is required for Cas9 DNA binding. This corresponds to NGG for the most commonly used Cas9 system, which is derived from *Streptococcus pyogenes* (SpCas9).

A CRISPR homing gene drive, at least in its simplest form, is composed of a single cassette that is inserted into a genomic locus of interest. The cassette contains the following elements: A CRISPR nuclease gene (typically SpCas9) driven by a heterologous promoter, a ubiquitous gRNA expression cassette, and an optional cargo element containing, for example, a gene to be spread through the population. The gRNA in the gene drive cassette targets the homologous WT chromosome at the same locus as the gene drive construct. As a result, expression of Cas9 and the gRNA results in

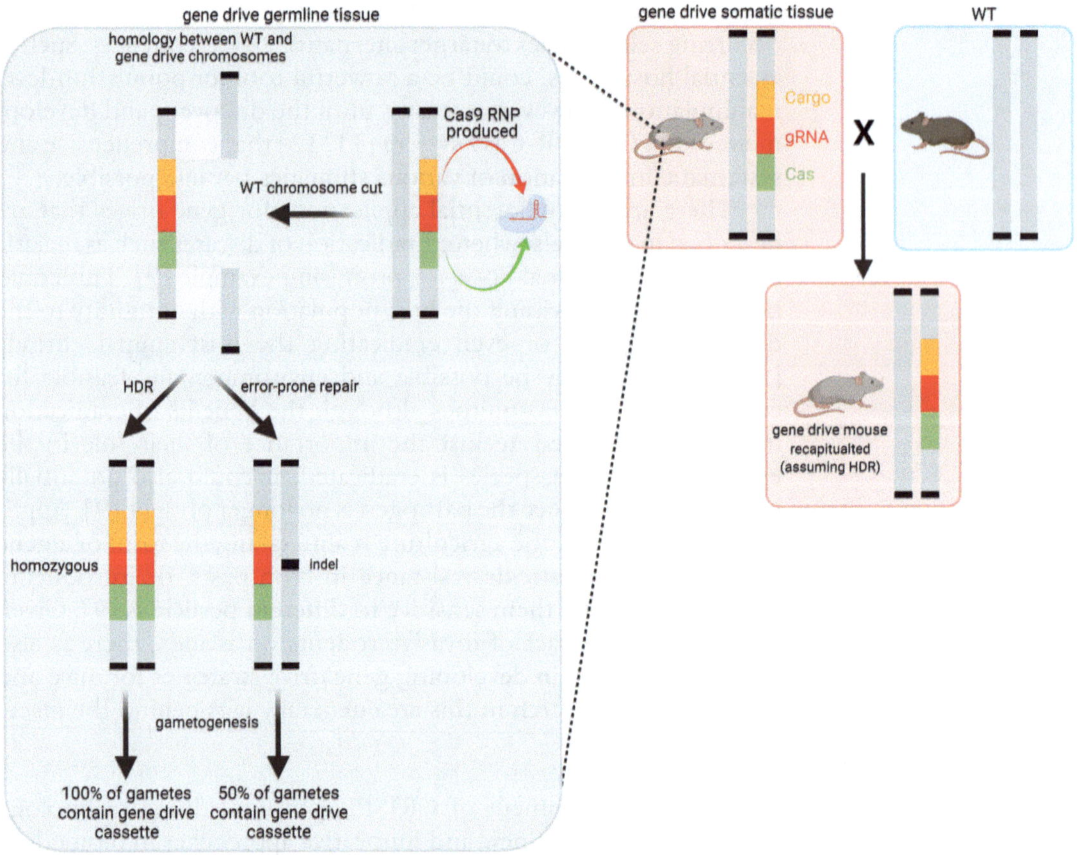

Fig. 2 Germline-homing gene drive in mice. A mouse carrying the gene drive is shown in the red box (top middle). In the somatic tissue, one chromosome is unaltered (WT) and the other contains the gene drive comprising the Cas9 gene (green), a gRNA expression element (red), and a cargo element (yellow). No Cas9 expression (and therefore no homing activity) occurs in the somatic tissue. The germline tissue is also shown (gray box). If homing occurs (via HDR), both chromosomes will carry the gene drive. Alternately, if error-prone repair occurs, an indel will be generated at the target site and the gene drive does not replicate. If homing occurs, the gene drive mouse is recapitulated in all offspring (red box, middle right) after mating to a WT mouse (blue box). Image created with BioRender

the formation of a CRISPR complex that cleaves the WT chromosome. In contrast, the chromosome containing the gene drive cassette is not cut as the gRNA target site is interrupted by the transgene. Repair of the DNA break via homologous recombination will result in the gene drive being copied onto the WT chromosome in a process referred to as "homing." This results in both chromosomal homologs carrying the gene drive cassette, ensuring that it is transmitted to all offspring (Fig. 2).

Suppression gene drives can be designed by insertion of the transgene into a haplosufficient gene that causes recessive lethality or infertility (i.e., a phenotype occurs only when both copies of the gene are inactivated) [10, 13]. By restricting the homing event to the germline, for example, by using germline-specific promoters to drive Cas9 expression, individuals that inherit the gene drive from only one parent will be unaffected as the somatic tissue remains hemizygous [13]. This allows the gene drive to spread through the population without a detrimental effect until it reaches saturation. As the proportion of carriers increase in the population, they will mate with each other resulting in inviable or infertile offspring, causing a population crash [13].

1.1.1 Gene Drive Technology Development in Insects

Gene drive technology has been developed most extensively in insects, in part due to their relative ease of manipulation and their short generation time which enables hundreds of progeny to be rapidly screened for drive activity. In addition to *Drosophila melanogaster* (fruit fly), the traditional model system for genetic research in insects, gene drives have been developed in vectors of significant pathogens including the mosquito species *Anopheles* (malaria) and *Aedes* (dengue, zika, and yellow fever viruses). Strategies for mosquito population control to limit malaria spread include sex-ratio distortion to drive a population toward male-only [14–16], targeting of female fertility genes [17], and sex chromosome shredding [18]. A population modification gene drive with an anti-malaria cargo gene has also been developed in the laboratory [19]. In the majority of studies published thus far, gene drive homing in insects has achieved efficiencies of at least 70%. Remarkably, in *Anopheles*, very high homing rates are consistently achieved (95–99%) which have enabled impressive population suppression in cage trials [15, 16].

1.1.2 Development of Gene Drives in Mice

In silico modeling has been used to assess various gene drive strategies for population modification or suppression in rodents [13, 20]. This work is critical as developing, constructing, and testing of various gene drive designs in mice is a time-consuming process and provides a means to select the most promising strategies. Modeling also allows researchers to incorporate ecological and environmental parameters that influence breeding success in species such as the impact of food availability, landscape, release site, migration range of offspring, efficiency of drive homing, and resistance allele formation, among other variables. Analysis of gene drive strategies targeting female fertility and embryo viability genes suggests these approaches are viable and could lead to efficient population suppression [13].

Proof-of-concept experiments for two types of CRISPR gene drives in mice have been published: zygotic- and germline-homing. The former is predicated on ubiquitous (and therefore zygotic) expression of Cas9 and aims to activate the homing event in the first embryonic cell—thus ensuring that all cells of the developing embryo (including the germline) carry the gene drive on both chromosomes. Data from two independent studies indicate that homing does not occur at a detectable rate in the early embryo [21, 22], and thus zygotic-homing drives are not considered further here. The second strategy, the germline-homing gene drive, differs in the promoter used to express the CRISPR nuclease. Based on successful gene drive experiments in insects, germline-homing drives in mice have used the murine germline-specific Vasa promoter [23] to express Cas9 in the developing germ cells. Two studies investigating the efficacy of germline-homing gene drives in mice have been published. The first report looked at homing of a gene drive construct targeting the tyrosinase gene which would provide a visual indication that homing has occurred due to a change of coat color. Cas9 was driven by a CAG flox-stop system and activity was restricted to the germline by crossing these mice with a Vasa-Cre line. Although no homing was detected in males, females exhibited variable homing rates of between 0% and 72% [21]. A second study investigated whether homing could occur in the germline when Cas9 expression was driven directly by the Vasa promoter [22]. While qPCR analysis showed Cas9 expression in testes and ovaries, no gene drive homing was seen in either sex. It was suggested that in males, Vasa promoter activity is too early to coincide with meiosis in primordial germ cells and, while expression in ovaries was detected, it appears to have been too low to induce homing [22].

A significant barrier to gene drive development that is highlighted by these studies is the generation of indels at the target site. This occurs when the CRISPR-induced DNA break is repaired by the non-homologous end-joining pathway (as opposed to homing which uses homology directed repair). Indels will typically alter the gRNA-binding sequence so that it is no longer susceptible to CRISPR cleavage, thus generating "resistant" alleles that prevent spread of the gene drive [24]. Several strategies to counter resistant alleles are currently under investigation. These include targeting of functionally important coding sequences such that the presence of an indel will invariably cause loss-of-function and gene drives with multiple gRNAs targeting different regions of the same gene [25–27].

1.1.3 X-Shredder

Meiotic gene drives offer an alternative approach to CRISPR homing drives for population suppression. These drives, which may be synthetic or naturally occurring (see *t* haplotype below), usually operate in males and typically function by reducing the fitness of gametes that don't contain the drive element [28]. One promising example is the "X-shredder," also known as the driving Y [29]. The aim of this approach is to generate a male-only population which will rapidly decline due to a lack of reproductively active females. The X-shredder drive uses a specific endonuclease such as CRISPR-Cas9 to target repeat elements on the X chromosome of developing sperm such that the chromosome is "shredded" causing loss of fitness. Spermatogenesis-specific promoters restrict Cas9 expression to the sperm, thereby removing or "shredding" the X chromosome, leaving only Y-bearing sperm to fertilize the egg. Insertion of the gene drive element into the Y chromosome ensures that it will be transmitted to all progeny, enabling gene drive spread and male population bias. This approach has already been partly developed in insects [14, 15, 30] although the precise mechanism by which the X-bearing sperm are inactivated remains unresolved. One limitation of this approach that is yet to be overcome is that silencing of sex chromosomes during meiosis hinders Cas9 and gRNA expression. To date, X-shredder gene drives have not been developed in mice although gRNA that are capable of selective deletion of the X chromosome have been published [31]. Y shredding has also been modeled and studied in mouse embryonic stem cells in vitro and in embryos in vivo. Y chromosome elimination in mice was found to convert XY males into XO females which are viable but sub-fertile [32]. A gene drive strategy based on Y chromosome shredding termed Y-CHOPE has been modeled which, although potentially useful for population suppression, appears to be less potent than the X-shredder approach [20, 32].

1.1.4 The *t* Haplotype: A Naturally Occurring Meiotic Drive in Mice

A naturally occurring meiotic drive in mice known as the *t* haplotype is composed of multiple large, non-overlapping inversions on chromosome 17. Over approximately 1.5 million years the *t* haplotype, which spans 40 Mb, has accrued a range of mutations [33], some of which alter the function of proteins involved in sperm motility, commonly referred to as "distorter" proteins. These mutant proteins diffuse throughout the syncytial cytoplasm via intercellular bridges that connect clonal spermatocytes as they develop into mature sperm. Sperm containing the *t* haplotype, however, produce a mutant "responder" protein, Smok, which does not diffuse throughout the syncytium, thereby rescuing the motility of all sperm containing the *t* haplotype.

Male heterozygotes transmit the *t* haplotype to >90% of progeny due to the enhanced motility of the rescued *t* sperm in competition with the retarded non-*t* haplotype bearing sperm. The inheritance of the *t* haplotype in heterozygous females is unaffected. Male or female homozygosity generally results in embryonic lethality although the exact phenotype of homozygotes depends on the specific *t* variant.

1.1.5 Daughterless Mice Approach

The existence of a natural gene drive in mice has inspired various strategies attempting to utilize this system to suppress invasive rodent populations. A sex-biasing strategy known as the "daughterless" mouse approach aims to suppress wild populations by reducing the proportion of females. This strategy is based on the integration of the male sex-determining gene *Sry* [34] into the *t* haplotype. The drive of the *t* haplotype, coupled with expression of *Sry*, will slowly spread through a population while preventing the development of females, instead producing sterile males with an XX karyotype. Because this strategy does not involve integration of any foreign (i.e., non-murine) genetic material, the resulting mice may not be considered by regulators to be "genetically modified," which may be an advantage with respect to public acceptance and potential deployment. This strategy was originally modeled by Backus and Gross who demonstrated that it may be useful for population suppression of islands with relatively small populations [35]. However, there are many variables that need further consideration such as the impact of polyandry and sperm competition, two factors that have been demonstrated to significantly restrict the spread of the *t* haplotype in wild populations [36, 37].

1.1.6 Alternative t Haplotype Strategies

Alternative *t* haplotype-based gene drive strategies involve the integration of a Cas9-gRNA transgene driven by a germline-specific promoter into the *t* haplotype. Unlike the "daughterless" mouse approach described above, these strategies would undoubtedly be classified as "genetically modified." The *t* haplotype mechanisms drive this element while the gRNA can be programmed to enable various population suppression strategies. For example, this strategy could be used to target haplosufficient embryo viability or female-specific fertility genes, thereby suppressing populations via a similar mechanism to a homing gene drive. The gRNA could also be designed to target X chromosome-specific repeats, shredding it, thereby acting similar to an X-shredder gene drive. These approaches provide an alternative to homing gene drives which are proving challenging to develop in mice, as outlined above.

1.2 Safeguards

With any genetic engineering technology that is designed to spread in a population, implementation of safeguards during development and testing of these systems is paramount. To this end, groups researching gene drive technology have used numerous safeguard systems aimed at preventing spread of drives in the wild following unintentional release of gene drive organisms. A split homing drive is a simple approach that separates the gRNA (homing) construct and the Cas9 construct on different chromosomes such that the latter is inherited in normal Mendelian fashion. Cas9 will be diluted from the population, resulting in the inability of the gRNA to home [38]. Expanding on this concept is the daisy-drive system which requires >2 genetic elements that drive the next element in sequence. This can result in the spread of a genetic element locally but due to loss through successive generations of the nondriving element, the drive does not propagate indefinitely [39]. Some studies have also used synthetic target sites that are completely absent in WT animals and thus, if a gene drive transgenic animal were to be accidently released, the gRNA would not be able to target any of the wild population [22]. In the case of isolated populations, alleles can become locally fixed due to genetic drift generating unique and stable DNA changes that could potentially be targeted by gene drives with reduced off-target potential [40]. A recent approach leveraging the development of Cas9 alternatives is to use a version of Cas9 that is only active in the presence of a small molecule [41]. Together, these approaches provide options for safeguarding the wild populations during gene drive development and avoiding undesirable spread into non-target populations after potential deployment.

1.3 Generating Mice with Gene Drive Components

As outlined above, generation of mice for gene drive experiments requires precise insertion of relatively large DNA constructs into specific genomic locations. Non-targeted transgenesis is also useful for the generation of split drives in which the Cas9 and gRNA-expressing components are not linked. The advent of CRISPR-Cas9 technology has enabled generation of transgenic mice through direct modification of the zygotic genome, circumventing the time-consuming, labour-intensive, and expensive embryonic stem cell route to transgenesis (referred to as "gene targeting"). As part of our efforts to generate gene drive mice, we have developed and adapted methods to incorporate plasmid-based transgenes in a targeted and non-targeted manner. Importantly, these protocols can be used for generating transgenic mice for any project that requires insertion of kilobase-scale transgenes such as knock-in of fluorescent reporters, gene swaps, overexpression/ectopic expression studies, and conditional "floxed" alleles. Animal experiments should only be performed after approval by the local animal ethics committee.

2 Materials

2.1 Mouse Vasectomy (for Generating Pseudopregnant Embryo Transfer Recipients)

1. Inhalation anesthetic (e.g., Isoflurane).
2. Analgesic for injection (e.g., Buprenorphine).
3. 1 mL sterile syringe.
4. 27G ½ (0.4 mm × 13 mm) needles.
5. Protecta pads (Kimberley Clark #KC-2705).
6. Supercut iris scissors ×2 (Coherent Scientific #14218).
7. Dumont tweezers, style 5A, oblique tips x2 (Pro Sci Tech # T05A-811).
8. Graefe Iris forceps ×1 (ProSciTech #T131).

9. Autoclip Wound Clip Researcher's Kit (#427638—Autoclip wound clip applier, autoclip wound clip remover and 100 mm × 9 mm autoclip wound clips—BD Life Sciences).
10. Reusable cautery unit (World Precision Instruments #500389).
11. Glass bead sterilizer (FST 250 #18000-45).
12. Surgical microscope (Olympus SZ61).
13. Cold light source.
14. Small animal anesthetic machine (Vet Tech Solutions) with scavenging system and oxygen source.
15. Warming pad ×2 (Able Scientific—10 watt animal cosy heat pad #ASCHP-RP).
16. Clean IVC recovery cage with bedding, nesting material, and wet food.
17. 4–10 male mice (CD1 or F1 hybrid @ 5–8 weeks of age).

2.2 Preparation of Zygotes for Microinjection

1. Donor C57BL/6 J females aged 3–6 weeks (10–12 per session) (*see* **Note 1**).
2. Stud C57BL/6 J males.
3. Pregnant Mares Serum Gonadotrophin (PMSG) @ 50 IU/mL (Prospec Bio 5000 IU diluted to 500 IU and 50 IU with saline and stored at −20 °C).
4. Human Chorionic Gonadotrophin (HCG) @ 50 IU/mL (Folligon by Intervet—purchased from Lyppard).
5. 0.20 μm syringe filters (Sartorius Stedim Minisart High-flow #16532).
6. 27G ½ (0.4 mm × 13 mm) needles.
7. Hyaluronidase @ 10 mg/mL (Sigma #H3506-30MG hyaluronidase type IV-s from bovine testes diluted with 3 mL sterile water and stored at −20 °C in 20 μL aliquots).
8. M2 handling media (Sigma #M7167-50ML filtered before use).
9. mHTF culture media (United Bioresearch (CosmoBio supplier) #KYD-008-02EX-X5 5 × 2 mL).
10. Embryo tested mineral oil (Sigma #M8410-1 L).
11. Embryo culture dishes (Eppendorf #0030700015 Cell Culture Dish, 35 mm × 10 mm).
12. Glass capillaries (SDR Scientific http://www.sdr.com.au/glass.php (Harvard apparatus supplier) #300036 GC100T-15 Clark borosilicate Thin Wall ID 0.78 mm, OD 1 mm, length 150 mm) pulled over a flame to create embryo handling pipettes.

13. Aspirator tube assembly (Sigma #A5177-5EA) modified to add a syringe filter.
14. Supercut iris scissors (Coherent Scientific #14218).
15. Dumont tweezers, style 5A, oblique tips ×2 (Pro Sci Tech # T05A-811).
16. Nikon dissection microscope SMZ800 1× objective.

2.3 Preparation of DNA for Microinjection

1. Restriction enzyme (New England Biolabs).
2. 10× Restriction digest buffer (New England Biolabs).
3. Agarose low EEO (PanReac Applichem #A2114).
4. 1× TBE buffer: make 10× TBE stock, 108 g Tris base, 55 g Boric acid, 7.4 g EDTA, 800 mL MilliQ H_2O, dissolve components then top up to 1000 mL with MilliQ H_2O. Dilute to 1× with MilliQ H_2O.
5. Invitrogen 1 kb + DNA ladder (ThermoFisher Scientific #10787018).
6. Zymoclean Gel DNA Recovery Kit (Zymo Research #D4001).
7. 0.5 M EDTA pH 8.0.
8. 1.0 M Tris pH 7.5.
9. Temperature stable heat block or thermocycler.
10. Gel casting apparatus.
11. Gel comb.
12. Electrophoresis gel tank.
13. Power Pac™ Basic Power Supply (BioRad #1645050).
14. GelDoc XR+ Gel Documentation System (BioRad #1708195).
15. Ethidium bromide Green Bag Disposal Kit (Mp Biomedicals #112350200).
16. Eppendorf® Centrifuge 5424 (Merck #Ep5405000042).
17. Nanodrop One (ThermoFisher Scientific #ND-ONE-W4).
18. 10x microinjection buffer: mix the following solutions in a sterile container inside a laminar flow cabinet. 10 μL 0.5 M EDTA pH 8.0 (final concentration 0.1 mM), 500 μL 1.0 M Tris pH 7.5 (final concentration 10 mM), 4.49 mL MilliQ H_2O to make up to 5 mL. Attach filter to 5 mL 0.22 μm syringe with plunger detached and decant the buffer solution into the open end of syringe. Insert plunger and eject ~500 μL before collecting aliquots of buffer in Eppendorf tubes, approximately 20 μL per aliquot. Store aliquots at −20 °C.

2.4 Microinjection (Cytoplasmic and Pronuclear)

1. Cultured embryos in CO_2 incubator (prepared earlier).
2. Microscope Cover Glasses 22 mm × 60 mm (Deckglaser #01011 52).
3. Dow Corning high vacuum grease (Sigma #Z273554-1EA).
4. Aluminum custom slide cover.
5. M2 handling media (Sigma #M7167-50ML filtered before use).
6. Embryo tested mineral oil (Sigma #M8410-1L).
7. GC100T-15 Clark borosilicate Thin Wall ID 0.78 mm, OD 1 mm, length 150 mm capillaries.
8. Harvard borosilicate glass capillaries #GC100TF-10 (for preparation of injection pipettes) http://www.sdr.com.au/glass.php.
9. Sutter borosilicate glass capillaries #B100-58-15 (for preparation of holding pipettes) https://www.sutter.com/index.html
10. Eppendorf Microloader pipettes #5242956003.
11. 0.20 μm syringe filters (Sartorius Stedim Minisart High-flow #16532).
12. Butane/propane gas and Portagaz burner.
13. Nikon Eclipse Ti inverted microscope.
14. Eppendorf TransferMan NK2 micromanipulator ×2.
15. Eppendorf Femtojet 4i (settings 90 pi, 20 pc hPa).
16. Eppendorf CellTram Air.
17. Sutter Micropipette Puller Model P-97 https://www.sutter.com/MICROPIPETTE/index.html
18. Microforge (MicroData Instruments, model no. MF-5) http://www.sdr.com.au/index.php (*see* **Note 2**).

2.5 Embryo Transfer

1. M2 handling media (Sigma #M7167-50 mL filtered before use).
2. Embryo culture dishes (Eppendorf #0030700015 Cell Culture Dish, 35 mm × 10 mm).
3. Post-injection embryos in culture.
4. Inhalation anesthetic (e.g., Isoflurane).
5. Analgesic for injection (e.g., Buprenorphine).
6. 0.20 μm syringe filters (Sartorius Stedim Minisart High-flow #16532).
7. 27G ½ (0.4 mm × 13 mm) needles.
8. Sterile disposable drapes (1 × 2 cm squares of paper towel, sterilized in small batches).

9. Embryo handling pipettes made as previously described (GC100T-15 Clark borosilicate Thin Wall ID 0.78 mm, OD 1 mm, length 150 mm capillaries).
10. Protecta pads (Kimberley Clark #KC-2705).
11. Butane/propane gas and Portagaz burner.
12. Aspirator tube assembly (Sigma #A5177-5EA) modified to add a syringe filter.
13. Supercut iris scissors ×2 (Coherent Scientific #14218).
14. Dumont tweezers, style 5A, oblique tips ×2 (Pro Sci Tech # T05A-811).
15. Graefe Iris forceps ×2 (ProSciTech #T131).
16. Bulldog Serrefine Clamp—Straight/35 mm (Fine Science Tools #18050-35).
17. Autoclip Wound Clip Researcher's Kit (#427638—Autoclip wound clip applier, autoclip wound clip remover and 100 mm × 9 mm autoclip wound clips—BD Life Sciences).
18. Glass bead sterilizer (FST 250 #18000-45).
19. Nikon dissection microscope SMZ800 1× objective.
20. Surgical microscope (Olympus SZ61).
21. Cold light source.
22. Small animal anesthetic machine (Vet Tech Solutions) with scavenging system and oxygen source.
23. Warming pad ×2 (Able Scientific—10 watt animal cosy heat pad #ASCHP-RP).
24. Clean IVC recovery cage with bedding, nesting material, and wet food.
25. Pseudopregnant embryo recipients 0.5 dpc (CD1 outbred females 6–20 weeks old).

2.6 Genotyping

1. High Pure PCR Template Preparation Kit (Roche #11796828001).
2. Sigma-Aldrich Standard DNA oligos (Merck).
3. Roche Taq DNA Polymerase (Merck #11146165001).
4. FailSafe™ PCR 2× PreMix (Lucigen #FSP995A-L).
5. Phusion® High-Fidelity DNA Polymerase (New England Biolabs #M0530S).
6. Agarose low EEO (PanReac Applichem #A2114).
7. 1× TBE buffer.
8. Gel Casting Apparatus.
9. Gel Comb.
10. Electrophoresis Gel Tank.

11. Power Pac™ Basic Power Supply (BioRad #1645050).
12. Invitrogen 1 kb + DNA ladder (ThermoFisher Scientific #10787018).
13. GelDoc XR+ Gel Documentation System (BioRad #1708195).
14. QIAquick® PCR purification kit (Qiagen #28106).
15. Monarch® DNA Gel Extraction Kit (New England Biolabs #T1020L).
16. pGEM®-T Easy Vector System I or II (Promega #A1360 or #A1380).
17. Nanodrop One (ThermoFisher Scientific #ND-ONE-W4).

3 Methods

3.1 Mouse Vasectomy

1. Turn on both heat pads—one for surgery and one for recovery cage.
2. Put the first male for surgery in the anesthetic induction box and induce anesthesia.
3. Place animal on a protecta pad on the warming pad and check for pedal reflex by pinching toes. If no response, inject subcutaneously with analgesia and proceed with surgery.
4. Push both testes down into the scrotal sac by gently applying pressure to the abdomen.
5. Spray the scrotal sac and surrounding area with 80% ethanol using a spray bottle. This wetting helps to prevent any loose hair.
6. Make a 5–8 mm incision through the skin along the midline of the scrotal sac.
7. Using scissors widen the incision using blunt dissection if more access is required.
8. Using supercut iris scissors and fine forceps separate the connective tissue between the scrotum and membrane surrounding the testes (*tunica vaginalis*).
9. Visualize the blood vessel running along the vas deferens through the *tunica vaginalis* and make a small opening in the tunica using two pairs of fine forceps.
10. Separate the vas deferens away from the left testis using two pairs of fine forceps (being careful not to puncture the testis) and excise a loop of vas deferens using a fine cautery unit.
11. Ensure that the testis is safely enclosed within the tunica before moving to the right side.

12. Repeat above on the other testis. It is advisable to keep the removed portions of each vas deferens on a paper towel next to the mouse to make sure that both sides were removed.
13. Ensure that the testis is safely enclosed within the tunica.
14. Grasp the two edges of the skin incision with blunt forceps, place the autoclip applicator over the incision, and squeeze to staple the wound. Check that clip is firmly in place. (*see* **Note 3**).
15. Record the weight of the mouse and place in the recovery cage on the warming pad.
16. Clean instruments with ethanol and sterilize in bead sterilizer in preparation for the next surgery.
17. Repeat **steps 1–15** for all vasectomies.
18. House all males individually.
19. Observe post-surgical males daily for 7–10 days before weighing and removing wound clips.
20. 7 days later test the males for sterility by housing with single adult females of any strain and checking daily for copulation plugs.
21. Females can be observed for pregnancy after 2 weeks by visual inspection and, if not pregnant, can remain with the males as constant companions.

3.2 Preparation of Zygotes for Microinjection

1. Day 1—Inject donor females with 0.15 mL of PMSG (7.5 IU) intraperitoneally at 2 pm.
2. Day 3—Inject donor females with 0.15 mL of PMSG (7.5 IU) intraperitoneally at 1.30 pm and mate with stud males (*see* **Note 4**).
3. Prepare embryo handling pipettes by pulling glass capillaries over a flame to create 10–20 pipettes with a diameter of 100–130 μm and a square cut end. Flame-polish the blunt ends of the embryo handling pipettes before storing in a sterile 50 mL Falcon tube.
4. Day 4—Prepare embryo culture dishes by pipetting 9 × 40 μL drops of embryo culture media onto the base of a culture dish and covering with 3 mL of embryo tested mineral oil. Prepare two culture dishes in this way and place into the 5% CO_2 incubator for equilibration for a minimum of 30 min.
5. Check donor females for the presence of copulation plugs as evidence of mating. Humanely kill all females in batches of 4–6 by cervical dislocation. Dissect out oviducts into a dish with 60 μL of M2 handling media at room temperature.

6. Add 15 μL of hyaluronidase solution to 1 mL of M2 (a final concentration of 150 μg/mL) for digestion of cumulus cells from cumulus oocyte complexes (COCs).
7. Place 5 × 40 μL drops of M2 media onto the lid of a culture dish and cover with the base of the dish to prevent evaporation (1 in the center and 4 evenly spaced around the center drop).
8. Pipette 450 μL of hyaluronidase/M2 solution onto the lid of another culture dish and place the oviducts into the drop. Place an additional drop of 50 μL next to the large drop.
9. Tear the swollen ampullas with sharp forceps allowing clutches of COCs to be released into the drop.
10. When all clutches are released remove the tissue from the drop with fine forceps and return it to the M2 collection drop.
11. Aspirate the COCs from the large hyaluronidase drop with a glass capillary and mouth pipette and place into the smaller drop. As the cumulus cells disperse, move the zygotes into a clean drop of M2 (from **step 4**) and wash into another clean drop to remove cumulus cells and the hyaluronidase solution to prevent further digestion of the zygotes.
12. Return the tissue to the hyaluronidase/M2 drop and check the M2 only drop and the hyaluronidase/M2 drop for COCs which may have been attached to the tissue.
13. Move the zygotes into the center drop of culture media in the first dish and return dish to the incubator.
14. Humanely kill the second batch of donors and perform the same COC collection procedure. Move the clean zygotes into the culture dish with the first batch.
15. Assess zygotes for fertilization by visualization of two pronuclei and a polar body. Separate groups of 20–30 embryos into the surrounding drops in the culture dish (*see* **Note 5**).

All animals are housed in IVC cages from Tecniplast with food and water ad libitum and a 12/12 light cycle.

3.3 Preparation of DNA for Microinjection

1. Digest plasmid by mixing the following components: 10 μg plasmid, 10 U (*see* **Note 6**) restriction enzyme, 10 μL 10× NEBuffer, up to 100 μL H_2O.
2. Incubate at enzyme-specific temperature for 1 h.
3. Check digest by making 2× 0.8% agarose gels by mixing 40 g agarose powder with 50 mL 1× TBE buffer for each gel before microwaving to dissolve agarose powder. Once melted, add 2 μL ethidium bromide to one of the gels (*see* **Note** 7).
4. Pour melted and slightly cooled agarose into gel casting apparatus with well comb in place and allow to set (*see* **Note 8**). Once set, remove gel comb and place gel with ethidium

bromide into electrophoresis gel tank with 1× TBE running buffer.

5. Load the gel as follows: lane 1—2 μL 1 kb + DNA ladder, lane 2—1 μL digested plasmid (100 ng), lane 3—100 ng undigested plasmid.
6. Run at 100 V for 30 min.
7. Remove gel from tank and image using the BioRad GelDoc. Confirm that plasmid has been digested and that the time and voltage ethidium gel has run for will be sufficient for ethidium-free gel to separate fragment containing the transgene from the plasmid backbone (*see* **Note 9**). If digestion and separation are sufficient, safely discard 1× TBE buffer from electrophoresis tank before rinsing and refilling with fresh 1× TBE and submerging ethidium bromide-free gel (*see* **Note 10**).
8. Load the gel as follows: lane 1—2 μL 1 kb + DNA ladder, lane 2—blank, lane 3 (large well)—94 μL digested plasmid (10 μg), lane 4—5 μL digested plasmid.
9. Run for appropriate time, as determined by previous step.
10. Remove gel from tank and use scalpel blade to isolate gel slice containing lane 4. Immerse gel slice in 1× TBE containing ethidium bromide (final concentration of 0.15 μg/mL) for 20 min. Place post-stained gel slice in gel imager alongside ruler. Image under UV light to determine how far the transgene fragment has migrated.
11. Place ruler alongside well 3 and determine the position of the transgene fragment using the distance from **step 6**. Use a new scalpel blade to dissect the transgene fragment from the ethidium bromide-free gel. Post-stain with ethidium bromide and image the remaining gel to confirm that fragment was excised.
12. Place excised band into 5 mL tube. Tare balance with empty 5 mL tube before weighing tube containing gel slice.
13. Purify DNA using Zymoclean Gel DNA Recovery Kit. Elute in 6 μL of 1× microinjection buffer (*see* **Note 11**).
14. Measure the concentration of purified DNA using a nanodrop with 1× microinjection buffer as the blanking solution. If the injection mix is to be made up at a later date, store DNA at −20 °C.

3.4 Microinjection (Cytoplasmic and Nuclear)

1. On day of microinjection, dilute the DNA fragment to 3 ng/μ L in 15 μL 1× microinjection buffer. Keep injection mix on ice until ready for microinjection.
2. Prepare embryo handling pipettes by pulling glass capillaries over a flame to create 10–20 pipettes with a diameter of 100–130 μm and a square cut end. Flame-polish the blunt

ends of the embryo handling pipettes before storing in a sterile 50 mL Falcon tube.

3. Prepare microinjection pipettes by pulling Harvard borosilicate glass capillaries #GC100TF-10 using a Sutter P-97 puller. Settings: Pressure = 200, Heat = 557, Pull = 0, Velocity = 150, Time = 175 (RAMP = 518). Lightly flame-polish the blunt ends of the capillaries with butane burner flame to prevent damage to the o-rings of the Eppendorf pipette holders.
4. Using the microforge, create a bend of 30–40° near the end of the microinjection needle.
5. Store 10 freshly made injection capillaries in a capillary storage box to prevent dust accumulating on or in the needles.
6. A minimum of 30 min before beginning to microinject, preload 6 microinjection capillaries with 1.5 μL of microinjection reagents. Return to the storage box and place in a 4 °C refrigerator with the box angled to allow release of air bubbles from the blunt ends of the needles and prevent evaporation of reagents.
7. Prepare holding pipettes by pulling Sutter borosilicate glass capillaries #B100-58-15 using a Sutter P-97 puller. Settings: Pressure = 200, Heat = 540, Pull = 120, Velocity = 80, Time = 170 (RAMP = 518). Lightly flame-polish the blunt ends of the capillaries to prevent damage to the o-rings of the Eppendorf pipette holders.
8. Using the microforge, create a bend of 30–40° near the end of the holding pipette.
9. Store in a capillary storage box (*see* **Note 12**).
10. Prepare the injection slide by removing the plunger of a 2 mL syringe and dispense vacuum grease into the barrel. Replace the plunger and depress until an even thread of grease comes from the tip of the syringe.
11. Dispense a thin line of vacuum grease around the very edge of the custom slide cover and gently press onto a clean cover glass until it is sealed onto the glass.
12. Carefully wipe any excess grease from the edge of the glass and ensure that there is none on the bottom of the glass before placing onto the inverted microscope stage.
13. Turn on the microscope and select the lowest objective (4×).
14. Create an oblong 80 μL drop of M2 media onto the center of the glass and cover the drop with 700 μL of embryo tested mineral oil to prevent evaporation of the media.
15. Turn on the left- and right-hand micromanipulators.
16. Load a holding pipette into the pipette holder of the left-hand micromanipulator and, using the "coarse" setting, lower onto

the slide until it is visible in the drop of M2 and just touching the glass.

17. When the holding pipette is positioned centrally and slightly to the left (by observation at 4×) change the micromanipulator setting to "fine."
18. Using the CellTram Air, gently draw media into the holding pipette from the slide and adjust until media is stable and visible in the capillary. Ensure that media does not travel the length of the capillary into the pipette holder.
19. Turn on the Femtojet and allow it to reach pressure with the pipette tubing disconnected. Once pressure is reached (as indicated by the screen depicting the current settings), attach the pipette tubing and set to "Change capillary" in the menu.
20. Take the culture dish of embryos from the incubator, place on the dissection microscope stage and, using the aspirator tube and embryo handling capillary, aspirate 20–30 fertilized embryos from one of the microdrops and dispense onto the injection slide to the "north" of the holding pipette.
21. Return the culture dish to the incubator.
22. Load a microinjection pipette into the pipette holder of the right-hand micromanipulator and, on the "coarse" setting, lower onto the slide until it is visible in the drop of M2 and just touching the glass.
23. When the microinjection pipette is positioned centrally and slightly to the right, by observation at 4×, change the micromanipulator setting to "fine."
24. Return the Femtojet to the injection setting in the menu.
25. Using the right-hand manipulator, gently tap the tip of the microinjection needle on the holding pipette to open the tip of the needle.
26. Using the foot pedal, "inject" to confirm that the needle opening is patent, and the reagents are able to be dispensed.
27. Using the holding pipette, pick up an embryo and confirm that it is fertilized by visualizing the two pronuclei.
28. If injecting into the pronucleus, position the embryo so that the pronucleus is in the same plane as the holding pipette and injection pipette by adjusting the focus and the position of the holding and injection pipettes. Push the injection needle through the zona, oolemma, and pronuclear membrane and use the foot pedal to inject. The reagents should be seen leaving the tip of the needle and the pronuclear membrane should slightly expand. This will vary depending on the size

of the opening of the needle, the viscosity of the reagents, and the pressure settings on the Femtojet.

29. If injecting into the cytoplasm, position the embryo so that neither of the pronuclei are aligned with the injection needle. Push the needle through the zona and oolemma into the cytoplasm and use the foot pedal to inject. The reagents leaving the tip of the needle will be seen as a slight displacement of the cytoplasm.
30. Continue to inject (either pronuclear or cytoplasmic) the batch of embryos on the slide.
31. Once completed, aspirate the embryos into the middle wash drop of the unused culture dish (prepared earlier). Leave any lysed embryos in the wash drop and move the surviving embryos into a fresh drop.
32. Continue to inject in batches, using a fresh needle for each batch, until they are all injected.
33. Note that embryos that may have earlier been deemed not to be fertilized are worth checking as they may progressively present with two pronuclei and can thus be added to the pool of embryos for injection (*see* **Note 13**).

3.5 Embryo Transfer

1. Day 1—Set up CD1 adult females with vasectomized males 2: 1.
2. Day 2—Check all females for copulation plugs (indicating a successful mating) and group house the plugged females.
3. Place a recovery cage on the warming pad.
4. Allow 19–22 embryos per transfer and determine how many recipients can be used for the embryos that have survived the microinjection process.
5. Prepare a transfer dish by placing 5 × 40 μL drops of M2 on the lid of a culture dish (one in the center and four evenly spaced around the center drop) and placing the base on top to reduce evaporation.
6. Using the aspirator tube and embryo handling capillary, take out enough embryos from the culture dish for two transfers and place into the center wash drop. Wash into two drops in preparation for embryo transfer and return culture dish containing remaining embryos to the incubator.
7. Put the first female for surgery in the anesthetic induction box and induce anesthesia.
8. Load the embryo transfer pipette with a small volume of M2 media, followed by a small gap of air, then the embryos in media, ensuring that the embryos are not too close to the opening of the capillary.

9. Lay the mouse prone on a protecta pad on the warming pad with the face in the nose cone and inject with analgesic subcutaneously.
10. Check for pedal reflex by pinching toes. If no response, continue with the procedure.
11. Spray the back with 80% ethanol.
12. Lift the skin with blunt forceps and make a small transverse incision (1 cm) in the skin with iris scissors, above the spinal cord directly below the ribs.
13. Using a sterile gauze swab, wipe the incision from the anterior to posterior, removing any loose hairs in the process.
14. Place a square of drape moistened with saline above the skin incision.
15. Move the incision over to the left above the body wall. The orange-colored ovary and a white fat pad attached should be visible beneath the body wall.
16. Make a 3–5 mm incision through the body wall above the fat pad and stretch the incision using the blunt forceps.
17. Using the blunt forceps, pull out the fat pad joined to the ovary. The oviduct and uterus will be pulled out simultaneously.
18. Attach the serrefine clamp to the fat pad, taking care to avoid the ovary. The reproductive tract may then be held in position over the back of the animal by the clip-on top of the moist drape.
19. Rotate the clamp so that the coils of the oviduct are uppermost.
20. Follow the oviduct coils to reveal a recess that lies below the ovary and behind the coils of the oviduct. The opening of the oviduct (infundibulum; the target of the transfer procedure) is located within this recess behind a transparent membrane, the bursa that covers the oviduct and ovary.
21. Gently tear the bursa with two pairs of fine forceps to the right of the infundibulum. If the infundibulum is not visible through the bursa, it may be necessary to lift it up from behind the coils of the oviduct. However, be very careful not to damage it. (Note: if there is excessive bleeding, a small piece of sterile drape can be used to absorb the blood).
22. Test that the opening is patent by very gently inserting the tip of one end of the forceps into the opening.
23. Push the pipette into the infundibulum until it has entered all the way into the ampulla. Correct placement is important: The pipette tip must be far enough into the infundibulum so that it does not fall out when the embryos are expelled, but not so far in that the pipette tip is against the wall of the ampulla, so

restricting expulsion of the embryos into the reproductive tract.

24. Confirm expulsion of the embryos into the ampulla by visualizing the presence of the air bubble that follows the embryos into the reproductive tract.
25. Remove the serrefine clamp, grip the fat pad with a pair of blunt forceps, and return the reproductive tract back inside the body wall.
26. Close the skin incision with two wound clips.
27. Record the weight of the mouse and place in the recovery cage.
28. Clean instruments with ethanol and sterilize in bead sterilizer in preparation for the next surgery.
29. Repeat for **steps 6–27** all embryo transfers.
30. Females can be housed in pairs in order to share the care of their newborns.
31. Observe post-surgical females daily for 7–10 days before weighing and removing wound clips.

3.6 Designing PCR Primers

1. Design primers with online tool, Primer-BLAST (https://www.ncbi.nlm.nih.gov/tools/primer-blast/).
2. Input entire FASTA sequence to search for primers in "PCR template" field and restrict range of forward and reverse primer by filling the "From" and "To" range fields (*see* **Note 14**).
3. Increase the "Max" field for "PCR product size" to suit the length of template entered.
4. In Primer Pair Specificity Checking Parameters Select "Genomes for selected organisms (primary reference assembly only)" for Database field and enter "Mus musculus (taxid:10090)" for Organism field.
5. Adjust "Max target amplicon size" if required.
6. Select "Get Primers."
7. Select tick boxes for any intended or allowed sequences.
8. Assess primers for "Products on potentially unintended templates" if presented (*see* **Notes 15** and **16**).
9. Order as standard DNA oligos in tubes at 100 μM in water from Merck (sigmaaldrich.com).

3.7 Genotyping

1. Obtain tissue biopsy from mice to be screened (*see* **Note 17**).
2. Process tissue using the Roche High Pure PCR Template Preparation Kit following method for "Isolation of Nucleic Acids from Mouse Tail" (pg.9) and "Protocol for Washing and Elution"(pg.11) (*see* **Notes 18–20**).

3. Select PCR kit based on parameters of amplicon (*see* **Note 21**). Set up and run as per manufacturer's instructions (*see* **Note 22**).
4. Make agarose gel by mixing agarose powder with 1×TBE buffer before microwaving to dissolve powder. The percentage of agarose in the gel will depend on the size of the amplicon: 2.0% for <250 bp, 1.0% for 250 bp–1 kb, 0.8% for >1 kb.
5. Allow melted agarose to cool slightly before pouring into gel casting apparatus with comb in place and allow to set.
6. Submerge the agarose gel in TBE buffer in a gel tank and load 5–10 μL each sample of PCR + loading dye into its own well, flanked by DNA ladder loaded into the first and last well.
7. Run Gel at 100 V for an appropriate amount of time, using the following as a guide: amplicon <250 bp for 30 min, 250 bp–750 bp for 45 min, 750 bp–1.5 kb for 1 h, >1.5 kb for >1 h.
8. Remove gel from tank and image using the BioRad GelDoc.
9. For sequencing sample preparation and analysis (for targeted insertions only), purify the PCR sample (*see* **Note 23**).
10. Quantify purified PCR product using a Nanodrop.
11. Design a number of sequencing primers approximately 800 bp apart so that the sequencing reads cover the entirety of the PCR product (*see* **Note 24**).
12. Mix purified PCR product and primer and submit according to your local service provider's instructions.
13. When sequencing files are received, align sequencing chromatograms to a mutant sequence using bioinformatics platform Benchling.com (*see* **Note 25**).

4 Notes

1. Most protocols indicate that embryo donors must be of a particular age (with strain variation) for optimal embryo production via superovulation. This is true, but there is some flexibility. Of course, females must be over 3 weeks of age (i.e., weaned), and not too small to be paired with adult males, but in our experience, it is possible to have a productive microinjection session using animals 27–84 days of age.
2. Becoming familiar with the many complex (and expensive) pieces of equipment used for microinjection only really happens with repeated use of this equipment. Take time to learn the basics of how things actually work, read the manuals, experiment with settings, be prepared to troubleshoot and, where necessary, modify your standard protocol. Nobody

wants to be at the mercy of waiting for a technician to be available to deal with an issue that could be solved by yourself.

3. On occasion, the male will attempt to remove the wound clip from the scrotum and will sometimes succeed. For this reason, the animals must be observed frequently in the first couple of days. If they do manage to remove the clip, often the incision is barely visible in which case the clip will not need replacement. Where the clip does need replacement, the animal must be anesthetized and given another dose of analgesic to facilitate this.
4. Ideally, hormone injections should be performed at required times. However, there may be occasions when this is not able to be strictly adhered to. In our experience, late or early injections will always provide some fertilized embryos to work with. Even a small yield of embryos can be enough to make a mouse model if the CRISPR modification is efficient. Additionally, we do not alter the hormone injection times to account for daylight savings.
5. Regardless of the age of your donors, it is common to find that embryos do not all fertilize and develop at the same rate. On the day of microinjection, you will find that your initial assessment of 2PN embryos is far below the final number that you microinject. Keep checking for 2PNs as the day progresses to ensure the maximum number of injectable embryos is obtained.
6. Each restriction enzyme will have recommended concentration, time, and temperature required to digest the donor DNA. Check the protocol for the selected restriction enzyme at nebiolabs.com.
7. Ethidium bromide is suspected of causing genetic defects. Work in an area designated for use with ethidium bromide and be sure to wear gloves and eye protection. Before adding ethidium bromide to melted agarose, ensure that the agarose has cooled enough to not release steam which may carry ethidium bromide into the respiratory system.
8. If gel comb wells hold <100 μL volume, masking tape can be used to make a larger well by taping a number of comb teeth together (Fig. 3).
9. If transgene and backbone fragments are of similar size, electrophoresis may need to be performed for longer than 30 min.
10. Any solutions containing ethidium bromide should be decontaminated before disposal by using the Ethidium Bromide Green Bag Disposal Kit.
11. The pH of the buffered solution is optimal for elution and allows for easy dilution of the purified fragment for injection.

Fig. 3 Masking tape can be used to join multiple wells together in order to hold entire volume of plasmid digest reaction

12. We have had experience with using microinjection needles and holding pipettes from both large and small companies but prefer to generate our own in house using a Sutter P-97 puller and a microforge. In-house generation is considerably cheaper and enables us to make changes to the design (length, taper, tip size) without having to articulate specific measurements and then wait for a custom order to be delivered. Needles may occasionally become blocked due to viscous plasmid preparations or particulates in the solution. Sometimes, this issue can be resolved by simply "clearing" the needle. Pressing the clear button on the Femtojet causes a quick blast of high pressure to push out anything that may impeding the flow of solution. It is also useful to tap the needle on the holding pipette a few times to slightly widen the opening, thus allowing better flow from the needle. If neither of these approaches, clear the blockage, remove the needle, and replace with a freshly pre-filled one from the fridge. The preparation can also be briefly spun down in a mini centrifuge and a new batch of needles filled and rested for use.
13. Quite often we suffer from a lack of embryo transfer recipients on the day of microinjection (Day 1). If this occurs, we culture the 1-cell embryos in the incubator overnight and transfer 2-cell embryos into the oviducts of 0.5dpc pseudopregnant recipients the following day (Day 2). If there are no recipients available on Day 2, it is possible to "incubate" 2-cell embryos in a single 80ul drop of handling media (M2 or FHM) in a 35 mm dish overlaid with oil in the refrigerator at 4 °C overnight. The following day (Day 3), the 2-cell embryos are transferred into a pre-equilibrated culture dish for at least

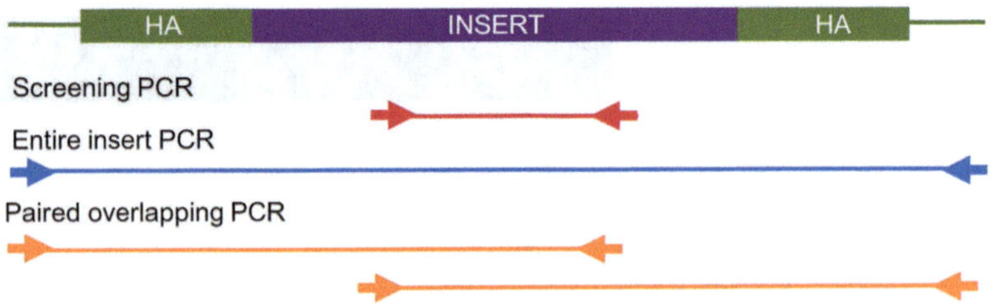

Fig. 4 Primer placement schematic for different genotyping PCR strategies

30 min prior to transferring into the oviducts of 0.5dpc pseudopregnant recipients.

14. Genotyping Assay Design (Fig. 4).

    Random integration.

    A simple screening PCR can be used when the transgene is randomly integrated into the genome. Primers should target within the insert and should be specific to the insert.

    Targeted insertion.

    The ideal screening strategy to detect a targeted insertion will amplify the entire insert with primers set outside of any homology arms that were used to insert the foreign DNA, so that a mutant allele is distinguishable from a WT allele by amplicon length. However, amplifying the entire insert can be challenging as homology arms are often large and PCR can favor the smaller WT fragment. For this reason, genotyping of large, targeted insertions by PCR is often done in stages.

    A simple screening PCR, amplifying a region within the insert can be used initially to screen the progeny for potential founders. Once potential founders have been identified, locus-specific PCRs can identify correctly targeted mice. The locus-specific PCR should be anchored by one primer outside of the homology arm and the other within the introduced DNA.

15. Mismatches at the 3′ end are less well tolerated, so will reduce off-target amplification, as will an increased number of mismatches throughout the primer sequence.

16. If one or both primers are located in the inserted sequence, a mismatch profile on unintended templates will not be shown.

17. An ear notch or any other tissue taken to identify the mouse can be used for the DNA extraction process. If another identification method such as tattooing or chipping is used, a tail snip will suffice.

18. The tissue lysis step (**step 1**) can be carried out overnight if required.

19. Tissue shearing (**step 2**) can be replaced with a brief vortex to save on consumables. Some tissues may remain visible after vortexing, but there will be enough digested material to generate a workable concentration of DNA.

20. The volume of elution buffer (Washing and Elution **step 5**) can be reduced to 100 μL, if necessary, to increase the DNA concentration.
21. Short amplicons that will be targeted for random integration identification or initial screening for targeted insertions can be performed with a standard Taq polymerase (Roche Taq DNA Polymerase) and buffer (FailSafe™ PCR 2x PreMix—Buffer D and J are efficient for many amplicons) whereas larger or more complex amplicons for locus specific confirmation in targeted insertion projects may require a high-fidelity polymerase with specialty buffers for complex or GC-rich regions (Phusion® High-Fidelity DNA Polymerase).
22. PCR controls—It is important to include negative controls (no template control and WT control) and if possible, to include a positive control (such as a repair template).
23. PCR cleanup—If the PCR generates a clean product of the correct size, the remaining PCR product that was not loaded to the gel can be purified using a PCR cleanup kit (QIAquick® PCR purification kit). If the PCR produces more than one band, i.e., mutant and WT band it is best to isolate the band of interest either by gel extraction (Monarch® DNA Gel Extraction Kit) or cloning (pGEM®-T Easy Vector System I or II).
24. The sequencing primers do not need to be as stringent as the PCR primers because the DNA used for sequencing will be amplified to enrich the targeted locus compared with genomic DNA used for PCR.
25. Benchling is a life sciences R&D cloud-based platform that is free to use for academics.

References

1. Pimentel D, Zuniga R, Morrison D (2005) Update on the environmental and economic costs associated with alien-invasive species in the United States. Ecol Econ 52(3):273–288
2. Stenseth NC et al (2003) Mice, rats, and people: the bio-economics of agricultural rodent pests. Front Ecol Environ 1(7):367–375
3. McLeod R (2004) Counting the cost: impact of invasive animals in Australia. Cooperative Research Centre for Pest Animal Control, Canberra
4. Blackburn TM et al (2004) Avian extinction and mammalian introductions on oceanic Islands. Science 305(5692):1955–1958
5. Harris DB (2008) Review of negative effects of introduced rodents on small mammals on islands. Biol Invasions 11(7):1611–1630
6. Doherty TS et al (2016) Invasive predators and global biodiversity loss. Proc Natl Acad Sci U S A 113(40):11261–11265
7. Piaggio AJ et al (2017) Is it time for synthetic biodiversity conservation? Trends Ecol Evol 32(2):97–107
8. Gregory S et al (2014) Eradications of vertebrate pests in Australia: a review and guidelines for future best practice. Invasive Animals Cooperative Research Centre, Canberra, Australia, pp 1–90
9. Esvelt KM et al (2014) Concerning RNA-guided gene drives for the alteration of wild populations. elife 3:e03401
10. Burt A (2003) Site-specific selfish genes as tools for the control and genetic engineering

of natural populations. Proc Biol Sci 270(1518):921–928
11. Cong L et al (2013) Multiplex genome engineering using CRISPR/Cas systems. Science 339(6121):819–823
12. Jinek M et al (2012) A Programmable dual-RNA—guided DNA endonuclease in adaptive bacterial immunity. Science 337(6096): 816–821
13. Prowse TAA et al (2017) Dodging silver bullets: good CRISPR gene-drive design is critical for eradicating exotic vertebrates. Proc Biol Sci 284(1860):20170799
14. Galizi R et al (2014) A synthetic sex ratio distortion system for the control of the human malaria mosquito. Nat Commun 5:3977
15. Simoni A et al (2020) A male-biased sex-distorter gene drive for the human malaria vector Anopheles gambiae. Nat Biotechnol 38(9): 1054–1060
16. Kyrou K et al (2018) A CRISPR-Cas9 gene drive targeting doublesex causes complete population suppression in caged Anopheles gambiae mosquitoes. Nat Biotechnol 36(11): 1062–1066
17. Hammond A et al (2016) A CRISPR-Cas9 gene drive system targeting female reproduction in the malaria mosquito vector Anopheles gambiae. Nat Biotechnol 34(1):78–83
18. Haghighat-Khah RE et al (2020) Cellular mechanisms regulating synthetic sex ratio distortion in the Anopheles gambiae germline. Pathog Glob Health 114(7):370–378
19. Gantz VM et al (2015) Highly efficient Cas9-mediated gene drive for population modification of the malaria vector mosquito Anopheles stephensi. Proc Natl Acad Sci U S A 112(49): E6736–E6743
20. Champer SE et al (2021) Modeling CRISPR gene drives for suppression of invasive rodents using a supervised machine learning framework. PLoS Comput Biol 17(12):e1009660
21. Grunwald HA et al (2019) Super-Mendelian inheritance mediated by CRISPR-Cas9 in the female mouse germline. Nature 566(7742): 105–109
22. Pfitzner C et al (2020) Progress toward zygotic and germline gene drives in mice. CRISPR J 3(5):388–397
23. Gallardo T et al (2007) Generation of a germ cell-specific mouse transgenic Cre line. Vasa-Cre Genesis 45(6):413–417
24. Champer J et al (2018) Reducing resistance allele formation in CRISPR gene drive. Proc Natl Acad Sci U S A 115(21):5522–5527
25. Unckless RL, Clark AG, Messer PW (2017) Evolution of resistance against CRISPR/Cas9 gene drive. Genetics 205(2):827–841
26. Noble C et al (2017) Evolutionary dynamics of CRISPR gene drives. Sci Adv 3(4):e1601964
27. Champer SE et al (2020) Computational and experimental performance of CRISPR homing gene drive strategies with multiplexed gRNAs. Sci Adv 6(10):eaaz0525
28. Lyttle TW (1991) Segregation distorters. Annu Rev Genet 1991:511–577
29. McFarlane GR, Whitelaw CBA, Lillico SG (2018) CRISPR-based gene drives for pest control. Trends Biotechnol 36(2):130–133
30. Burt A, Deredec A (2018) Self-limiting population genetic control with sex-linked genome editors. Proc Biol Sci 285(1883):20180776. https://doi.org/10.1098/rspb.2018.0776
31. Zuo E et al (2017) CRISPR/Cas9-mediated targeted chromosome elimination. Genome Biol 18(1):224
32. Prowse TA et al (2019) A Y-chromosome shredding gene drive for controlling pest vertebrate populations. elife 8
33. Kelemen RK, Vicoso B (2018) Complex history and differentiation patterns of the t-Haplotype, a mouse meiotic driver. Genetics 208(1):365–375
34. Koopman P et al (1991) Male development of chromosomally female mice transgenic for Sry. Nature 351(6322):117–121
35. Backus GA, Gross K (2016) Genetic engineering to eradicate invasive mice on islands: modeling the efficiency and ecological impacts. Ecosphere 7(12):e01589
36. Manser A et al (1909) Controlling invasive rodents via synthetic gene drive and the role of polyandry. Proc Biol Sci 2019(286): 20190852
37. Manser A, Konig B, Lindholm AK (2020) Polyandry blocks gene drive in a wild house mouse population. Nat Commun 11(1):5590
38. Kandul NP et al (2020) Assessment of a split homing based gene drive for efficient knockout of multiple genes. G3 (Bethesda) 10(2): 827–837
39. Noble C et al (2019) Daisy-chain gene drives for the alteration of local populations. Proc Natl Acad Sci U S A 116(17):8275–8282
40. Sudweeks J et al (2019) Locally fixed alleles: a method to localize gene drive to island populations. Sci Rep 9(1):15821
41. Lopez Del Amo V et al (2020) Small-molecule control of super-mendelian inheritance in gene drives. Cell Rep 31(13):107841

Part IV

Large Animal Models of Human Disease

Chapter 12

Generation of a Human Deafness Sheep Model Using the CRISPR/Cas System

Martina Crispo, Vanessa Chenouard, Pedro dos Santos-Neto, Laurent Tesson, Marcela Souza-Neves, Jean-Marie Heslan, Federico Cuadro, Ignacio Anegón, and Alejo Menchaca

Abstract

CRISPR/Cas9 system is a promising method for the generation of human disease models by genome editing in non-conventional experimental animals. Medium/large-sized animals like sheep have several advantages to study human diseases and medicine. Here, we present a protocol that describes the generation of an otoferlin edited sheep model via CRISPR-assisted single-stranded oligodinucleotide-mediated Homology-Directed Repair (HDR), through direct cytoplasmic microinjection in in vitro produced zygotes.

Otoferlin is a protein expressed in the cochlear inner hair cells, with different mutations at the *OTOF* gene being the major cause of nonsyndromic recessive auditory neuropathy spectrum disorder in humans. By using this protocol, we reported for the first time an *OTOF* KI model in sheep with 17.8% edited lambs showing indel mutations, and 61.5% of them bearing knock-in mutations by HDR. The reported method establishes the bases to produce a deafness model to test novel therapies in human disorders related to *OTOF* mutations.

Key words Genome editing, *OTOF*, ssODNs, In vitro fertilization, Embryo

1 Introduction

Genome editing in productive species has been largely postponed since the available methodology until few years ago was laborious and time consuming. These methods include the use of somatic cell nuclear transfer (SCNT) with gene targeted cells [1–3], or the use of embryonic stem (ES) cells [4], with very poor outcomes. Since the discovery of endonucleases and more specifically CRISPR/Cas system [5], more and more KO and KI large animal models are being published in several laboratories around the world [6, 7], due to its greater efficiency, easier of application, and lower cost [8]. Previously described endonucleases such as zinc finger nucleases

Paul John Verma, Huseyin Sumer and Jun Liu (eds.), *Applications of Genome Modulation and Editing*,
Methods in Molecular Biology, vol. 2495, https://doi.org/10.1007/978-1-0716-2301-5_12,

(ZFN) [9] or transcription activator-like effector nucleases (TALENs) [10] rely on the use of proteins that are somehow difficult and expensive to engineer in large animals. CRISPR/Cas approach allows a widespread application of the technology in almost all species in a simply manner.

The goal of this procedure is to generate an *OTOF* KO model in sheep with acceptable efficiencies, by using a KI strategy with single-stranded oligodinucleotide (ssODN) to induce two-stop codons in exon 5 or 6. We describe a simple and reliable approach to incorporate the CRISPR/Cas9 system through microinjection into the cytoplasm of sheep zygotes provided by in vitro production [11]. Otoferlin protein is produced by the *OTOF* gene, and it is mostly expressed in the cochlear inner hair cells. Mutations in this gene will induce a form of deafness that is the major cause of nonsyndromic recessive auditory neuropathy spectrum in humans. The method described in this report can be incorporated in any laboratory with expertise in molecular biology, genetic engineering, and mammalian embryo micromanipulation.

2 Materials

Prepare all solutions using ultrapure water and analytical grade reagents. For embryo culture and microinjection, use embryo-tested ultrapure water. Embryo culture media last 15 days at 4 °C. Prepare all reagents at room temperature unless otherwise indicated.

2.1 CRISPR Design and Validation

1. National Center Biotechnology Information (NCBI) database (https://www.ncbi.nlm.nih.gov).
2. Ensembl database (http://www.ensembl.org/index.html).
3. CRISPOR (http://crispor.tefor.net).
4. Specific primers.
5. GeneArt CRISPR nuclease vector kit.
6. HiScribe T7 RNA synthesis kit.
7. Sheep A15 astroglia cell line.
8. Lipofectamine.
9. Thermocycler.
10. Herculase II Fusion DNA polymerase kit.
11. Specific Exon 5 and 6 primers.
12. T7 Endonuclease I.
13. Capillary electrophoresis device.

2.2 In Vitro Embryo Production

1. Search medium: TCM199 medium supplemented with 0.5% bovine serum albumin (BSA), 25 mM HEPES, and 50 IU/mL–50 μg/mL penicillin-streptomycin.
2. Maturation medium: TCM 199 medium supplemented with 10% estrus goat/sheep serum, 10 μg/mL FSH, 10 μg/mL LH, 100 μM cysteamine, and 50 IU/mL–50 μg/mL of penicillin-streptomycin.
3. Fertilization media: Synthetic Oviductal Fluid (SOF) medium supplemented with 2% estrus sheep serum, 10 μg/mL heparin, and 1 μg/mL hypotaurine.
4. Estrus sheep serum: Produced by blood collection from females in estrus, which after centrifugation requires complement heat-inactivation at 56 °C for 30 min. Aliquoted and frozen at −80 °C for further use.
5. Embryo culture medium: SOF medium supplemented with 0.4% fat acid-free BSA, 5% essential and 5% non-essential amino acids.
6. Embryo-tested mineral oil.
7. Plastic Petri dishes: 35 mm, 60 mm, and 90 mm.
8. Automatic pipettes: 5, 20, 200, and 1000 μL.
9. CO_2 incubator.
10. Stereomicroscope: 60x magnification.
11. Inverted microscope: 400x magnification.
12. Liquid nitrogen dewar.

2.3 Zygote Microinjection

1. RNA microinjection buffer: 10 mM Tris pH 7.5, 0.1 mM EDTA.
2. Protein microinjection buffer: 20 mM HEPES, 150 mM KCL.
3. Eppendorf tubes: 0.5 mL.
4. Microinjection pulled capillary: thin-walled borosilicate glass capillary with internal filament, 1 mm outer diameter, and 0.78 mm inner diameter.
5. Holding pipette: Bended, tip with 25 μm inner diameter and 100 μm outer diameter.
6. Glass depression slide.
7. M2-buffered medium.
8. Microinjection station: Inverted microscope (200× magnification), two micromanipulators, and pressure injector.
9. CO_2 incubator.
10. Pipette puller.
11. Glass microforge.

2.4 Embryo Transfer and Pregnancy

1. Tomcat catheter: 1.0 mm in diameter.
2. Holding medium.
3. Shearing machine.
4. Stretcher.
5. Laparoscope.
6. Scalpel blade.
7. Atraumatic grasping forceps.
8. Paper clips.
9. Anesthesia: 5 mg/kg ketamine and 0.35 mg/kg diazepam.
10. Suture.
11. Antibiotic: 5 mg/kg ceftiofur.
12. Ultrasound equipment (5 MHz probe) for pregnancy diagnosis and monitoring of fetal growth.

2.5 Genotyping

1. Scalpel blade.
2. Scissors.
3. Eppendorf tubes.
4. DNA extraction kit.
5. OTOF-specific primers: OTOF-Ex5-Up (CACA CAGTTGTTGGGGGTCA) and OTOF-Ex5-Lo (CTCCTC CCACCGTACCTTTC); OTOF-Ex6-Up (GCTGAGTCTGTGCTTGTCAGT) and OTOF-Ex6-Lo (GA AACCCCTTCCCTCCTCTTG); potential big deletion with OTOF-Ex5-Up / OTOF-Ex6-Lo.
6. Restriction enzyme SpeI.
7. Microfluidic electrophoresis device.
8. PCR-based Sanger sequencing.
9. T7EI enzyme.

3 Methods

3.1 CRISPR Reagents

Sheep Otoferlin gene (Gene ID: 101111175), located on chromosome 3, has 47 exons.

3.1.1 Selection of Targeting Single Guide RNAs (sgRNAs)

1. Target exons 5 and 6 and run their genomic DNA sequences on the CRISPOR short guide design tool (Tefor).
2. Select two guides with the highest scores for both exons.

3.1.2 Construction of CRISPR/Cas9-Targeting Vector

1. Clone short guide DNAs into CRISPR Cas9 vector.
2. Amplify and purify recombinant plasmids.
3. Confirm proper integration of the sg oligos by sequencing the cloning site of plasmids.

3.1.3 Production of CRISPR/Cas9-Targeting Vector

1. Use PCR of the CRISPR plasmids with synthetic 5′ oligo harboring the T7 promoter upstream of the sgRNA sequence and a downstream 3′ Tracr oligonucleotide to generate full sgDNA matrix.
2. Purify PCR products and use them as matrix for in vitro transcription with the HiScribe T7 RNA synthesis kit.
3. Purify SgRNA on silica columns.
4. Produce 30–40 μg of sgRNA and control purity by MOPS-agarose gel electrophoresis.

   (oligo 5′T7prom-OTOF-Ex5-sg54: TTAATACGACT CACTATAGGTGGAGGAGAACCACGTGG,

   oligo 5′T7prom-OTOF-Ex5-sg95: TTAATACGACTCAC TATATGGGGGAGGAGTCCCTTCAC.

   oligo 3′T7-sgRNATracr: TTTAAAAGCACCGACTCGGTGCC).

3.1.4 In Vitro Validation of System Efficiency

1. Asses CRISPR activity by transfecting 1–5 μg of the plasmid onto sheep A15 cell line along with Lipofectamine.
2. Harvest A15 cells 72 h later and purify their genomic DNA.
3. Amplify 50 ng of genomic DNA by PCR for each targeted exon followed by heat denaturation/slow renaturation of PCR products and treat by T7 Endonuclease I.
4. Separate PCR products on microfluidic electrophoresis device to reveal gDNA edition of targeted exons. For that, load 5 K or 1 K microfluidic chip with gel and marker in the specified containers. Centrifuge PCR tube strips or plates for 5 min at 447 × g and place without caps or adhesive film in the electrophoresis system. Prime the chip for 5 min for the 5 K or 10 min for the 1 K. Inject PCR samples into the chip's microcapillaries connected to electrodes. Run each sample, mixed automatically with an internal marker, through these microcapillaries according to their sizes and nucleotidic compositions. DNA fluorescence measurement is recorded as a function of time on an electrophoregram, and the DNA ladder allow to define the product sizes. Perform the analysis at 28 s/sample for the 5 K chip or 1 min/sample for the 1 K chip on a limited sample volume (<1 mL).
5. Estimate CRISPR activity as ratio between the amounts of PCR products cleaved to the uncleaved WT product.

3.2 In Vitro Production of Zygotes for Microinjection

1. Recover oocytes from follicles (2–8 mm) of ovaries retrieved from a nearby slaughterhouse. About 250–300 ovaries can be processed in a single day.
2. Load recovered follicular fluid with cumulus-oocyte complexes (COCs) into a 90 mm Petri dish and add pre-warmed search medium to complete the entire Petri dish surface.

3. Recover COCs using a glass capillary pipette under a stereomicroscope at 20 to 40× magnification and transfer them to a 35 mm Petri dish (35 mm) containing search medium.
4. Select Grade 1 and 2 COCs (*see* **Note 1**) and wash in drops of search medium.

3.2.1 In Vitro Maturation (IVM)

1. Perform IVM into 35 mm Petri dishes, placing 20–25 selected COCs into 100 μL drops of maturation medium (1 oocyte/2–5 μL of medium) under mineral oil, for 24–27 h at 39 °C in a humidified atmosphere containing 5% CO_2.

3.2.2 In Vitro Fertilization (IVF)

1. Fertilize expanded COCs using frozen-thawed semen. In our laboratory, motile sperm are selected by swim-up. Place a sample of thawed semen into a pre-warmed 1.5 mL plastic tube. Evaluate an aliquot of 5–10 μL sperm under microscope at 100× magnification to check viability (*see* **Note 2**). For swim-up, trespass the straw content in 1 mL of capacitation/fertilization medium into a 15 mL Falcon conical tube and incubate in 45° angle position for 15 min at 39 °C in a humidified atmosphere with 5% CO_2. After incubation, trespass 840 μL of the supernatant medium into a pre-warmed 1.5 mL plastic tube (*see* **Note 3**).
2. After sperm selection, evaluate the percentage of viable sperm and individual progressive sperm motility under inverted microscope at 400× magnification. Determine sperm concentration in a Neubauer chamber and subsequently calculate the insemination dose.
3. Perform IVF in 35 mm Petri dishes by co-incubating the matured COCs with 1×10^6 sperm into 100 μL of fertilization medium drops, covered by mineral oil. Co-incubate for 16–18 h under mineral oil at 39 °C in a humidified atmosphere containing 5% CO_2.

3.3 Zygote Microinjection

3.3.1 Microinjection Mix Preparation

1. For Cas9 RNA, mix 10 ng/μL of each guide, 50 ng/μL of Cas9 RNA, and 20 ng/μL of ssODN in RNA microinjection buffer (*see* **Note 4**). Prepare at list 10 μL of mix for a standard microinjection day (up to 500 zygotes) (*see* **Note 5**).
2. For Cas9 RNP, mix 200 ng/μL of each guide, 500 ng/μL of Cas RNP, and 200 ng/μL of ssODN in protein microinjection buffer. Prepare at list 10 μL of mix for a standard microinjection day (up to 500 zygotes). Incubate the mix at room temperature for 10 min to allow formation of RNP complexes and then kept at 4 °C until microinjection (within 1 h) (*see* **Notes 5** and **6**).

3.3.2 Microinjection

1. Sixteen hours post fertilization, denude presumptive zygotes by gently pipetting the cumulus-oocyte complexes (COCs) with a 1000 μL micropipette and a sterile plastic tip in a 500 μL drop of buffered SOF medium until all cumulus cells have been released (*see* **Note 7**).
2. Produce injection pipettes by using a pipette puller and thin-walled borosilicate glass capillary. The tip of pulled pipettes may be bended at an angle of 10–15° with the microforge, especially when using cavity slides as injection chambers. Prepare at least 20 pulled pipettes for a typical microinjection day (*see* **Note 8**).
3. Glass holding pipettes with angle can be purchased at a commercial vendor or be homemade.
4. Switch on the microinjection pressure device to stabilize pressure at list 1 h before starting the first microinjection session. Set the injection pressure at 500–1000 hPa and the compensation pressure at 30–40 hPa.
5. Load 20 μL of M2 medium on the depression of the glass slide injection chamber, cover with mineral oil, and place it on the microscope stage. Use low magnification (25×) to get a full view of the microinjection drop.
6. Insert the holding pipette into the holder of the left-hand-side micromanipulator. Low the pipette into the M2 drop through the oil and adjust it so that the tip is parallel to the bottom of the chamber.
7. Backfill the injection pipette by dipping the back into the 0.5 mL tube containing the CRISPR/Cas9 solution and let the mix enter the tip of the pipette by capillary force. Clean the back of the pipette with paper towel and assemble into the holder of the right-hand-side micromanipulator. Low it into the microinjection drop and adjust the pipette in the same focus than the holding pipette, parallel to the bottom of the injection chamber. Change the magnification to 200× to have a better view of the tips of both pipettes.
8. Carefully tap the tip of the injection pipette against the tip of the holding pipette to break it up. Test the microinjection mix flow by applying pressure.
9. Wash a group of 100–150 denuded zygotes in M2 medium drops. Transfer to the upper region of the injection chamber.
10. Attach one zygote to the tip of the holding pipette by applying negative pressure with the air device attached to the holder of the left hand micromanipulator. Focus the zona pellucida (ZP) and the tip of the injection pipette on the same level.
11. Slowly insert the injection pipette through the ZP and the cell membrane into the cytoplasm. When the tip of the pipette appears to be inside, apply injection pressure through the

injector (*see* **Notes 9** and **10**). When the cytoplasm swells, quickly pull the pipette out of the zygote's cytoplasm (*see* **Note 11**). Place the injected zygote on the lower region of the M2 drop, far away from the non-injected zygotes. Inject the 100–150 zygotes one by one within 30 min until the whole group is ready (*see* **Note 12**).

12. Transfer the injected zygotes into a fresh drop of warm culture medium for washing and then move to a 100 μL drop of SOFaa in a 35 mm plastic culture dish under mineral oil for further incubation (*see* **Note 13**). Load a new group of embryos to the chamber and proceed to inject them with a new loaded injection pipette.

3.4 Embryo Culture

1. Incubate injected zygotes in 100 μL of culture medium drops in a tri-gas chamber with 5% CO_2, 5% O_2, and 90% N_2 in humidified atmosphere at 39 °C until DNA extraction for genotyping and sequencing, cryopreservation, or embryo transfer to recipients (*see* **Note 14**).

3.5 Embryo Transfer to Recipients

1. Synchronize the estrous cycle of the recipients with the stage of the embryos to be transferred. The day of ovulation of the recipient should be synchronized with the day of IVF (*see* **Notes 15** and **16**).
2. Shear and clip the recipients cranial to the udder, wash, and sanitize the skin.
3. Administer antibiotics (e.g., ceftiofur 5 mg/kg, im) and anesthesia (ketamine 5 mg/kg and diazepam 0.35 mg/kg iv) prior to the transfer procedure.
4. Make an incision (~3 cm) in the midventral region just cranial to the udder and introduce atraumatic grasping forceps into the abdominal cavity.
5. Using the laparoscope and with careful manipulation with the forceps, examine both ovaries to find at least one corpus hemorrhagicum denoting recent ovulation. No large follicles should be present. Those females bearing either a small or pale corpus luteum should be rejected.
6. Carefully exteriorize the horn through the incision in the right paramedian region. Hold the horn between the thumb and index finger using powder-free gloves.
7. Perform a puncture through the uterine wall by using a sterilized small diameter instrument (e.g., a paper clip or blunted 20 g needle).
8. Load the embryos into a Tomcat catheter (i.e., 1.0 mm in diameter) with a small volume of holding medium and 1 or

2 air bubbles positioned on both sides of the embryo (i.e., air-medium-air-embryo-air-medium-air).

9. Insert the tip of the Tomcat catheter at least 1–2 cm into the uterine tube and direct it cranially into the horn. Slowly inject the contents of the catheter into the lumen, ensuring no bubbling reflux of fluid occurs (*see* **Note 17**).
10. After the catheter is removed, push the oviduct and uterine horn back into the abdomen.
11. Using 4 metric absorbable suture material, with a half-round swaged on needle, close the body cavity by carefully picking up the peritoneum and the abdominal wall by the *linea alba* as one layer, and using a continuous interlocking suture pattern. The skin incision may be closed with a non-absorbable suture material using a continuous suture pattern or metal clips.

3.6 Pregnancy and Birth

1. Check pregnancy and fetal development using B-mode transabdominal ultrasonography with a 5 MHz probe at Day 30. Check single, twin, or triplet pregnancy and register.
2. Birth assistance is required in order to avoid lamb complications and death.
3. Register offspring survival during the first month of life.

3.7 Genotyping

1. When the lambs reached one week of life, recover tissue samples from skin and muscle with a scalpel and sterile scissors, identify correctly and freeze at −20 °C until genotyping.
2. Withdraw blood from the jugular vein, and store until DNA extraction.
3. Extract genomic DNA using blood and tissue DNA extraction kit.
4. Perform PCR amplification of OTOF target sites with the specific primers.
5. Digest PCR products for exon 5 to evaluate KI by restriction enzyme SpeI. Load PCR products on a microfluidic electrophoresis device.
6. Validate on-target activities using PCR-based Sanger sequencing, T7EI assay, and TA cloning-based Sanger sequencing.
7. To validate off-target activities, amplify potential off-target sites by PCR and follow the same steps of genomic validation of on-target activities.

4 Notes

1. Oocyte quality and competence is a key factor for IVEP success and the most used method for COC evaluation relies on its morphology under stereomicroscope at 40× magnification:
   - Grade 1: Good, completely surrounded by multiple and compact cumulus cells, homogeneous cytoplasm.
   - Grade 2: Fair, partially surrounded by cumulus cells, homogeneous cytoplasm with dark peripheral areas.
   - Grade 3: Naked, no cumulus cells, irregular cytoplasm.
   - Grade 4: Non-viable, oocytes surrounded by fibrin, aged expanded cells, irregular cytoplasm with dark areas.

     Grade 3 COCs result in low embryo production and most of the laboratories discard them. Grade 4 COCs are death and always are discarded. Normally, about 90% of good quality oocytes achieve metaphase II stage.
2. Variability in sperm fertility occurs between bucks or rams, and also between ejaculates from the same male. Sperm evaluation by motility and morphology should be conducted according to the standardized procedure. In practice, because there is no single in vitro measurement that reflects the real fertility of the sperm, it is recommended to test each batch in one IVEP session by comparing it with other sperm batch of known fertility previously used.
3. In vitro sperm capacitation is achieved adding capacitating agents during sperm washing and incubation, and also during sperm co-incubation in IVF. Chemical capacitating agents include heparin, hypotaurine, ionomycin, or a mixture of penicillamine, hypotaurine, and epinephrine.
4. Use RNAse-free materials for the CRISPR mix preparation and manipulate the mix in ice during the whole microinjection session using powder-free disposable gloves.
5. Aliquots of Cas9 RNA microinjection mix ready to inject can be frozen at −80° for several months. RNP microinjection mix should be prepared fresh immediately before starting the first microinjection session.
6. Lower RNPs concentrations can be used, with better embryo survival but lower editing efficiencies.
7. Zygotes should be microinjected around 18 h post fertilization.
8. A good injection pipette should have a tip with a diameter less than 1 μm. This is critical for good embryo survival. Pull the pipettes the same day to avoid dust accumulation.

9. Adjust the injection pressure according to the flow of microinjection mix when injecting an embryo.
10. If the tip of the injection pipette becomes too large, discard the pipette and use a new one. Wide openings can damage the zygote and produce high lysis rates.
11. If the cytoplasm swells visibly, it has been successfully injected (approx. 5–10 pl). If the cytoplasm does not swell, the pipette is clogged or has not entered the plasma membrane. Unclog the pipette and reinject the zygote.
12. Injection pipette should be changed before every microinjection session since mRNA will degrade after 30 min.
13. Culture medium must be equilibrated in the incubator at 39 °C and 5% CO_2 for at least 2 h prior to use. Replace 80% of the culture medium per drop on Day 3 and Day 6 [12].
14. Cleavage rate is recorded on Day 2 after fertilization (2–4 cell embryos/total oocytes). Development rate of morulae and blastocysts stages are recorded on Day 6. As reference, the expected blastocyst rate applying this procedure is around 30–40%.
15. Ideally, female recipients should be multiparous, preferably with a recent history of having successfully delivered and raised offspring. Select healthy females, in good body condition score, with good teeth and feet, and pay particular attention to condition of the udder (e.g., mastitis, damaged teats during shearing). Since for IVEP usually two or more embryos are transferred per recipient, body size should consider the possibility of pregnancies bearing twins.
16. Fixed-time embryo transfer is performed in a single day in those females that showed estrus behavior within 48 h after device removal, which requires the use of paint-marked vasectomized males from 24 to 48 h (or alternatively, estrus detection).
17. Embryos can be transferred at the blastocyst stage (i.e., 6–7 days after IVF) into the uterus (late embryo transfer), or at 1 to 4-cell stage (i.e., within 2 days after IVF) into the oviduct (early embryo transfer). Often 3–5 embryos are placed per recipient that usually results in singleton or twin pregnancies.

Acknowledgments

The technique described here was supported by *Agencia Nacional de Investigación e Innovación* (ANII), PEDECIBA, *Institut Pasteur de Montevideo* (*IPMontevideo*) and *Instituto de Reproducción*

Animal de Uruguay (IRAUy). Some activities were funded by "TEFOR" project («Investissements d'Avenir» French Government program, managed by the French National Research Agency, ANRII-INSB-0014) and the IHU-Cesti project (ANR-10-IBHU-005) which both are part of the «Investissements d'Avenir» French Government program managed by the ANR, the IHU-Cesti project is also supported by Nantes Métropole and Région Pays de la Loire. Part of this work was realized in the context of the Labex IGO project (ANR-11-LABX-0016-01).

References

1. Schnieke AE, Kind AJ, Ritchie WA, Mycock K, Scott AR, Ritchie M, Wilmut I, Colman A, Campbell KH (1997) Human factor IX transgenic sheep produced by transfer of nuclei from transfected fetal fibroblasts. Science (New York, NY) 278(5346):2130–2133
2. Cibelli JB, Stice SL, Golueke PJ, Kane JJ, Jerry J, Blackwell C, Ponce de Leon FA, Robl JM (1998) Cloned transgenic calves produced from nonquiescent fetal fibroblasts. Science (New York, NY), 280 (5367):1256-1258
3. Keefer CL, Baldassarre H, Keyston R, Wang B, Bhatia B, Bilodeau AS, Zhou JF, Leduc M, Downey BR, Lazaris A, Karatzas CN (2001) Generation of dwarf goat (Capra hircus) clones following nuclear transfer with transfected and nontransfected fetal fibroblasts and in vitro-matured oocytes. Biol Reprod 64(3):849–856
4. Soto DA, Ross PJ (2016) Pluripotent stem cells and livestock genetic engineering. Transgenic Res 25(3):289–306. https://doi.org/10.1007/s11248-016-9929-5
5. Wang H, Yang H, Shivalila CS, Dawlaty MM, Cheng AW, Zhang F, Jaenisch R (2013) One-step generation of mice carrying mutations in multiple genes by CRISPR/Cas-mediated genome engineering. Cell 153(4):910–918. https://doi.org/10.1016/j.cell.2013.04.025
6. Tait-Burkard C, Doeschl-Wilson A, McGrew MJ, Archibald AL, Sang HM, Houston RD, Whitelaw CB, Watson M (2018) Livestock 2.0 - genome editing for fitter, healthier, and more productive farmed animals. Genome Biol 19(1):204. https://doi.org/10.1186/s13059-018-1583-1
7. Menchaca A, Dos Santos-Neto PC, Mulet AP, Crispo M (2020) CRISPR in livestock: from editing to printing. Theriogenology. https://doi.org/10.1016/j.theriogenology.2020.01.063
8. Menchaca A, Anegon I, Whitelaw CB, Baldassarre H, Crispo M (2016) New insights and current tools for genetically engineered (GE) sheep and goats. Theriogenology. https://doi.org/10.1016/j.theriogenology.2016.04.028
9. Remy S, Tesson L, Menoret S, Usal C, Scharenberg AM, Anegon I (2010) Zinc-finger nucleases: a powerful tool for genetic engineering of animals. Transgenic Res 19(3):363–371
10. Boch J (2011) TALEs of genome targeting. Nat Biotech 29(2):135–136
11. Menchaca A, Dos Santos-Neto PC, Souza-Neves M, Cuadro F, Mulet AP, Tesson L, Chenouard V, Guiffes A, Heslan JM, Gantier M, Anegon I, Crispo M (2020) Otoferlin gene editing in sheep via CRISPR-assisted ssODN-mediated homology directed repair. Sci Rep 10(1):5995. https://doi.org/10.1038/s41598-020-62879-y
12. Vilariño M, Crispo M, dos Santos-Neto PC, Wijma R, Menchaca A (2012) The effect of culture medium changes on in vitro production of sheep embryos. Reproduct Domestic Anim = Zuchthygiene 47:1806

Chapter 13

Targeted Gene Editing in Porcine Germ Cells

Taylor Goldsmith, Alla Bondareva, Dennis Webster, Anna Laura Voigt, Lin Su, Daniel F. Carlson, and Ina Dobrinski

Abstract

As the genetic mutations driving human disease are identified, there is an increasing need for a biomedical model that can accurately represent the disease of interest and provide a platform for potential therapeutic testing. Pigs are a better model for human disease than rodents because of their genetic and physiological similarities to humans. However, current methods to generate porcine models are both technically challenging and expensive. Germline genetic modification through gene edited spermatogonia provides an effective alternative to how these models are developed. Here, we report an improved technique of gene editing in spermatogonia of pigs using CRISPR-Cas9 to generate different edits that reflect the genotypes of human diseases.

Key words CRISPR/Cas9, Spermatogonia, Gene editing, Pig

1 Introduction

Historically, the mouse is considered a foundational animal model across the majority of scientific fields, driving many pharmacological and human disease-related studies. While mouse models offer advantages such as cost-effectiveness and simplicity of care, they often do not fully recapitulate human physiology and in some cases do not even manifest the disease of interest. Pigs, by contrast, are genetically, physiologically, and metabolically more similar to humans than rodent models and are quickly becoming an invaluable platform for the research of human disease [1]. The use of pigs as biomedical models does, however, present a unique set of challenges. Currently, generation of gene edited pigs relies heavily on the inefficient processes of somatic cell nuclear transfer (SCNT) or zygote microinjection or electroporation of genetic editing systems such as zinc finger nucleases (ZFNs), transcription activator-like effector nucleases (TALENs), and clustered regularly interspaced short palindromic repeats and CRISPR-associated-9

Paul John Verma, Huseyin Sumer and Jun Liu (eds.), *Applications of Genome Modulation and Editing*,
Methods in Molecular Biology, vol. 2495, https://doi.org/10.1007/978-1-0716-2301-5_13,

(CRISPR-Cas9) [2–5]. The limitations of these techniques point to the need for an alternative approach, making the genetic editing and subsequent transplantation of spermatogonia a promising candidate for the generation of porcine biomedical models [6, 7].

Spermatogonia is a broad term used to describe a population of male germ cells including cells at earliest stages of differentiation and the undifferentiated unipotent spermatogonial stem cell (SSC) [8]. These cells can be transplanted into a germ cell depleted donor testis and restore full spermatogenesis [9, 10]. Their ability to recolonize and proliferate within the recipient testis niche make spermatogonia ideal targets for genetic editing. Targeted gene knock-out and the correction of disease alleles has been demonstrated in mouse and rat spermatogonia through the use of engineered nucleases and viral-mediated transgenesis, at the expense of spermatogonial viability and off-target effects, respectively [11–14]. In the approach outlined below, we used CRISPR-Cas9 gene editing technology to introduce custom indels and single nucleotide polymorphisms (SNPs) into specific loci pertinent to human disease modeling in porcine spermatogonia. CRISPR-Cas9 complexes were delivered by electroporation in ribonucleoprotein (RNP) format alongside a single strand oligo donor (ssODN) repair template. We demonstrated higher viability post-transfection and increased editing efficiency when using CRISPR-Cas9 compared to TALENs in isolated spermatogonia [14, 15].

2 Materials

All cell culture work and resuspension of oligos was carried out in a biological safety cabinet to prevent contamination. Analysis of editing was done on the bench top. Digestion solutions are filter sterilized prior to use. All tissue culture reagents are warmed to 37 °C prior to use unless otherwise noted.

2.1 Spermatogonial Isolation and Differential Plating

1. 1× Dulbecco's Phosphate-Buffered Saline (DPBS) without calcium and magnesium, with 1× Penicillin-Streptomycin (100 U/mL Penicillin and 100 μg/mL Streptomycin; P/S).
2. High-glucose DMEM medium.
3. Solution A: 20 mg of Sigma Collagenase IV, 40 mg Worthington Collagenase IV in 25 mL of DMEM. 1 mL of FBS added after resuspension.
4. Solution B: 80 mg Worthington Collagenase IV, 40 mg Hyaluronidase in 35 mL of DMEM.
5. 7 mg/mL Deoxyribonuclease I (DNase I) solution in DMEM.
6. DMEM/F12 with 5% FBS and 1× P/S.
7. Sterilized surgical tools: forceps, scissors, scoopula.

8. 0.22 μm filter units and syringe filters.
9. 40 μm cell strainers.
10. 70 μm cell strainers.
11. 0.25% Trypsin-1 mM EDTA solution.
12. 10 cm cell culture plates.
13. Centrifuge.
14. Water bath at 37 °C.
15. Hemocytometer or calibrated electronic cell counter.
16. Cell culture incubator at 37 °C and 5% CO_2.
17. Beaker.
18. Trypan blue viability dye.
19. 50 mL conical tubes.

2.2 Fluorescence-Activated Cell Sorting of Spermatogonia

1. Alpha Mem Advanced with 10 μg/mL holo-transferrin, 10 μg/mL insulin, 1× non-essential amino acids, 1 mM Na-pyruvate, 1% FBS, 10 μM β-mercaptoethanol, 25 ng/mL GFRα1 recombinant rat, 40 ng/mL GDNF recombinant human, 20 ng/mL EGF recombinant human, 1× P/S, 1% BSA, 1× L-glutamine, and 15 mM Hepes.
2. Alpha MEM Advanced with 1% BSA, 15 mM Hepes, 10 μM β-mercaptoethanol, and 1× P/S.
3. Fetal Bovine Serum with 10 μM β-mercaptoethanol.
4. 40 μm filter.
5. 15 mL conical tubes.
6. 130 μm nozzle.
7. Sorter.
8. Centrifuge.
9. 15 mL tube rack.
10. Ice bucket.
11. Trypan blue viability dye.

2.3 CRISPR-Cas9 and Nucleofection Reagents

1. sgRNA or crRNA for gene of interest.
2. tracrRNA.
3. Cas9 protein.
4. Nuclease-free water.
5. DPBS without calcium or magnesium.
6. PCR strip tubes.
7. Thermocycler.
8. Lonza Cell Line Nucleofector Kit V (VCA-1003).

9. Lonza Nucleofector 2b Device.
10. ssODN template for gene of interest.
11. Electroporation enhancer.
12. 6-well tissue culture plates.
13. 0.2% Gelatin in PBS.
14. Spermatogonia Media: αMEM Advanced culture medium, 1% FBS, 0.1% BSA, 1× non-essential amino acids, 1 mM sodium pyruvate, 15 mM HEPES, 2 mM L-glutamine, 10 μM beta-mercaptoethanol, 100 U/mL Penicillin-100 μg/mL Streptomycin, 10 ng/mL glial cell line-derived neurotrophic factor.
15. Trypsin 0.25% EDTA.
16. PBS without calcium and magnesium.

2.4 Detection and Analysis of Editing

1. Lysis buffer: 10 mM Tris–Cl pH 8.0, 2 mM EDTA, 2.5% Tween20, 2.5% Triton-X 100, 100 mg/mL Proteinase K.
2. PCR strip tubes.
3. Thermocycler.
4. Forward and reverse primers flanking region of interest.
5. Hi-Fi Taq polymerase.
6. Surveyor mutation detection kit.
7. Restriction endonuclease(s).
8. Appropriate enzyme buffer.
9. Nuclease-free water.
10. 10% TBE polyacrylamide gel.
11. 10× TBE: 500 mL of distilled water, 121.1 g Tris Base, 55.0 g Boric acid, 40.0 mL of .5 M EDTA.
12. 1× TBE: diluted 10× TBE with ddH_2O.
13. 5× Ficoll Buffer: 2.5 g Ficoll-400, 1 mL 1 M Tris–HCl, pH 7.4, 2 mL 0.5 M EDTA, 20 mL distilled water, 25 mg orange G, and xylene cyanol.
14. Electrophoresis rig.
15. Staining Buffer: 130 mL 1× TBE buffer, 6.5 μL ethidium bromide.
16. Gel imaging machine/software.
17. Adobe Photoshop or similar.
18. Image J.

3 Methods

The workflow is schematically outlined in Fig. 1. Subheadings 3.1 and 3.2 are described in detail in Sakib et al. (2019) [16].

3.1 Spermatogonial Cell Isolation

1. Dissolve the components of Solutions A, B, and DNase I in 25 mL, 35 mL, and 10 mL of DMEM, respectively. Filter-sterilize each solution through a 0.22 μm filter (*see* **Note 1**).

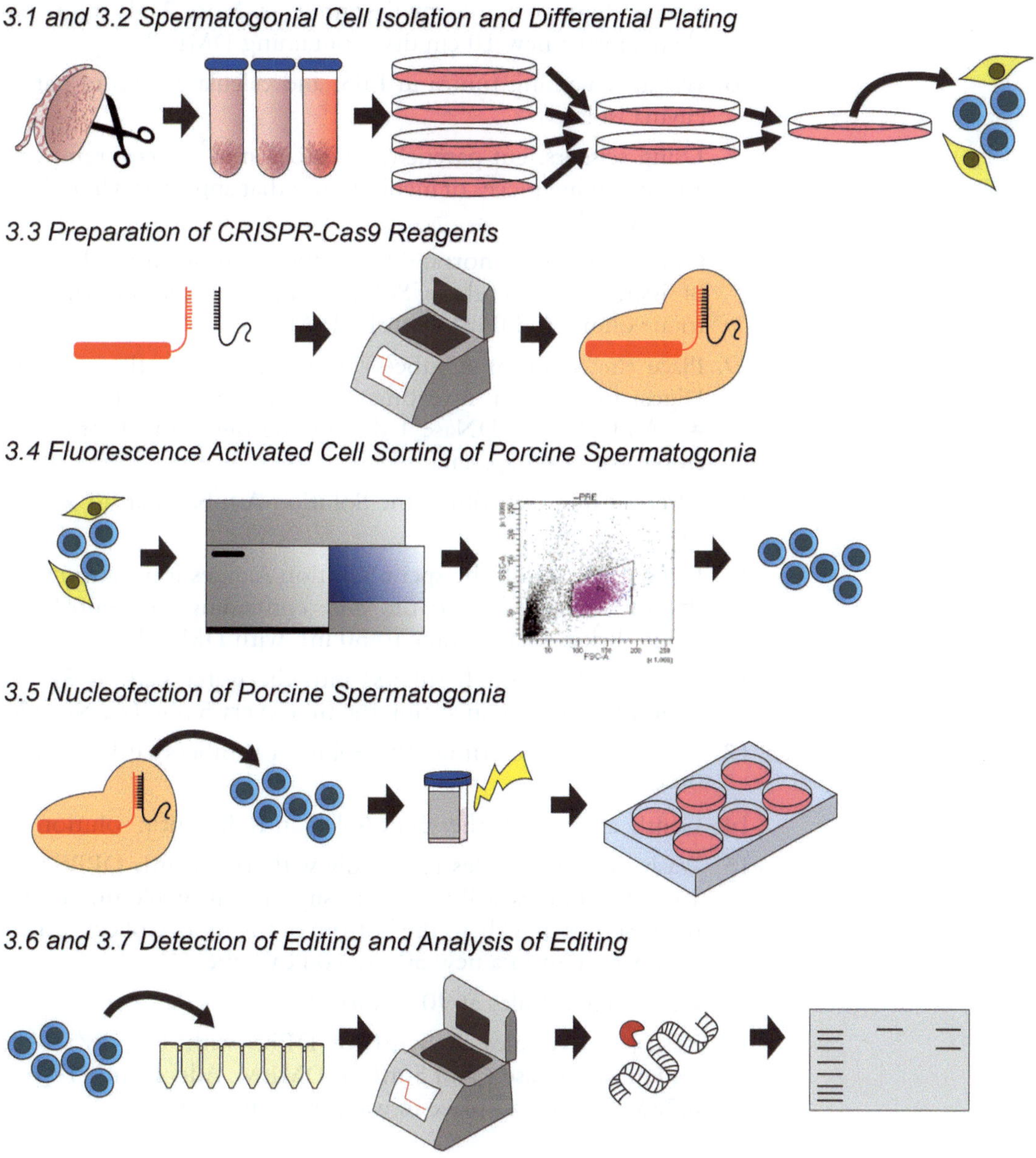

Fig. 1 Schematic representation of the workflow involved in targeted gene editing of porcine spermatogonia

2. Wash testes in a beaker containing DPBS with 1 × P/S. Remove testes and place into a 10 cm dish containing DPBS with 1 × P/S.
3. Using sterile scissors and forceps, remove and discard tunica vaginalis and epididymis. Transfer testes to another 10 cm dish containing DPBS with 1 × P/S.
4. Weigh dish containing isolated testes and record weight.
5. Make a longitudinal cut in the tunica albuginea and using scissors and forceps peel and scrape the parenchyma from the tunica. Remove visible mediastinum (*see* **Note 2**). Place parenchyma into a new 10 cm dish containing DMEM.
6. Weigh remaining tunica in PBS, and calculate mass of parenchyma isolated.
7. Using scissors, cut parenchyma into 1 mm^3 pieces, removing any remaining pieces of mediastinum that appear as white tissue (*see* **Note 3**).
8. Once the tissue is thoroughly chopped, transfer around 7–10 g of tissue into each tube of Solution A using the scoopula. Bring total volume to 50 mL with DMEM.
9. Place the Solution A tube(s) into the water bath at 37 °C. Digest for 30 min, inverting the tube every 5 min (*see* **Note 4**). Add 100 μL DNase I during the digestion if strings or clumps of DNA appear.
10. After 30 min, centrifuge the Solution A tube(s) at 90 × *g* for 90 s.
11. Gently aspirate off the spent Solution A digestion solution and decant Solution B into the tube containing the centrifuged tissue. Bring total volume to 50 mL with DMEM.
12. Place the Solution B tube(s) into the water bath at 37 °C. Digest for 30 min, inverting the tube every 5 min (*see* **Note 5**).
13. After 30 min, centrifuge the Solution B tube(s) at 90 × *g* for 90 s.
14. Gently aspirate off the spent Solution B digestion solution.
15. Wash isolated tubules repeatedly with 10–15 mL DPBS with 1×P/S. Tubules will remain in supernatant while undigested tissue settles quickly to the bottom of the tube. Collect tubules in supernatant in a new 50 mL conical tube.
16. Centrifuge tubules at 90 × *g* for 90 s.
17. Resuspend tubules in 5 mL of DPBS with 1 × P/S. Add 500 μL of DNase I and 5 mL of DMEM. Add 15 mL Trypsin 0.25% EDTA, and place in water bath at 37 °C.

18. Check progress of digestion every 5 min by removing a 10 μL of solution and examining under microscope (*see* **Notes 6** and 7).
19. Once the digestion is approaching finality, strain solution through a 70 μm mesh filter into 5 mL of FBS.
20. Filter once more through a 40 μm mesh filter.
21. Centrifuge isolated cells at 500 × *g* for 5 min. Once finished decant and discard solution and resuspend cells in 50 mL of DMEM.
22. Count viable cells using hemocytometer and trypan blue exclusion.

3.2 Differential Plating

1. Plate 25 million cells per 10 cm tissue culture dish, in a final media volume of 8 mL of DMEM/F12 with 5% FBS and 1×P/S.
2. Allow cells to sit in a cell culture incubator at 37 °C and 5% for 1.5 h (*see* **Note 8**).
3. Consolidate the supernatant from two plates into one new 10 cm dish.
4. Allow cells to sit in a cell culture incubator at 37 °C and 5% CO_2 for 1 h.
5. Consolidate the supernatant from two plates once more into one new 10 cm dish.
6. Allow cells to sit in a cell culture incubator at 37 °C and 5% CO_2 overnight.
7. The following day collect two fractions of cells: Non-adherent fraction of cells in the supernatant and the slightly adhered fraction. Collect the former by just aspirating the supernatant into a 50 mL conical tube. Collect the latter first by using a 1: 20 dilution of 0.25% Trypsin-EDTA in PBS for 5 min at room temperature. Collect the cells that detach into a new 50 mL conical containing 5 mL FBS to stop the trypsin. Then followed by a 1:5 dilution of 0.25% Trypsin-EDTA in PBS for 2.5 min at 37 °C and 2.5 min at room temperature. Collect the detached cells same as mentioned above.
8. Centrifuge both fractions at 500 × *g* for 5 min, resuspend in PBS, and count using a hemocytometer and trypan blue.
9. Move forward with the fraction that has the highest percentage of spermatogonia, or combine the fractions if necessary, to achieve the appropriate cell number for transfection.

3.3 Preparation of CRISPR-Cas9 Reagents

For CRISPR-Cas9 experiments, either a sgRNA or a crRNA: tracrRNA complex can be used. If using sgRNA, skip **steps 1** and **2**. The following CRISPR design tool was used to create the guide RNAs used: http://www.rgenome.net/cas-designer/

1. To create the crRNA:tracrRNA complex, combine equimolar concentrations of crRNA for the locus of interest with tracrRNA per reaction in PCR strip tubes. In this experiment, we used 200 pmol.
2. Using a thermal cycler, heat the reaction components at 95 °C. Then cool the reaction to 22 °C at a ramp rate of 0.1 °C/s.
3. Incubate complex or sgRNA with 175 pmol of Cas9 protein and 3.5 μL of DPBS at room temperature for 15 min. If the complex is not being used immediately, store on ice. Total volume is 8.33 μL.

3.4 Fluorescence-Activated Cell Sorting of Porcine Spermatogonia

1. Prepare FACS collection tubes with 3 mL of fetal bovine serum with 10 μM β-mercaptoethanol.
2. Place rack for collection tubes on ice (*see* **Note 9**).
3. Filter cells through 40 μm filter.
4. Resuspend spermatogonia to a concentration of 5×10^6/mL in Alpha Mem Advanced with 15 mM Hepes, 1% BSA, 10 μM β-mercaptoethanol, and 1× P/S in 15 mL tube.
5. Set up sorter with 2.0 ND and a 130 μm nozzle with 10 psi at lowest flow rate.
6. Place tube with enriched cells into sample collection chamber at room temperature and set up sample spin to 200 rpm.
7. Open sorter software workspace with FSC-A/SSC-A dot plot, SSC-A/SSC-H dot plot, FSC-A/FSC-H dot plot, and GFP-A PERCP-Cy5–5-A dot plot.
8. Adjust laser intensity for FSC and SSC to receive cell population cloud appearance as shown.
9. Gate "germ cell cloud" according to distinct FSC-A/SSC-A tightly (Fig. 1, P1).
10. Remove doublets from sort from cell population P1 by gating out events falling under a linear line in an FSC-A/FSH-H dot plot and create gate P2.
11. Remove doublets from sort from cell population P2 by gating out events falling under a linear line in SSC-A/SSC-H dot plot and create gate P3.
12. Gate out dying cells from sort by plotting events on GFP-A/PERCP-CY5–5-A dot plot and gating out events that appear on a linear diagonal line in the plot, create gate P4.
13. Place 15 mL collection tubes with filled with 3 mL FBS with 10 μM β-mercaptoethanol into sample collection chamber.
14. Place ice packs into sample collection chamber.
15. Sort cells at lowest flow rate out of gate P4 in 15 mL collection tube filled with 3 mL FBS with 10 μM β-mercaptoethanol.

16. After obtaining required cell number spin down cells at 500G for 5 min immediately or store short term on the rack on ice.
17. Test purity of post sort cell population with running a sample in FACS.
18. Optional: Assess viability with trypan blue.
19. Resuspend cells at required concentration in Alpha Mem Advanced with 10 μg/mL holo-transferrin, 10 μg/mL insulin, 1× non-essential amino acids, 1 mM Na-pyruvate, 1% FBS, 10 μM β-mercaptoethanol, 25 ng/mL GFR*α*1 recombinant rat, 40 ng/mL GDNF recombinant human, 20 ng/mL EGF recombinant human, 1× P/S, 1%BSA, 2× L-glutamine, and 15 mM Hepes.

3.5 Nucleofection of Porcine Spermatogonia

Prior to nucleofection, coat 6-well tissue culture plates with 0.2% gelatin for 1–3 h at room temperature. Aspirate off gelatin and allow plates to dry. Once dry, add spermatogonial culture medium and allow to equilibrate prior to transfection in a cell culture incubator set to 37 °C and 5% CO_2. Allow nucleofection reagents to come to room temperature and set aside appropriate number of cuvettes and transfer pipettes for the reactions.

1. Centrifuge isolated spermatogonia in 15 mL tubes at one million cells per tube for 5 min at 500 × *g*.
2. While cells are centrifuging, mix an appropriate volume of Lonza nucleofection solution V and supplement 1. For one reaction, 90 μL of solution V and 20 μL of supplement 1 are used.
3. Gently aspirate off PBS wash.
4. Resuspend the cell pellets in 100 μL of the transfection solution just before transfection. Gently pipette the cells up and down a few times (*see* **Note 10**).
5. Immediately add 168 pmol of ssODN repair template (if using homology-directed repair) and 0.1 nmol electroporation enhancer to the RNP complex. Skip this step if you are not using a repair template.
6. Directly add the RNP complex (with ssODN if applicable) to the cells in transfection solution.
7. Gently pipette up and down and transfer solution to the Lonza nucleofection cuvettes. Gently tap the cuvette to ensure the solution has reached the bottom of the cuvette. Avoid introducing air bubbles.
8. Insert cuvette into the Amaxa nucleofector machine, and electroporate on setting X-005.
9. Remove cuvettes and collect 6-well plates that were equilibrated. Using the provided Lonza transfer pipettes, add

0.5 mL of prewarmed media to the nucleofection cuvettes, then aspirate the transfected cells gently out of the cuvette (*see* **Note 11**).

10. Transfer cells to the appropriate wells and return the plate to the incubator and leave cells overnight.
11. The following morning, change media by collecting the media from the wells and centrifuging for 5 min at 500 × *g*. Add 1 mL of fresh media back to the wells during centrifugation to ensure slightly adhered cells do not desiccate. Aspirate and resuspend pellet in 0.5 mL fresh media and add back to the wells.
12. Change half of the media every other day following transfection by following **step 11**.
13. Five days after transfection, collect non-adherent and slightly adhered cells with 1:5 diluted 0.25% Trypsin-EDTA in DPBS for 5 min at room temperature. Stop the trypsin reaction with 500 μL of FBS per 1 mL of trypsin.
14. Centrifuge cells and then wash once with DPBS. Proceed to analysis of editing. If analysis is not occurring right away, store cell pellets at −80 °C.

3.6 Detection of Editing

1. Lyse cell pellets using 50–100 μL of lysis buffer and move lysis over to PCR strip tubes.
2. Place in thermocycler and incubate at 50 °C for 60 min and 95 °C for 15 min.
3. Using primers that were created to flank the gRNA target site, amplify the region of interest using Hi-Fi Taq polymerase (*see* **Note 12**).
4. To examine the frequency of mutations caused by non-homologous end joining (NHEJ), use 10 μL of the PCR product in a Surveyor mutation detection reaction.
5. To examine the frequency of mutations caused by homology-directed repair (HDR), digest 10 μL of PCR product with required restriction endonuclease in appropriate enzyme buffer at temperature and time dictated by the endonuclease (*see* **Note 13**).
6. Remove and place a 10% TBE polyacrylamide gel into an electrophoresis rig, and fill the rig with 1× TBE. Place the rig on ice or in a cold room.
7. Add 5× Ficoll to the reactions to a final 1× concentration to ensure they sink to the bottom of the gel wells.
8. Load 3 μL of an appropriate ladder for the product size.
9. Load 6 μL of a Surveyor reaction and/or 10 μL of an endonuclease reaction.

10. Run the gel per manufacturer's recommendations to ensure sufficient band separation. In these experiments 200 V for 1 h was used.
11. Once the time is up, remove the gel from the gel mold by cracking it with a gel removal tool and a razor blade.
12. Fully submerge gel in staining buffer in appropriate container.
13. Place on a rocker and stain for 10 min at room temperature.
14. Dispose of ethidium bromide solution and wash three times in DI water (*see* **Note 14**).
15. Image gel at an exposure time less than 2 s as not to over expose the images.

3.7 Analysis of Editing

For more information on the theory behind analysis see Guschin et al. (2010) [17].

1. Open the image of the gel in photoshop or a similar editing program.
2. Remove the color of the image, and then invert the black and white image.
3. Change the Shadow input level to 100.
4. Save the image as a .jpg.
5. Open the editing image in Image J.
6. Using the subtract background function, set pixels to 20.
7. Select the rectangle tool and draw a box around your first lane of interest. Press the "1" key. Make sure to include all bands in that lane.
8. Click and drag the rectangle to your next lane of interest, and press the "2" key. Continue this until you reach your last lane. On the last lane, press the "3" key.
9. Another window will appear containing the graphs of the band densities. The left peak represents the density of the wildtype band, while other peaks represent other products from either the NHEJ or HDR.
10. Select the line tool, and while holding "Shift" drag a line across the graph, starting at the lower left of the wildtype peak, where the line begins to ascend.
11. Select the wand tool, and click on the area under the peak, above the recently drawn line. Do this for all peaks.
12. Another window will appear containing the area of the selection. Copy these numbers into an Excel file and label the numbers with their corresponding sample names and which band they represent (WT, edited, etc.).

13. For Surveyor reactions, calculate the percent editing using the following formula:

$$\%\text{Editing} = 100 \times \left(1 - \left(1 - \left(\frac{\Sigma \text{ edited bands}}{\Sigma \text{ all bands}}\right)^{0.5}\right)\right)$$

For endonuclease reactions, calculate the HDR rate using the following formula:

$$\%\text{HDR} = ((\Sigma \text{ edited bands})/(\Sigma \text{ all bands})) \times 100$$

4 Notes

1. It is recommended to only filter-sterilize 50 mL at a time as larger volume filters can get clogged by the enzymes.
2. It is easiest to hold the tunica with forceps and pull as you scrape along the tunica with scissors to collect the parenchyma.
3. It is important to remove as much as the mediastinum as possible. Because it is mainly connective tissue, it can use up much of the enzyme in solution, making the digestion of the parenchyma less efficient.
4. If floating DNA is noticed at this step, up to 500 μL of DNase I can be added to the digestion solution.
5. As per **Note 4**, DNase I can also be added at this step if required.
6. DNA will be noticed upon trypsinization. Add 1 mL of DNase I along with 3 mL of DMEM media without FBS. More DNase I can be added throughout this step if required.
7. At the beginning of the digestion, an aliquot of digestion solution will appear mainly as tubules. As the digestion progresses, the tubules break apart into single cells. Ideally, the digestion should be stopped when the majority of cells are in single cell format to avoid over-digestion. One can also triturate during this step to help break up tubules faster.
8. This step is effective because the somatic cells of the testis will adhere rapidly to a tissue culture dish and will remain adhered while the spermatogonial remain in suspension.
9. A cell population sorted for subsequent culture should be maintained at room temperature to avoid stress caused by drastic temperature changes. Cells sorted for RNA extraction however, should be sorted at 4 °C and kept on ice until further processing.
10. Cells should not remain in the nucleofection solution for longer than 15 min. It is recommended to only do 5–6 transfections at once to limit the time spent in this solution.

11. When aspirating the cells out of the cuvette, avoid taking the "popcorn" like bubbles/DNA that accumulate near the top of the solution in the cuvette.
12. Hi-Fi Taq polymerase is used so that there is a lower likelihood of mismatches being introduced by the polymerase, as these would be detected by the Surveyor assay and provide a false result.
13. The restriction site should be designed as part of the ssODN that was introduced upon transfection. This site should be specific to the edit.
14. Ensure to wash the gel well after ethidium bromide, as failure to do so can result in high background when imaging the gel.

Acknowledgments

This work was supported by NIH ORIP 9 R01 OD016575-17A1.

References

1. Aigner B, Renner S, Kessler B, Klymiuk N, Kurome M, Wünsch A et al (2010) Transgenic pigs as models for translational biomedical research. J Mol Med 88(7):653–664
2. Tan WS, Carlson DF, Walton MW, Fahrenkrug SC, Hackett PB (2012) Precision editing large animal genomes. Adv Genet 80:37–97
3. De Sousa PA, King T, Harkness L, Young LE, Walker SK, Wilmut I (2001) Evaluation of gestational deficiencies in cloned sheep fetuses and placentae. Biol Reprod 65:23–30
4. Bonk AJ, Li R, Lai L, Ho Y, Liu Z, Samuel M et al (2008) Aberrant DNA methylation in porcine in vitro-, parthenogenetic-, and somatic cell nuclear transfer-produced blastocysts. Mol Reprod Dev 75:250–264
5. Tian XC, Park J, Bruno R, French R, Jiang L, Prather RS (2009) Altered gene expression in cloned piglets. Reprod Fertil Dev 21:60–66
6. Hamra FK, Gatlin J, Chapman KM, Grellhesl DM, Garcia JV, Hammer RE, Garbers DL (2002) Production of transgenic rats by lentiviral transduction of male germ-line stem cells. Proc Natl Acad Sci U S A 99:14931–14936
7. Orwig KE, Avarbock MR, Brinster RL (2002) Retrovirus-mediated modification of male germline stem cells in rats. Biol Reprod 67: 874–879
8. De Rooij DG (2001) Proliferation and differentiation of spermatogonial stem cells. Reproduction 121:347–354
9. Brinster RL, Zimmermann JW (1994) Spermatogenesis following male germ-cell transplantation. Proc Natl Acad Sci U S A 91: 11298–11302
10. Honaramooz A, Megee SO, Dobrinski I (2002) Germ cell transplantation in pigs. Biol Reprod 66(1):21–28
11. Chapman KM, Medrano GA, Jaichander P, Chaudhary J, Waits AE, Nobrega MA et al (2015) Targeted germline modifications in rats using CRISPR/Cas9 and spermatogonial stem cells. Cell Rep 10:1828–1835
12. Sato T, Sakuma T, Yokonishi T, Katagiri K, Kamimura S, Ogonuki N et al (2015) Genome editing in mouse spermatogonial stem cell lines using TALEN and double-nicking CRISPR/Cas9. Stem Cell Rep 5:75–82
13. Zeng W, Tang L, Bondareva A, Honaramooz A, Tanco V, Dores C et al (2013) Viral transduction of male germline stem cells results in transgene transmission after germ cell transplantation in pigs. Biol Reprod 88(1):27. 1–9
14. Tang L, Bondareva A, Gonzalez R, Rodriguez-Sosa JR, Carlson DF, Webster D et al (2018) TALEN-mediated gene targeting in porcine spermatogonia. Mol Reprod Dev 85:250–261
15. Webster D, Bondareva A, Solin S, Goldsmith T, Su L, De Lima e Martins Lara N et al (2021) Targeted gene editing in porcine spermatogonia. Front Genet 11:27673

16. Sakib S, Yu Y, Voigt A, Ungrin M, Dobrinski I (2019) Generation of porcine testicular organoids with testis specific architecture using microwell culture. J Vis Exp (152). https://doi.org/10.3791/60387

17. Guschin DY, Waite AJ, Katibah GE, Miller JC, Holmes MC, Rebar EJ (2010) A rapid and general assay for monitoring endogenous gene modification. Meth Mol Biol 649:247–256

Chapter 14

Generating a Heat-Tolerance Mouse Model

Jun Liu and Paul J. Verma

Abstract

Creating mouse models of human genetic disease (Gurumurthy and Lloyd, Dis Models Mech 12(1): dmm029462, 2019) and livestock trait (Schering et al. Arch Physiol Biochem 121(5):194–205, 2015; Habiela et al. J Gen Virol 95 (Pt 11):2329–2345, 2014) have been proven to be a useful tool for understanding the mechanism behind the phenotypes and fundamental and applied research in livestock. A single base pair deletion of prolactin receptor (*PRLR*) has an impact on hair morphology phenotypes beyond its classical roles in lactation in cattle, the so-called slick cattle (Littlejohn et al. Nat Commun 5: 5861, 2014). Here, we generate a knock-in mouse model by targeting the specific locus of *PRLR* gene using Cas9-mediated genome editing via homology-directed repair (HDR) in mouse zygotes. The mouse model carrying the identical *PRLR* mutation in slick cattle may provide a useful animal model to study the pathway of thermoregulation and the mechanism of heat-tolerance in the livestock.

Key words CRISPR, Genome editing, Prolactin receptor, Zygote microinjection, Homology-directed repair, Thermoregulation, Heat-tolerance

1 Introduction

A single homozygous frameshift mutation was identified in purebred Senepol sires, consisting of a single base deletion of chr20: 39136558GC>G (UMD3.1) in exon 10 of prolactin receptor (*PRLR*) gene that frameshifts the long form of PRLR (NM_001039726.2) which corresponds to an p.Ala461Val substitution. Subsequently, a premature stop codon (p.Leu462*) and loss of 120 C-terminal amino acids were introduced [4]. Association analysis suggested that the *PRLR* p.Leu462* mutation in four Senepol sires was responsible for the slick-coat phenotype in the cow. Studies also suggested a correlation of sweating rate with coat length in cow [5, 6], and an impact on hair follicle growth cycles in prolactin receptor knockout mice [7]. Therefore, the impact of the *PRLR* p.Leu642* mutation on the heat-tolerance is of great industrial significance in cow as well as other livestock as a whole. In this project, the objective was to generate a mouse *PRLR* mutation

Paul John Verma, Huseyin Sumer and Jun Liu (eds.), *Applications of Genome Modulation and Editing*, Methods in Molecular Biology, vol. 2495, https://doi.org/10.1007/978-1-0716-2301-5_14,

model carrying the equivalent mutation in cow for a further assessment animal behavior and phenotypes under a hot environment. We first aligned mouse and cow PRLR protein sequences and generate a frameshift mutation to replace Leu450 with a stop codon (TAA, p.Leu450*) in mouse *PRLR* gene using gene editing strategies (Fig. 1), [8, 9].

2 Materials

2.1 Animal

1. C57BL/6, age 3–4 weeks, for zygote collection.
2. F1 (C57BL/6 × CBA) mice for pseudo-pregnant mother and vasectomized males.

2.2 Plasmids and Oligos

1. pX330-Cas9 plasmid (Addgene), used for construction of pX330-PRLR-sgRNA vector.
2. A small guide RNA (sgRNA) sequence (CTAAAGTTGTCT GAAGATGC) is designed (*see* **Note 1**) using an online CRISPR Design Tool (https://zlab.bio/guide-design-resources) (*see* **Note 2**). Two oligonucleotides (PRLR-sgRNA_top and PRLR-sgRNA_bottom) used for cloning sgRNA sequence into expression vector pX330 are listed in Table 1.
3. The constructed bicistronic expression vector pX330 containing the *PRLR*-sgRNA is used for IVT of the PRLR-sgRNA.
4. Custom-designed single-stranded DNA (ssDNA) as a donor for point mutation is listed in Table 1 (*see* **Note 3**).
5. Gene-specific primers PRLR_F and PRLR_R for genotyping are listed in Table 1 (*see* **Note 4**).

2.3 Reagents and Medium

1. Order two synthesized oligos PRLR-sgRNA_top and PRLR-sgRNA_bottom. The sequences are listed in Table 1.
2. Order primers PRLR_F and PRLR_R. The sequences are listed in Table 1.
3. Preparation of Cas9 mRNA: resuspend and dilute Cas9 mRNA in nuclease-free water at a final concentration of 500 ng/μL (*see* **Note 5**). Store the stock at −80 °C.
4. Preparation of ssDNA stock: resuspend and dilute ssDNA in nuclease-free water at a final concentration of 1 μg/μL. Store the stock at −20 °C for at least 3 months.
5. LB medium: add 20 g of LB powder to 1 L of distilled water. Autoclave and store the medium at room temperature.
6. LB-Ampicillin plates: Add 7.5 g of agar into 500 mL of LB medium, autoclave, and store it at 4 °C. Liquefy the medium by boiling it in a microwave oven, add ampicillin stock solution to

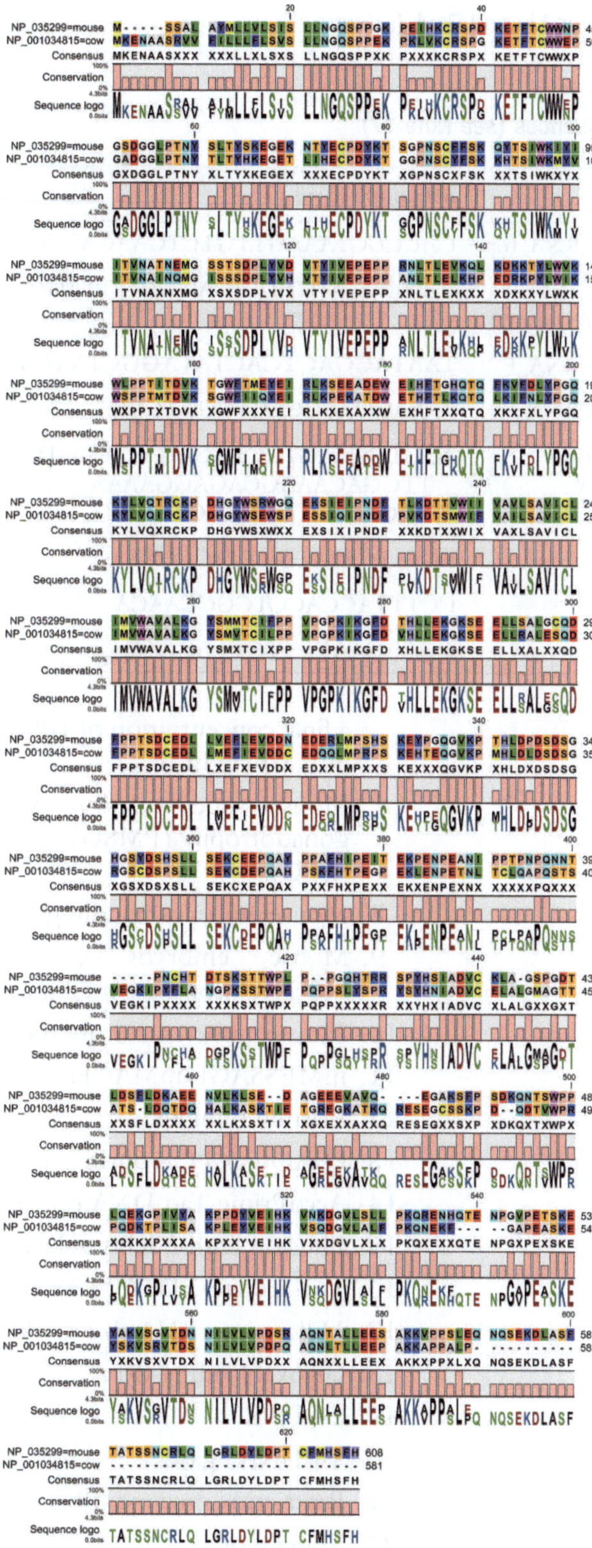

Fig. 1 Cow-mouse PRLR protein sequence alignment using Qiagen CLC Main workbench software. Alignment analysis indicates the equivalent residue for a stop codon is Leu450

Table 1
Oligo sequences (*see* Note 17)

Name	Sequence (5′–3′)
PRLR-sgRNA_top	*CACCG*CTAAAGTTGTCTGAAGATGC
PRLR-sgRNA_bottom	*AAAC*GCATCTTCAGACAACTTTAG*C*
PRLR-sgRNA_F	**TAATACGACTCACTATAGG**CTAAAGTTGTCTGAAGATGC
PRLR-sgRNA_R	AAAAGCACCGACTCGGTGCC
PRLR donor ssDNA	AGCTGGAAGTCCTGGAGATACACTGGACTCTTTC TTGGACAAAGCAGAGGAAAATGTTTAAAAGTTGTCTGAAGATGC TGGAGAGGAAGAAGTGGCTGTGCAAGAAGGGGCCAAAAGCT
hU6 seq_F	ACTATCATATGCTTACCGTAAC
PRLR_F	CCTTTACCACCTGGCCAACA
PRLR_R	GGTGGAACCTTCTTGGCTGA

a final concentration of 100 μg/mL to the medium and pour it into 10-cm dishes.

7. Mouse superovulation hormones: Pregnant mare serum gonadotropin (PMSG) and human chorionic gonadotropin (hCG).
8. Mouse embryos handling medium: M2 medium.
9. Mouse embryos cumulus cell removing medium: M2-Hyaluronidase medium.
10. Mouse embryos culture medium: KSOM medium.
11. mMESSAGE mMACHINE T7 ULTRA kit for IVT.
12. Plasmid mini kit.
13. QIAQuick PCR purification kit.
14. AccuPrime Taq DNA polymerase (high fidelity).
15. Phusion high-fidelity DNA polymerase.
16. Tris-acetate-EDTA (TAE) buffer.
17. Millennium RNA size marker.
18. Glyoxal loading dye.
19. T4 PNK.
20. T4 DNA ligase.
21. Restriction enzyme: BbsI and buffer.
22. Mineral oil.
23. Nuclease-free water.

2.4 Equipment

1. Microinjection holding pipette.
2. Microinjection injection pipette.
3. Transfer pipette.
4. Falcon 10-cm dishes, bottoms are suitable for oocyte/embryo collection and lids are suitable for micromanipulation.
5. Falcon 6-cm dishes used for embryo culture.
6. 26-G needles.
7. Surgical instruments: fine forceps, serrefine clamp, fine iris scissors, needle holder, suture needles, suture silk, wound clips, and applicator.
8. Inverted microscope with Hoffman optics.
9. Micromanipulator set.
10. Microinjector device.
11. Stereomicroscope.
12. 5% CO_2 incubator.

3 Methods

3.1 Generation of pX330-sgRNAs Expressing Vector

Digest bicistronic expression vector pX330 with BbsI, treat the digested product with Antarctic Phosphatase, and then purify the linearized pX330 vector. Anneal the pair of PRLR-sgRNA_top and PRLR-sgRNA_bottom oligos, phosphorylate the annealed oligos, and ligate it with the linearized pX330 vector.

1. Order two synthesized oligos PRLR-sgRNA_top and PRLR-sgRNA_bottom from IDT
2. Phosphorylate and anneal the two oligos

1.0 μL	PRLR-sgRNA_top oligo (100 μM)
1.0 μL	PRLR-sgRNA_bottom oligo (100 μM)
1.0 μL	10× T4 ligation buffer
6.5 μL	nuclease-free H_2O
0.5 μL	T4 PNK
10.0 μL	Total volume

Anneal in thermocycler using the following parameters:

37 °C	30 min
95 °C	5 min and then ramp down to 25 °C at rate of 5 °C/min
	Dilute the annealed oligo 1:250 (250-fold).

3. Setup digestion-ligation reaction

2.0 μL	pX330 backbone vector (100 ng)
2.0 μL	phosphorylated and annealed oligo duplex (1:250 dilution)
2.0 μL	10× FastDigest Buffer
1.0 μL	DTT (10mM to a final concentration of 1mM)
1.0 μL	ATP (10mM to a final concentration of 1mM)
1.0 μL	FastDigest BbsI
0.5 μL	T7 DNA ligase
10.5 μL	nuclease-free H_2O
20.0 μL	Total volume

Incubate the ligation reaction in a thermocycler:

37 °C	5 min
23 °C	5 min

Cycle the two steps for 6 cycles (total running time 1 h)

4 °C	hold until ready to proceed

4. Treat ligation reaction with exonuclease to prevent unwanted recombination products.

11.0 μL	ligation reaction from **step 3**
1.5 μL	10× PlasmidSafe buffer
1.5 μL	10 mM ATP
1.0 μL	PlasmidSafe exonuclease
15.0 μL	Total volume

Incubate reaction at 37 °C for 30 min.

5. Transformation with 1–2 μL of the final product into competent cells.
6. Pickup colonies for mini prep and sequencing using human U6 promotor primer (Table 1) to verify the clones.

3.2 PCR Amplification of PRLR-sgRNA Sequence

T7 promoter sequence is added to sgRNAs PCR product by PCR amplification using primers PRLR-sgRNA_F and PRLR-sgRNA_R.

1. PCR setup

2.0 μL	pX330-sgRNA (5ng/μL to a final amount of 10 ng)
10.0 μL	5× GC buffer
1.5 μL	DMSO
2.0 μL	primer PRLR-sgRNA_F (to final concentration of 0.2 μM)
2.0 μL	primer PRLR-sgRNA_R (to final concentration of 0.2 μM)
2.0 μL	dNTP
0.5 μL	Phusion high-fidelity DNA polymerase
30.0 μL	nuclease-free H_2O
50.0 μL	Total volume

2. Perform PCR in a thermocycler with one initiation denaturation cycle of 98 °C for 5 min, 35 cycles of 98 °C for 30 s, annealing temperature of 60 °C for 30 s, and 72 °C for 30 s, and a final extension step of 72 °C for 5 min.

3.3 PCR Product Purification and IVT of T7-PRLR-sgRNA

T7-PRLR-sgRNA PCR products are purified and used as the template for IVT using mMESSAGE mMACHINE T7 ULTRA kit.

1. Run the PCR product (~100 bp) on a 2% agarose gel.
2. Gel-purify the T7-PRLR-sgRNA PCR product using QiaQuick PCR purification kit (*see* **Note 6**).
3. Spectrum the purified PCR product.
4. IVT to generate PRLR-sgRNA mRNA using mMESSAGE mMACHINE T7 ULTRA kit with the purified PCR products as templates. (*see* **Note** 7).
5. IVT reaction setup:

10 μL	T7 2× NTP/Arca
2.0 μL	10× T7 buffer
5.0 μL	Purified T7-PRLR-sgRNA PCR products (0.2 μg)
2.0 μL	T7 Enzyme mix
1.0 μL	nuclease-free H_2O
20.0 μL	Total volume

6. Incubate the IVT reaction at 37 °C for 2 h.
7. Purify the IVT mRNA using the MEGAclear kit and elute it with 50 μL of nuclease-free water at room temperature (*see* **Note 8**).

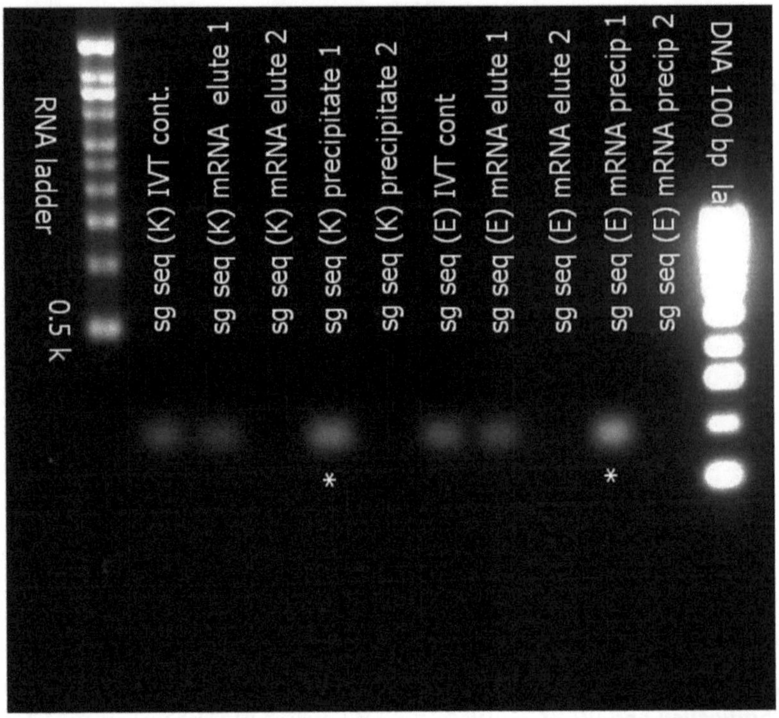

Fig. 2 Agarose gel image of electrophoresis of PRLR-sgRNA (bands with *) after in vitro transcription (IVT) and purification using the MEGAclear kit (K) or ethanol precipitation (E). The products with * are used in the next embryo microinjection experiments

8. Dilute the PRLR-sgRNA mRNA to 250 ng/μL in nuclease-free water and check its quality on a 2% agarose gel (Fig. 2).

2.5 μL	mRNA sample
7.5 μL	Formaldehyde load dye
Incubate at 75 °C for 10 min.	
Run 2% agarose gel in NortherMax gel buffer.	

9. Store the mRNA stock at −80 °C for up to 3 months.

3.4 Mouse zygote Preparation

1. Inject 12–15 C57BL/6 (3–4 weeks old) mice with PMSG (5 IU) at 3:00–4:00 pm.
2. After 48 h, inject the mice with hCG (5 IU), and house the females with C57BL/6 male mice overnight.
3. In the morning of next day, pick up female mice with vagina plugs into a new cage.
4. Prepare the KSOM medium for embryo culture. Place several drops (30–50 μL/drop) on 6-cm dishes covered with mineral oil and incubate in a 37 °C incubator for at least 30 min before use.

5. At 20–21 h after hCG injection, euthanize the vagina-plugged mice and collect zygote-cumulus complexes from oviducts and transfer to M2 medium.
6. Transfer the zygote-cumulus complexes into M2-+Hyaluronidase medium to remove cumulus cells and then transfer the zygotes into KSOM medium and incubate at 37 °C in a 5% CO_2 incubator.

3.5 Microinjection and Embryo Transfer of Mouse Zygotes

1. Prepare 20 μL of CRISPR reagent mix (20 ng/μL Cas9 mRNA + 10 ng/μL PRLR-sgRNA + 10 ng/μL ssDNA donor oligos) (*see* **Note 9**).

2.0 μL	Cas9 mRNA stock
2.0 μL	PRLR-sgRNA mRNA stock
0.5 μL	ssDNA stock
15.5 μL	nuclease-free H_2O
20.0 μL	Total volume

2. Inject CRISPR reagent mix into male pronucleus of each zygote. Approximately 2–5 pico liters was injected into the pronucleus (*see* **Note 10**).
3. Culture the injected zygotes in KSOM medium at 37 °C in a 5% CO_2 incubator till develop to blastocyst stage embryos.
4. Prepare pseudo-pregnant foster mice by mating estrous Fl (C57BL/6 × CBA) female mice with vasectomized male mice at the same day of embryo injection.
5. Transfer the blastocyst stage embryos into uteri of day-3.5 post-coitus (*dpc*) pseudo-pregnant recipient mice (*see* **Note 11**).
6. Separate male and female offspring into different cages around 3 weeks after birth.

3.6 Genotyping of Founder Mice

1. Tattoo the offspring and collect tail tips.
2. Extract genomic DNA from tail tip samples using DNeasy kit (Qiagen) (*see* **Note 12**).
3. Setup PCR reactions:

5.0 μL	gDNA (final amount of ~500 ng)
5.0 μL	10× buffer II
1.0 μL	primer PRLR_F (to final concentration of 0.2 μM)
1.0 μL	primer PRLR_R (to final concentration of 0.2 μM)
0.5 μL	AccuPrime H-F Taq DNA polymerase (*see* **Note 13**)
37.5 μL	nuclease-free H_2O
50.0 μL	Total volume

4. Perform PCR in a thermocycler with one initiation denaturation cycle of 95 °C for 2 min, 35 cycles of 95 °C for 30 s, annealing temperature of 60°C for 30 s, and 68°C for 40 s, and a final extension step of 68°C for 5 min.
5. Purify the PCR products using Qiaquick PCR purification kit.

3.7 T7E1 Assay

T7 endonuclease I (T7E1) assay is to detect non-homologous end joining (NHEJ) mutations, which result in random insertion or deletion at the double-strand break (DSB).

1. Setup assay:

10 μL	PCR products
2 μL	NEB buffer 2
7 μL	nuclease-free H_2O
19 μL	Total volume

2. Perform the assay in a thermocycler with initiation denaturation at 95 °C for 5 min, then ramp down from 95 °C to 85 °C at rate of −1.5 °C/s, then ramp down from 85 °C to 25 °C at rate of −0.1 °C/s, then cool down to 4 °C.
3. Add 1 μL of T7 endonuclease I.
4. Perform incubation of the reaction at 37 °C for 20 min.
5. Run 2% agarose gel electrophoresis (Fig. 3).

3.8 pGEM-T Assay, Transformation and Sanger Sequencing

1. Setup reaction

5 μL	2× T4 ligation buffer
1 μL	pGEM-T Easy vector
1 μL	purified PCR products from above reaction (*see* **Note 14**)
1 μL	T4 ligase
2 μL	nuclease-free H_2O
10 μL	Total volume

2. Incubate the reaction at room temperature for 1 h.
3. Setup transformation reaction: add 5 μL of pGEM-T ligation from the above step to 50 μL of DH5α competent cells; incubate 10 min on ice; Heat shock in water bath at 42 °C for 1 min, and then incubate 3 min on ice; add 950 μL of SOC medium, and then incubate at 37 °C for 60 min in a shaking incubator with shaking rate of 225 rpm.

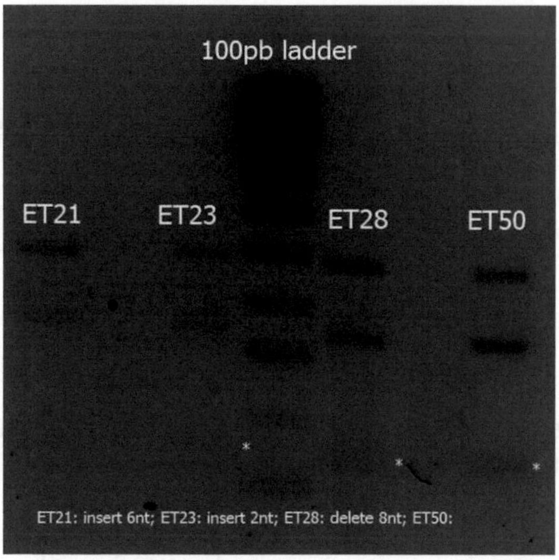

Fig. 3 Agarose gel image of electrophoresis of T7E1 assay. Three founder mice (ET23, ET28, and ET50) were analyzed. In each assay, three bands can be identified: the full length of PCR product (the top band), and two digested products (the middle band and the bottom band labeled with *)

4. Add 40 μL of 100mM IPTG and 120 μL of 20 mg/mL X-Gal to the surface of each LB agar (with 100 μg/mL ampicillin) plate and spread over entire surface. Dry the surface at room temperature.
5. After 1 h shaking culture, take 20 μL of transformed bacteria culture and spread over entire surface of the LB agar plate prepared from the above **step 4**.
6. Incubate the plates at 37 °C incubator overnight.
7. Remove plates from the 37 °C incubator next morning and check for colonies.
8. Pick up individual white color colony into 5 mL of LB medium with 100 μg/mL ampicillin. Pick up a total of five colonies from each plate.
9. Incubate at 37 °C overnight in a shaking incubator at a shaking rate of 225 rpm.
10. At next day, perform mini-pre to extract plasmids from each bacterium cultures.
11. Setup Sanger sequencing reaction.

X μL	Plasmid DNA (400 ng)
1 μL	5 μM PRLR_F primer
Y μL	nuclease-free H_2O
16.0 μL	Total volume

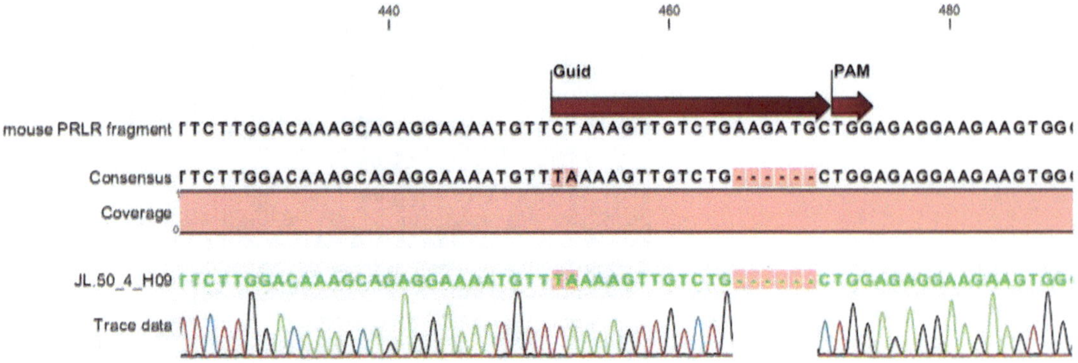

Fig. 4 Sanger sequencing analysis of founder mouse ET50. A single nucleotide deletion (CT>T) results in frameshift mutation to generate a stop codon (TAA) subsequently

12. Perform standard Sanger sequencing (*see* **Note 15**). Sequencing results confirm the homology-directed repair (HDR) mutation in the desired location of PRLR gene (Fig. 4) (*see* **Note 16**).

4 Notes

1. The double strand break (DSB) introduced by the sgRNA is ideally less than 10–15bp away from the mutation site.
2. Here are some online CRISPR design tools the authors recommend:

   https://zlab.bio/guide-design-resources
   http://wwwuser.cnb.csic.es/~montoliu/CRISPR/
   http://crispor.tefor.net/
   http://www.e-crisp.org/E-CRISP/designcrispr.html
   http://crispr.dbcls.jp/
3. For introducing less than 50bp small insertion or deletion or a single-point mutation, the best HR template is probably a single-stranded DNA (ssDNA) oligo which usually works better than plasmids. For ssDNA oligo design, we typically use around 100–150 bp total homology. The mutation is introduced in the middle, giving about 50–70 bp on left and right sides that are homologous to the target region.
4. We typically use single-stranded DNA oligos in a format of PAGE purified ultramer oligos.
5. We use commercial Cas9 mRNA from System Biosciences (SBI). Cas9 mRNA can be transcribed in vitro (IVT) in house using pX330-Cas9 plasmids as templates.
6. PCR product purification can be achieved using ethanol precipitation: add 600 μL ice-cold 100% ethanol and keep it at

−20 °C for 1 h, then spin 16,000 × g for 15 min at room temperature; remove supernatant, and then add 300 μL ice-cold 70% ethanol, spin 16,000 × *g* for 5 min at room temperature, and then remove supernatant; air-dry 10 min and resuspend in 50 μL nuclease-free water.

7. Keep 10× T7 buffer at room temperature while making the reaction mix.
8. We elute the mRNA with nuclease-free water instead of elution buffer provided in the kit. After adding the water, we keep them at room temperature for 5–10 min, instead of 65–70 °C for 5–10 min that is recommended on the manufacturing instructions.
9. We recommend using the freshly prepared reagent mix before the microinjection experiments.
10. For pronuclear injection usually 2–5 pico liters is the volume to double the size of the pronucleus.
11. The injected embryos can be transferred at 2-cell stage to oviducts of day-0.5 *dpc* pseudo-pregnant recipient mice next day after microinjection.
12. We also use other DNA extraction kits such as Extract-N-Amp from Sigma-Aldrich and QuickExtract DNA Extraction Solution from Epicenter.
13. Taq DNA polymerase creates the A-overhangs that is required for the subsequent ligations.
14. Calculate the ratio of vector:PCR product at molar ratio of 3:1.

    ng of PCR product = (ng of vector × 0.5 kb size of PCR product/3.5 kb size of vector) × 1/3.
15. We use a commercial service facility to perform all Sanger sequencing experiments.
16. We use Qiagen CLC Main Workbench software for sequence analysis.
17. Sequence highlighted in italic is to create a BbsI overhang. Sequence highlighted in bold is T7 promoter sequence.

Acknowledgment

This work was supported by Livestock Improvement Corporation Limited (LIC), New Zealand (grant code: LICNZ-CA14). The pX330 plasmid was a gift from Feng Zhang (Addgene plasmid # 42230; http://n2t.net/addgene:42230; RRID: Addgene_42230).

References

1. Gurumurthy CB, Lloyd KCK (2019) Generating mouse models for biomedical research: technological advances. Dis Models Mech 12(1): dmm029462. https://doi.org/10.1242/dmm.029462
2. Schering L, Hoene M, Kanzleiter T, Jahnert M, Wimmers K, Klaus S, Eckel J, Weigert C, Schurmann A, Maak S, Jonas W, Sell H (2015) Identification of novel putative adipomyokines by a cross-species annotation of secretomes and expression profiles. Arch Physiol Biochem 121(5):194–205. https://doi.org/10.3109/13813455.2015.1092044
3. Habiela M, Seago J, Perez-Martin E, Waters R, Windsor M, Salguero FJ, Wood J, Charleston B, Juleff N (2014) Laboratory animal models to study foot-and-mouth disease: a review with emphasis on natural and vaccine-induced immunity. J Gen Virol 95(Pt 11):2329–2345. https://doi.org/10.1099/vir.0.068270-0
4. Littlejohn MD, Henty KM, Tiplady K, Johnson T, Harland C, Lopdell T, Sherlock RG, Li W, Lukefahr SD, Shanks BC, Garrick DJ, Snell RG, Spelman RJ, Davis SR (2014) Functionally reciprocal mutations of the prolactin signalling pathway define hairy and slick cattle. Nat Commun 5:5861. https://doi.org/10.1038/ncomms6861
5. Dikmen S, Alava E, Pontes E, Fear JM, Dikmen BY, Olson TA, Hansen PJ (2008) Differences in thermoregulatory ability between slick-haired and wild-type lactating Holstein cows in response to acute heat stress. J Dairy Sci 91(9): 3395–3402. https://doi.org/10.3168/jds.2008-1072
6. Dikmen S, Khan FA, Huson HJ, Sonstegard TS, Moss JI, Dahl GE, Hansen PJ (2014) The SLICK hair locus derived from Senepol cattle confers thermotolerance to intensively managed lactating Holstein cows. J Dairy Sci 97(9): 5508–5520. https://doi.org/10.3168/jds.2014-8087
7. Craven AJ, Ormandy CJ, Robertson FG, Wilkins RJ, Kelly PA, Nixon AJ, Pearson AJ (2001) Prolactin signaling influences the timing mechanism of the hair follicle: analysis of hair growth cycles in prolactin receptor knockout mice. Endocrinology 142(6):2533–2539. https://doi.org/10.1210/endo.142.6.8179
8. Cong L, Zhang F (2015) Genome engineering using CRISPR-Cas9 system. Methods Mol Biol 1239:197–217. https://doi.org/10.1007/978-1-4939-1862-1_10
9. Cong L, Ran FA, Cox D, Lin S, Barretto R, Habib N, Hsu PD, Wu X, Jiang W, Marraffini LA, Zhang F (2013) Multiplex genome engineering using CRISPR/Cas systems. Science 339(6121):819–823. https://doi.org/10.1126/science.1231143

Part V

Large Animal Welfare and Production Outcomes

Chapter 15

Generation of Pigs that Produce Single Sex Progeny

Björn Petersen and Stefanie Kurtz

Abstract

In livestock industry, one sex is usually preferred over the other due to its impact on the production (e.g., milk from cows, eggs from laying hens, or meat from bulls). Boar taint, to which most of the consumers are susceptible, is a major challenge for the pork industry in the light of the enacted ban of castration without anesthesia from 2021 in Germany. Consequently, a shift towards an increased female ratio would be of great benefit for the pork production. We recently described that a CRISPR/Cas9-mediated knockout of the porcine SRY gene by intracytoplasmic microinjection or SCNT resulted in genetically male pigs with a female phenotype. This sex reversal study in pigs revealed a pivotal role of the SRY gene in male sex determination and might pave the way for the generation of boars that produce only female offspring.

Key words Pig, SRY-knockout, CRISPR/Cas9 system, Intracytoplasmic microinjection, SCNT

1 Introduction

In livestock, the pre-determination of the sex is of great economic importance [1]. Altering the ratio of male to female offspring is profitable, mainly if only one sex provides the desired product (e.g., eggs from laying hens or milk from cows) or one sex has a higher production efficiency (e.g., meat from bulls). In pork production, males are undesirable due to the male-specific boar taint that occurs in 5–10% of the boars and which is characterized by an unpleasant smell released during the heating process [2]. Boar taint originates from two components, androstenone, a steroid, functioning as a pheromone, which is produced in the testes and skatole, produced by microbial L-tryptophan degradation that accumulates in the adipose tissue when boars reach sexual maturity [3, 4]. A well-established method to prevent boar taint is the surgical castration of piglets, usually done without anesthesia within the first 7 days after birth. A heated discussion concerning animal welfare resulted in the ban of this surgical intervention without anesthesia within most countries of the EU. Currently, the following alternative methods for surgical castration such as castration with anesthesia,

Paul John Verma, Huseyin Sumer and Jun Liu (eds.), *Applications of Genome Modulation and Editing*, Methods in Molecular Biology, vol. 2495, https://doi.org/10.1007/978-1-0716-2301-5_15,

boar fattening [5, 6], genetic selection for "low-taint" pigs [7, 8], feeding management strategies [9, 10], immunocastration [11, 12] and slaughter at a younger age [13, 14] are discussed. However, at present, all alternatives suffer from either lack of efficiency to reliably preventing boar taint or lack of broad acceptance by farmers, consumers, and the pork industry [1, 15]. The development of the CRISPR/Cas9 system to edit the mammalian genome offers new opportunities to address current problems in livestock farming [16]. Therefore, we employed the CRISPR/Cas9 system to target the porcine sex determining region on the Y-chromosome (SRY gene). The SRY gene is a small, intron-less sequence located on the short arm of the Y-chromosome. The centrally located HMG domain of the SRY gene is considered the main functional domain for SRY protein synthesis [17, 18]. During embryogenesis, the SRY expression in Sertoli cells causes the formation of primary precursor cells of the tubuli seminiferi and the development of testicles from the undifferentiated gonads [19]. Recently, we produced SRY-knockout pigs by intracytoplasmic microinjection of CRISPR/Cas9 RNPs or somatic cell nuclear transfer using edited donor cells [20]. The knockout of the porcine SRY gene resulted in sex reversal in genetically male pigs. These pigs developed a complete set of internal and external female genitalia, which underlined the critical role of the SRY gene as the master gene that initiates male sex development during embryogenesis. The SRY-knockout pigs form the basis for the generation of boars that only produce female offspring, rendering surgical castration obsolete.

2 Material

2.1 Preparation of the CRISPR Vector

1. pX330-U6-Chimeric_BB-CBh-hSpCas9 (Addgene).
2. Restriction enzyme: BbsI with supplied buffer.
3. Sense and antisense gRNA oligonucleotides.
4. Annealing buffer: 10 mM Tris–HCL (pH 7.5), 1 mM EDTA, 50 mM NaCl.
5. T4 ligase with supplied ligation buffer.
6. NEB® 5-alpha competent *E. coli* with supplied SOC outgrowth medium.
7. LB medium: 5 g NaCl, 5 g bacto tryptone, 2.5 g yeast extract in 500 mL ddH_2O.
8. Ampicillin agar plate: 3 g Agar, 200 μL ampicillin in 200 mL autoclaved LB medium.
9. PCR master mix: GoTaq®G2 Hot Start (Promega), 10 mM dNTPs, 20 nM forward and reverse primer for SRY locus.
10. Plasmid Miniprep Kit.

11. Vacuum Elution Device.
12. Plasmid Maxiprep System Kit.
13. 80% glycerin.
14. Oligo sequences (BbsI overhang is indicated in blue):

    SRY_1 duplex: 5′- CACCGATTGTCCGTCGGAAA TAGTGGTTT-3′

    SRY_3 duplex: 5′- CACCGAAATACCGACCTCGTCG CAAGTTT-3′

2.2 Efficiency Testing of gRNA

1. Neon™ Transfection System with Neon™ 100 μL Kit Transfection System.
2. Dulbecco's phosphate-buffered saline (PBS).
3. 10× EDTA/Trypsin.
4. Dulbecco's modified Eagle's medium stock solution (DMEM-stock): 2 mM L-Glutamine, 0.1 mM ß-Mercaptoethanol in 500 mL high glucose DMEM.
5. Electroporation medium: DMEM-stock with 1% non-essential amino acids, 1% sodium pyruvate, 30% fetal calf serum.
6. Fibroblast culture medium: DMEM-stock with 1% penicillin/streptomycin, 1% non-essential amino acids, 1% sodium pyruvate, 30% fetal calf serum.
7. Cell lysis buffer: 10% SDS-Solution, 10 mg/μL proteinase K, 1 M Tris–HCL (pH 8.4) in ddH_2O
8. PCR master mix (Subheading 2.1, **item 9**)
9. TBE buffer: 108 g Tris–HCL (pH 7.5), 54 g EDTA, 7.4 g Borate in 10 L ddH_2O.
10. 0.8% agarose gel: 800 mg agarose gel powder in 100 mL TBE buffer.
11. High Performance Ultraviolet Transilluminator.
12. Ultraviolet light (Fusion-SL 3500.WT).
13. DNA Invisorb®Fragment CleanUp (Stratec) for DNA purification and DNA extraction from agarose gel.
14. Primer sequences:

    primSRY forward: 5′-CCCTTTTCAAATGGTGCAGT-3′
    primSRY reverse: 5′-CCTTGGCGACTGTGTATGTG-3′

2.3 Preparation of CRISPR/Cas RNP Complexes

1. Synthetic sgRNA fused to purified 2NLS-Cas9.
2. crRNA and tracRNA fused to Alt-R S.p. Cas9 nuclease 3NLS.
3. Microinjection buffer: 10 mM Tris–HCL (pH 7.47), 0.1 mM EDTA in ddH_2O.

2.4 Intracytoplasmic Microinjection of CRISPR/Cas RNPs

1. Porcine ovaries from slaughterhouse.
2. 0.9% NaCl solution, 60 mg/L penicillin G, 0.13 g/L streptomycin in ddH_2O.
3. Vacuum-based aspiration system and an 18 G cannula for follicle punctuation.
4. Washing buffer (PXM-Hepes Air): 108 mM NaCl, 10 mM KCl, 0.35 mM KH_2PO_4, 0.4 mM $MgSO_4$, 25 mM $NaHCO_3$, 25 mM HEPES, 0.2 mM Na-Pyruvate, 4 mM Ca-Lactate, 0.1 mL/L Gentamycin, 1 g/L bovine serum albumin, 1 mM Glucose in ddH_2O.
5. Oocyte maturation medium (FLI-medium): 40 ng/mL FGF-ß, 2000 U/mL LIF, 20 ng/mL IGF-1, 0.57 mM cysteine, 10 ng/mL EGF, 10 I.E. hCG and PMSG, 50 μg/mL Gentamycin, 22 μg/mL Na-Pyruvate, 1 mg/mL BSA, 2.2 mg/mL $NaHCO_3$, 14.7 mg/mL TCM 199 in ddH_2O.
6. Frozen boar semen.
7. Androhep®Plus.
8. FerTalp medium stock: 0.01 mM Polyvinyl alcohol (PVA), 0.071 mM Gentamycin, 0.003 mM Phenol red solution, 114 mM NaCl, 3.2 mM KCl, 0.5 mM $MgCl_2 \times 6\ H_2O$, 21.4 mM 60% Na-Lactate, 0.35 mM $NaH_2PO_4 \times H_2O$, 5 mM Glucose, 25 mM $NaHCO_3$, 2 mM Coffein, 8 mM Ca-Lactate $\times$ 5 H_2O in ddH_2O.
9. FerTalp medium: 0.6 g BSA, 200 μL Na-Pyruvate in 200 mL FerTalp stock.
10. Microinjection medium (TL-Hepes 296 Ca^{2+} stock): 114 mM NaCl, 3.2 mM KCl, 2 mM $CaCl_2 \times 2\ H_2O$, 0.4 mM $NaH_2PO_4 \times H_2O$, 0.5 mM $MgCl_2 \times 6\ H_2O$, 2 mM $NaHCO_3$, 10 mM HEPES, 10 mM 60% Na-Lactate, 100 U/L Penicillin G, 50 mg/L Streptomycin in ddH_2O.
11. Culture medium for microinjection: 0.25 mM Na-Pyruvate, 32 mM Sucrose, 0.4% BSA in TL-Hepes 296 Ca^{2+} stock
12. Microinjection mineral oil.
13. Micromanipulation unit including FemtoJet, CellTram Vario, InjectMan NI2 (all from Eppendorf).
14. 20 μL microloader tips.
15. Stereomicroscopes.
16. Warming plate.
17. 4-well cell culture dish.
18. 35 mm culture dish.

19. Glass needle diameters: transport needle for oocytes (Ø 225 μm), transport needle for blastocysts (Ø 300 μm), holding needle (Ø 100–125 μm).
20. Porcine zygote medium (PZM): 108 mM NaCl, 10 mM KCl, 0.35 mM KH_2PO_4, 0.4 mM $MgSO_4$ + 7 H_2O, 25.07 mM $NaHCO_3$, 0.2 mM Na-Pyruvate, 2 mM Ca-lactate + 5 H_2O, 5 mM Hypotaurin, 10% 50× BME, 1% MEM, 0.05 mg/mL Gentamicin sulfate, 3 mg/mL BSA in ddH_2O.
21. Synchronization of gilts: 20 mg/day/gilt Altrenogest, 1500 I.U. PMSG (pregnant mare serum gonadotropin), 500 I.U. hCG (human chorionic gonadotropin).

2.5 DNA Preparation of Tail Tissue for Genetically Analysis of Offspring

1. Tail tissue of piglets.
2. Tail lysis buffer: 100 mM Tris–HCl (pH 8), 100 mM NaCl, 100 mM EDTA, 1% SDS-Solution in ddH_2O.
3. 10 mg/mL Proteinase K.
4. Saturated NaCl solution.
5. 100% ethanol.
6. 70% ethanol.
7. Nuclease-free water or ddH_2O.

2.6 Somatic Cell Nuclear Transfer: Re-cloning of SRY-KO Pigs

1. Ear tissue from the SRY-KO piglets.
2. PBS with 2% penicillin/streptomycin.
3. Serum-reduced medium: DMEM-stock with 1% penicillin/ streptomycin, 1% non-essential amino acids, 1% sodium pyruvate, 0.5% fetal calf serum.
4. Calcium-free TL-Hepes stock solution: 114 mM NaCl, 3.2 mM KCl, 0.4 mM $NaH_2PO_4 \times H_2O$, 0.5 mM $MgCl_2 \times 6\ H_2O$, 238.3 mM HEPES, 112.06 mM 60% Na-Lactate, 100 U/L Penicillin G, 50 mg/L streptomycin in ddH_2O.
5. Calcium-free TL-Hepes 296: 0.25 mM Na-Pyruvate, 32 mM Sucrose, 0.4% BSA in 25 mL Calcium-free TL-Hepes stock solution.
6. Ca^{2+} TL-Hepes stock solution: 114 mM NaCl, 3.2 mM KCl, 2 mM $CaCl_2 \times H_2O$, 0.4 mM $NaH_2PO_4 \times H_2O$, 0.5 mM MgCl2 × 6 H_2O, 238.3 mM HEPES, 112.06 mM 60% Na-Lactate, 100 U/L Penicillin G, 50 mg/L streptomycin in ddH_2O.
7. Ca^{2+} TL-Hepes 296: 0.25 mM Na-Pyruvate, 32 mM Sucrose, 0.4% BSA in 25 mL Ca^{2+} TL-Hepes stock solution.
8. Multiporator with 0.2 mm micro fusion chamber (Eppendorf).
9. Calcium-free Sor2 medium for fusion: 0.25 mM Sorbitol, 0.5 mM Mg-Acetate, 0.1% BSA in ddH_2O.

10. Ca^{2+} Sor2 medium for electrical activation: 0.25 mM Sorbitol, 0.5 mM Mg-Acetate, 0.1 mM Ca-Acetate, 0.1% BSA in ddH_2O.
11. Medium for chemical activation: PZM with 500 nM Skriptaid.

2.7 General Equipment

1. 15 mL and 50 mL Corning falcons.
2. 14 mL polystyrene tubes.
3. Humidified incubator at 37 °C and 5% CO_2 for cell culture.
4. Humidified incubator at 38.8 °C and 5% CO_2 for oocyte maturation.
5. Humidified incubator at 39 °C and 5% O_2 and 5% CO_2 oocyte cultivation.
6. Freezer (−20 °C and −80°C) and refrigerator (4 °C).
7. Thermo-shaker at 37 °C.
8. Thermomixer (up to 95 °C).
9. Pipette filter tips (10/20 μL, 100 μL, and 1000 μL).
10. Pipetting aid.
11. T75 and T25 cell culture flasks.
12. Vortex mixer.
13. 200 mL and 500 mL Erlenmeyer flasks.
14. Laminar air flow.
15. Centrifuge.
16. Thermocycler.
17. PCR tubes, 0.6 and 1.5 mL tubes.
18. Agarose gel electrophoresis chamber.
19. Spectrophotometer.

3 Methods

3.1 Design and Cloning of gRNA Targeting *SRY* Gene

The CRISPR/Cas9 system was employed to induce an approx. 300 bp deletion encompassing the complete HMG domain of the porcine SRY gene (Fig. 1).

1. Design and evaluate the guide RNAs (gRNAs) targeting the SRY gene using a suitable CRISPR design tool (Table 1). The porcine SRY sequence (Sus scrofa 11.1) can be obtained from genome databases such as NCBI (https://www.ncbi.nlm.nih.gov/gene/407740) (*see* **Note 1**).
2. Digest 1 μg of the pX330 vector for 2.5–3 h at 37 °C with 1 μL of BbsI, 2 μL of supplied buffer solution, and add ddH_2O to a total volume of 20 μL.

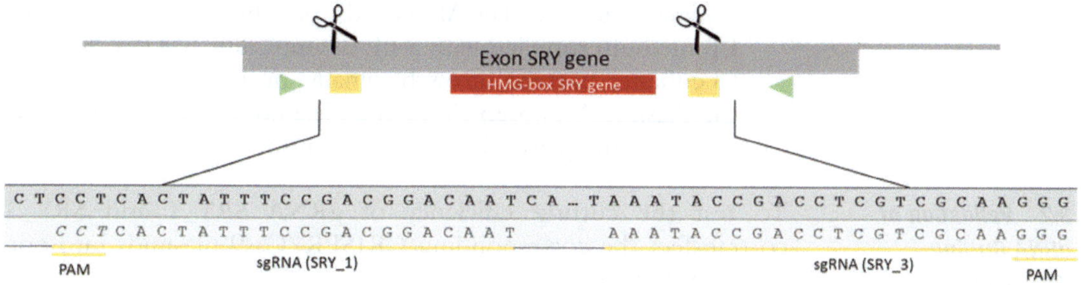

Fig. 1 Location of the two gRNAs (yellow) encompassing the HMG domain (red) of the porcine SRY gene (grey). The SRY primers are indicated in green

Table 1
List of web-based tools for guide RNA design

Tools for guide design	
CRISPOR	http://crispor.tefor.net/
Benchling	https://www.benchling.com/crispr/
Horizon discovery	https://horizondiscovery.com/en/products/tools/CRISPR-Design-Tool
CHOPCHOP	https://chopchop.cbu.uib.no/
Deskgen	https://www.deskgen.com/landing/cloud.html#/
E-CRISP	http://www.e-crisp.org/E-CRISP/designcrispr.html
IDT	https://eu.idtdna.com/site/order/designtool/index/CRISPR_CUSTOM
Synthego	https://www.synthego.com/products/bioinformatics/crispr-design-too

3. Design both SRY_1 and SRY_3 duplex oligonucleotide pairs with a BbsI overhang and anneal 10 μL of each oligo at 37 °C for 30 min with 80 μL of annealing buffer, heat inactivate the mixture at 95 °C for 5 min, and then cool it down with a ramp of 5 °C per minute to 25 °C to form gRNA duplexes.
4. Clone each gRNA duplex into the linearized CRISPR/Cas9 vector by 2 h ligation of 50 ng plasmid with 1 μL annealed oligonucleotide (diluted 1:200 with ddH_2O), 1 μL T4 ligase, 2 μL supplied ligase buffer and fill it up to a total volume of 20 μL with ddH_2O at 37 °C in a PCR tube (*see* **Note 2**).
5. For amplification of the SRY vector, transfer 3 μL of the ligation product into commercially available competent bacteria according to the manufacturer's recommendation (e.g., NEB® 5-alpha competent *E.coli* [high efficiency]) and plate the transformed bacteria onto an agar plate with ampicillin for overnight incubation at 37 °C. Pick single colonies from agar plate with a pipette tip and dip the colonies into the PCR mixture for insert amplification using an U6 sequencing primer and an SRY

reverse oligo next day. Afterwards, put the pipette tip in 3 mL LB medium supplied with 3 μL of ampicillin at 37 °C for 24 h to enhance bacterial growth. Finally, purify the plasmids with the Plasmid Miniprep Kit or the Plasmid Maxiprep System Kit and store the plasmids at - 20 °C until further use.

3.2 Validation of gRNA Cutting Efficiency

To test the cutting efficiency of gRNA SRY_1 and SRY_3, co-transfect the corresponding CRISPR/Cas9 plasmids into porcine fetal fibroblasts.

1. Obtain porcine fetal fibroblasts from 24- to 28-day-old fetuses. Prepare the fetuses from uterus. Decapitate and eviscerate the fetus and wash the corpus three times with PBS containing 2% penicillin and streptomycin in a petri dish. Cut the corpus tissue into 1 × 1 mm pieces using a scalpel. Transfer the tissue pieces to a 1.5 mL Eppendorf tube and digest the tissue with 500 μL EDTA/Trypsin at 37 °C for 20 min to separate cells. After incubation, transfer the tissue-cell solution into a T75 cell culture flask with fibroblasts culture medium and culture the solution in a humidified chamber at 37 °C and 5% CO_2 to provide porcine fetal fibroblast proliferation. Exchange fibroblast culture medium after 48–72 h.
2. After the cells reach 80–90 % confluency in T75 cell culture flask (~2.5 × 10^6 cells), detach the cells from the flask with 2 mL of EDTA/Trypsin solution for 5–10 min at a 37 °C warming plate. After centrifugation of trypzinized cells for 4 min at 200 g, resuspend the cell pellet in 200 μL resuspension buffer R (supplied in Neon™ 100 μL kit) and add 10 μL of each plasmid (SRY_1 and SRY_3) to the cell solution (maximum DNA concentration of 5 μg).
3. For electroporation, transfer 100 μL of the resuspended cell solution into a supplied glass tube with 3 mL electroporation buffer E2 (Neon™ 100 μL kit) using an electroporation cuvette. Apply the following electroporation settings for porcine fibroblasts: 1,350 V and 2 impulses for 20 s to allow entry of the plasmids into the cells. Transfer the transfected cells directly into 5 mL antibiotic-free fibroblast culture medium in a T25 cell culture flask until the next day and further culture the cells in fibroblast culture medium supplied with 1 % penicillin and streptomycin at 37 °C and 5 % CO_2 in a humidified chamber.
4. For genotyping of the transfected cells, wash the confluent cells with 5 mL PBS and lyse them overnight in 1 mL lysis buffer at 37 °C (*see* **Note 3**). On the next day, heat inactivate the proteinase K at 95 °C for 12 min and use the lysate for PCR amplification.

Table 2
List of web-based tools for primer design

Tools for primer design	
Primer3	http://primer3.ut.ee/
Eurofins Genomics	https://www.eurofinsgenomics.eu/de/ecom/tools/pcr-primer-design/
Primer-BLAST	https://www.ncbi.nlm.nih.gov/tools/primer-blast/
IDT	https://eu.idtdna.com/Primerquest/Home/Inde

5. To amplify the SRY sequence, design site-specific primers using your preferred primer design tool (Table 2). The porcine SRY sequence (Sus scrofa 11.1) can be obtained from a genome database (NCBI) (https://www.ncbi.nlm.nih.gov/gene/40 7740). The amplicon length should encompass both gRNAs. Unspecific binding of primers can be determined via BLAST (NCBI) (https://blast.ncbi.nlm.nih.gov/Blast.cgi). Primer sequences used in our study were listed in 2.2.14.
6. Perform PCR amplification as followed:

   Mix for the PCR master mix 10 μL of 5 × PCR buffer, 1.5 μL forward and reverse primer (concentration of 0.02 nM), 3 μL 25mM $MgCl_2$, 1 μL 10mM dNTPs, 0.25 μL Taq polymerase and 28 μL ddH_2O for one sample. Finally, add 5 μL of DNA sample to PCR master mix in a PCR tube. Start the PCR amplification in a thermocycler with an initial denaturation at 94 °C for 2 min, followed by 34 cycles of denaturation at 94 °C for 30 s, annealing at 59 °C for 45 s and extension at 72 °C for 30 s. Perform the final extension at 72 °C for 2 min and cool the PCR samples down to 4 °C (*see* **Note 4**).
7. After PCR amplification, separate the DNA fragments via gel electrophoresis in a 0.8 % agarose gel. Prepare the agarose gel by resuspension of 800 mg agarose gel powder in 100 mL TBE buffer when exposed to heat and finally add 20 μL of ethidium bromide to solution. The agarose solution forms gel after cooling for 30 min at room temperature. Set the standard conditions for agarose gel electrophoresis to 80 V, 400 mA, and 60 min. Visualize the DNA amplicons on the agarose gel under ultraviolet light (Fusion-SL 3500.WT).

   PCR-based detection reveals the edited cell population in the PCR product with an approx. 300 bp deletion within the SRY locus compared to male wild type control (Fig. 2).
8. Excise both DNA fragments from the co-transfection of CRISPR/Cas9 plasmids SRY_1 and SRY_3 from the 0.8 % agarose gel with a scalpel under UV light (High Performance Ultraviolet Transilluminator). Purify the agarose slices using DNA Invisorb®Fragment CleanUp to isolate DNA and to

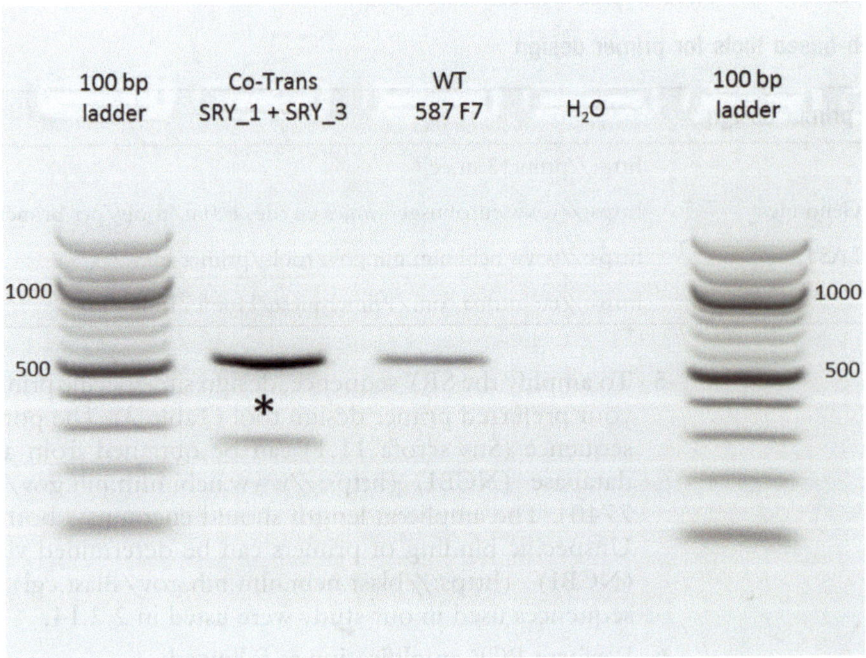

Fig. 2 PCR-based detection of the SRY gene displays two bands in porcine fibroblasts co-transfected with CRISPR/Cas plasmid SRY_1 and SRY_3. The upper band indicates the wild type cell population and the lower band a deletion of approx. 300 bp (black asterisk). The male wild type control only show the expected ~500 bp band (WT 587 F7)

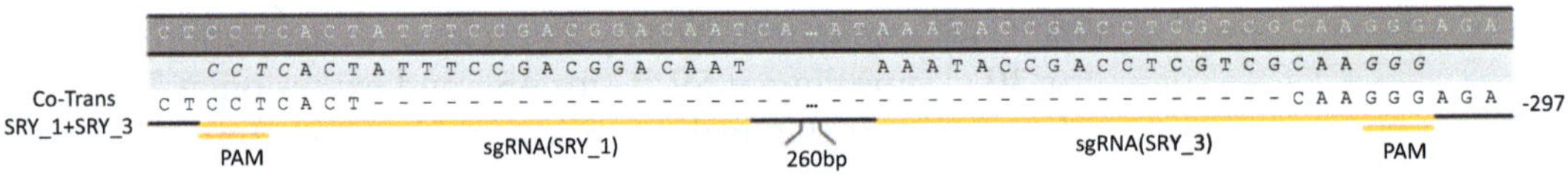

Fig. 3 Sanger sequencing of the lower band from agarose gel revealed a deletion of 297 bp within the SRY gene eliminating the entire HMG domain

confirm a large deletion within the SRY gene via Sanger sequencing. For Sanger sequencing, use the primSRY forward primer (Fig. 3).

The results of this section confirmed that a combination of both gRNAs is capable of inducing an efficiently large deletion encompassing the HMG domain of the porcine SRY gene and can be used further to generate genetically modified animals.

3.3 Preparation of RNP Complexes

CRISPR/Cas9 RNP complexes SRY_1 and SRY_3 are designed based on the previous tested gRNAs and used for intracytoplasmic microinjection into IVF-produced zygotes. In our study, one RNP complex was ordered from IDT (USA) and the other one from Synthego (USA). The RNP complexes were resuspended and diluted in microinjection buffer.

1. The first RNP complex from Synthego (USA) included the individually designed synthetic single-guide RNA that is based on the previously tested gRNA SRY_1: 5′-ATTGTCCG TCGGAAATAGTG-3′. To form the RNP complex, mix 0.84 μL sgRNA (25 pmols) with 1.25 μL of purified 2NLS-Cas9 nuclease (25 pmols) and 10.4 μL of microinjection buffer. Incubate all components for 10 min at room temperature in a 0.6 mL tube. Afterwards, spin the mixture down at 8928 g and 4°C for 10 min and transfer the supernatant into a new tube for immediately use.
2. The Alt-R CRISPR/Cas9 system from IDT (USA) consists of two RNP components: crRNA and tracrRNA. The crRNA is individually designed based on the previously tested gRNA SRY_3: 5' – AAATACCGACCTCGTCGCAA – 3'. To prepare the active gRNA for the RNP complex, mix 5 μL of crRNA (1 μg/μL) with 10 μL of tracrRNA (1 μg/μL) in a PCR tube. Anneal both components in a thermocycler at 95 °C for 5 min and ramp it down to 25 °C at 5 °C/min. To assemble the RNP complex for microinjection, mix 25 μL of gRNA (40 ng/μL) with 25 μL of Alt-R S.p. Cas9 nuclease 3NLS (40 ng/μL) and incubate both for 10 min at room temperature. Spin the mixture down at 8928 g and 4 °C for 10 min and transfer the supernatant into a new tube for immediately use (*see* **Note 5**).
3. Combine both RNP complexes in a ratio of 1 (SRY_1) to 1.7 (SRY_3) and directly use the microinjection solution for intracytoplasmic microinjection.

3.4 In Vitro Fertilization (IVF) and Intracytoplasmic Microinjection

The RNP complexes targeting the HMG domain of the porcine SRY gene are intracytoplasmatically injected into IVF-produced zygotes (Fig. 4).

1. Porcine ovaries can be obtained from a regional slaughterhouse (*see* **Note 6**). Before starting follicle aspiration, wash ovaries with 0.9% NaCl solution with penicillin/streptomycin at 37 °C. To collect oocytes, punctate follicles with a size between 2 and 4 mm at the surface of the ovaries with an 18 G cannula that is connected to a vacuum-based aspiration system (*see* **Note 7**). Wash the collected oocytes with sterile-filtered PXM. Afterwards, verify and collect oocytes with "good" oocyte criteria such as several layers of cumulus cells and intact cytoplasm and zona pellucida in a petri dish with 8 to 10 mL PXM on a warming plate at 37 °C. Again, wash the oocytes during selection with 2 mL PXM in culture dishes.
2. Maturate oocytes as previously described [21]. Equilibrate the maturation medium (FLI-medium) prior to oocyte maturation at 39 °C and 5% CO_2. Afterwards, maturate 50 collected oocytes in 500 μL FLI-medium for 40–44 h in a 4-well culture

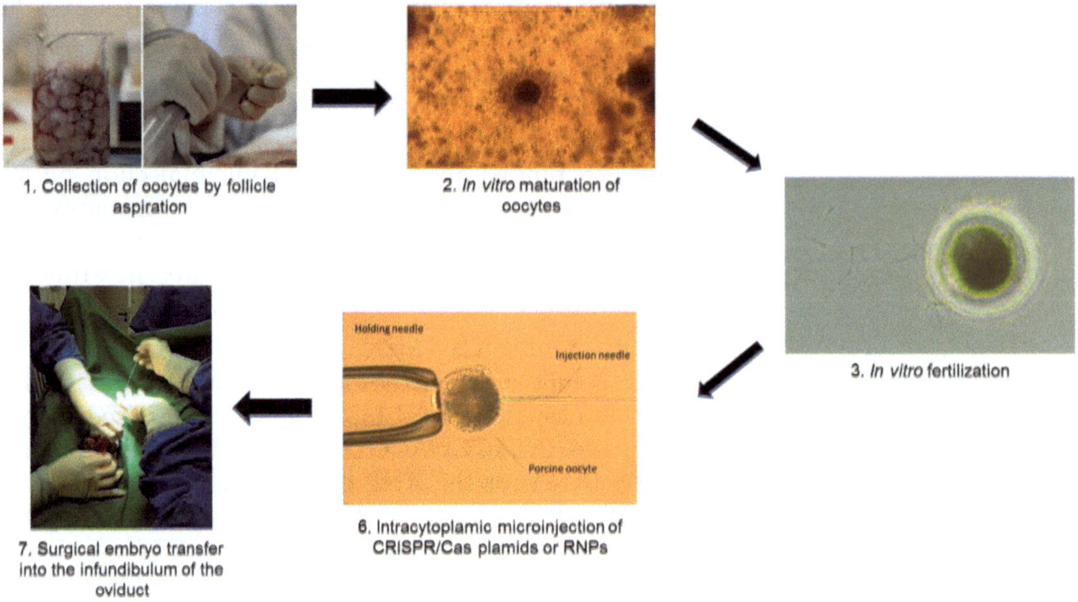

Fig. 4 Workflow for intracytoplasmic microinjection of CRISPR/Cas plasmids or RNPs into IVF-produced zygotes to generate genetically modified pigs

dish at 39 °C and 5% CO_2 in a humidified incubator. After incubation, select matured oocytes based on the presence of a visible polar body.

3. Prior to in vitro fertilization, equilibrate FerTalp with Na-pyruvate at 39 °C and 5% CO_2 overnight. The next day, add BSA to FerTalp medium and sterile-filtrate the solution. Transfer the maturated oocytes 42 to 46 after onset of maturation in 100 μL FerTalp medium in culture dishes. Remove the cumulus cells by vigorous pipetting the oocytes with a 1000 μL pipette. Wash the denuded oocytes with 2 mL FerTalp medium and collect them in groups of 50 oocytes per culture dish in 500 μL FerTalp medium (*see* **Note 8**).
4. For in vitro fertilization of matured oocytes, thaw frozen boar semen (50 million sperm cells in one straw) from a fertile boar for 30 s in a water bath at 37 °C. Analyze and estimate the motility of sperm on a microscope slide using a microscope (Olympus, BH-2) (*see* **Note 9**). Prior to fertilization, wash the boar semen with 3 mL Androhep®Plus in a 15 mL corning falcon and centrifuge it for 3 min at 867 g and 30 °C. Remove the supernatant and resuspend the sperm pellet in 3 mL Androhep®Plus. Centrifuge the solution for 3 min at 867 g and 30 °C, remove the supernatant and resuspend the sperm pellet in 0.5 mL FerTalp medium. Dilute the sperm in a ratio of 1:10 to calculate the sperm number using a Neubauer® counting chamber. Depending on sperm capacity, co-incubate 75 to

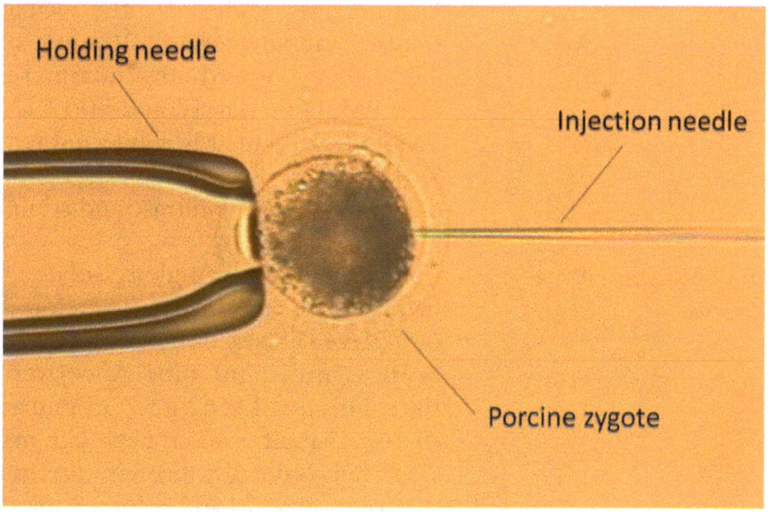

Fig. 5 Intracytoplasmic microinjection of CRISPR/Cas RNP using a glass needle (injection needle) into porcine oocyte fixated with a holding needle

100 sperm cells per oocyte in FerTalp for 2.5 h at 39 °C and 5% CO_2 in a humidified incubator (*see* **Note 10**). After fertilization, wash the IVF-produced zygotes in 2 mL of PZM and finally store the oocytes in groups of 50 oocytes per well in 500 μL PZM before microinjection (*see* **Note 11**).

5. Microinjection begins 20 h after fertilization. Before microinjection starts, fill the injection needle with 5 μL of microinjection solution including the RNP complexes using microloader tips (*see* **Note 12**). Connect the FemtoJet system with the injection needle to allow injection of RNP solution into oocytes. Break the injection needle with light pressure at the edge of the holding needle (*see* **Note 13**). Parameter settings on the FemtoJet are variable depending on the size of the needle's diameter. For intracytoplasmic microinjection, transfer oocytes with use of a transport pipette (225 μm) into 750 μL TL-Hepes 296 Ca^{2+} on a glass plate and warming plate at 30 °C. Fixate the oocytes with a holding pipette (100–125 μm) and inject the RNP solution carefully into the cytoplasm of the oocytes (*see* **Note 14**) (Fig. 5). After microinjection, wash oocytes with PZM and culture the injected oocytes in groups of 50 oocytes per well in 4-well culture dishes with 2 mL PZM at 39 °C, 5% CO_2, and 5% O_2.
6. After 5 to 6 days of incubation, embryos reach the blastocyst stage. Aspirate 30 to 35 blastocysts into a straw using the following pattern: medium, air, medium with embryos, air, medium, and surgically transfer them into one uterus horn, near the utero-tubal junction, of hormonally synchronized 7- to 9-month-old recipients (*see* **Note 15**).

7. Hormonally synchronize the recipient gilts prior to embryo transfer administering 20 mg/day/gilt Altrenogest for 12 days, followed by intramuscular injection of 1500 I.U. PMSG on day 13 and 500 I.U. hCG injection 78 h after PMSG treatment. Microinjected embryos are transferred 6 to 7 days later. Recipients can be pre-checked 14 days after embryo transfer by ultrasound scanning for pregnancy.

3.5 Genotyping of the Offspring

1. For genotyping the piglets, isolate DNA from tissue (tail, skin, hair, follicle, era punch). Therefore, digest 50 mg of tissue in 600 μL of tail lysis buffer with 45 μL of proteinase K overnight at 50 °C in a 1.5 mL tube. After overnight digestion, centrifuge the solution at 14462 g for 15 min and 15 °C. Transfer 500 μL of supernatant into a new 1.5 mL tube and add 700 μL saturated NaCl for protein precipitation. Again, centrifuge the solution at 14462 g for 15 min and 15 °C, before adding 700 μL of supernatant to 700 μL ethanol (100%) for DNA precipitation in a new 1.5 mL tube. Centrifuge the solution at 14,462 × *g* for 15 min at room temperature and wash the DNA pellet twice with 1 mL 70% ethanol. Finally, remove the ethanol and dry the DNA pellet at 37 °C for 1 h and eluate it in 50 μL ddH_2O (*see* **Note 16**). Determine the DNA concentration by using spectrophotometer and use the DNA sample for PCR amplification with SRY primers (Fig. 6). Sanger sequence the PCR product to reveal genetic modifications at the SRY locus.

3.6 Re-cloning

1. Donor cells for somatic cell nuclear transfer (SCNT) (Fig. 7) are isolated from ear tissue of piglets. Therefore, collect ear tissue (5 × 5 mm) and wash the tissue in PBS containing 2% penicillin/streptomycin. Remove the epidermis and cut the remaining ear tissue in small pieces using a scalpel and incubate them for 20 min at 37 °C with 500 μL EDTA/Trypsin in a 1.5 mL tube. Transfer the cell solution including solid tissue pieces into a T25 cell culture flask with fibroblast culture medium including 2% penicillin/streptomycin in a humidified chamber at 37 °C and 5% CO_2. Change the fibroblast culture medium 48–72 h later.
2. Use confluent cells on a 6-well plate (~0.3 × 10^6 cells) as donor cells for SCNT. Incubate these cells in serum-reduced medium (0.5 FCS) for 48 h before SCNT. At the day of SCNT, remove the medium from the cells, wash cells with 1 mL PBS, aspirate the PBS and detach the cells by adding 500 μL EDTA/Trypsin for 5–10 min at 37 °C. After detachment, resuspend the cells in 2.5 mL Ca-free TL-Hepes 296 medium plus 100 μL FCS in a 15 mL corning falcon tube and centrifuge the cell solution at room temperature for 4 min at 200 g to form a cell pellet. Remove the supernatant and resuspend the cell pellet in 3 mL

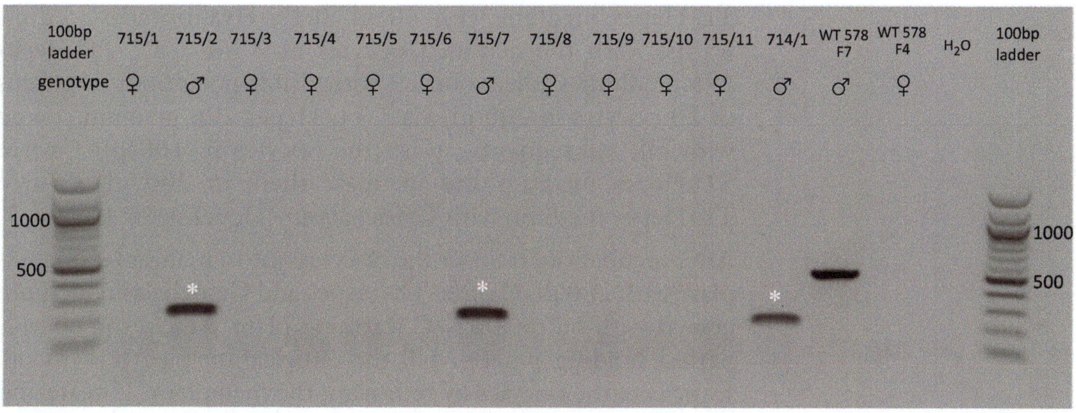

Fig. 6 PCR-based detection of the SRY gene in piglets generated by simulatneous intracytoplasmic microinjection of two CRISPR/Cas RNPs SRY_1 and SRY_3. An approx. 300 bp deletion within the SRY gene in piglet 714/1, 715/2 and 715/7 can be detected when compared to the genetically male wild type control (WT 578 F7)

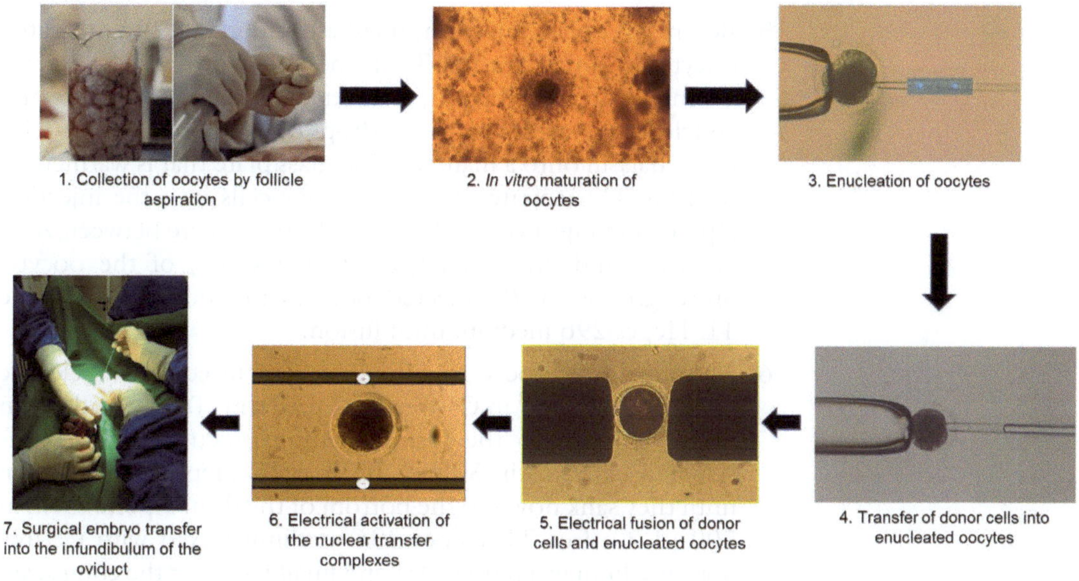

Fig. 7 Workflow of the somatic cell nuclear transfer (SCNT). Donor cells are isolated from ear tissue and electrically fused into enucleated oocytes resulting in cloned genetically modified offspring

of Ca-free TL-Hepes 296 medium for a second centrifugation run. Remove again the supernatant, resuspend the donor cells in 1 mL of Ca-free TL-Hepes 296 medium and store it in a 1.5 mL tube at room temperature until nuclear transfer.

3. In parallel, oocytes are prepared for nuclear transfer. Maturation of oocytes is similar to the one used for microinjection and is described in Subheading 3.4, **step 2**. In contrast to microinjection technique, remove the cumulus of oocytes by vigorous pipetting the oocytes with a 1000 μL pipette in 2 mL Ca^{2+}

TL-Hepes medium with 40 μL 0.1% Hyaluronidase. Afterwards, wash the denuded oocytes in 2 mL Ca^{2+} TL-Hepes 296 medium. Collect oocytes with visible polar body in groups of 20 oocytes in 200 μL Ca^{2+} TL-Hepes 296 medium coated with oil. Subsequently, wash the oocytes in 100 μL Ca-free TL-Hepes medium and incubate them in 200 μL Ca-free TL-Hepes medium with Cytochalasin/Hoechst for 10 min.

4. After incubation, transfer the oocytes onto a manipulation glass plate with 200 μL Ca-free TL-Hepes and Cytochalasin medium coated with oil on a 32 °C warming plate. Adjust the oocytes with a holding pipette. Fill the enucleation pipette with oil. Enucleate the oocytes by removing the metaphase-II-plate and polar body using the enucleation needle. Control the correct removal of the polar body and metaphase-II-plate under UV light (*see* **Note 17**). Store the enucleated oocytes in 75 μL Ca^{2+} TL-Hepes 296 on a 37 °C warming plate until donor cell injection.
5. Before fibroblast injection, incubate groups of 20 enucleated oocytes in 200 μL Ca-free TL-Hepes 296 with Cytochalasin for 10 min. After incubation, transfer the donor fibroblasts and enucleated oocytes in separate drops of Ca-free TL-Hepes with Cytochalasin onto a manipulation glass plate that is warmed up to 29.5 °C. Aspirate 20 to 25 donor cells into the injection pipette and inject one cell per enucleated oocyte between zona pellucida and cytoplasm (perivitelline space) of the oocyte. Store groups of 20 injected oocytes in 100 μL of Ca-free TL-Hepes 296 medium until fusion.
6. The enucleated oocyte and donor cell are fused with an electric pulse of 1 kV/cm^2 in Ca-free Sor2 medium. To allow adaption of the oocytes to the fusion medium, incubate the reconstructed embryos in 500 μL Ca-free TL-Hepes 296 medium until they sank down to the bottom of the dish. Pipette them in 500 μL Ca-free TL-Hepes 296 medium mixed with Ca-free Sor2 medium in a ratio of 1:1 and finally transfer the complexes into 500 μL Ca-free Sor2 medium. Place the reconstructed embryos between both electrodes of the fusion chamber in 100 μL Ca-free Sor2 medium and fuse them with an electrical pulse of 1 kV/cm^2. Incubate the nuclear transfer complexes for one minute in 500 μL Ca-free TL-Hepes 296 medium mixed with Ca-free Sor2 medium in a ratio of 1:1 and 500 μL Ca-free TL-Hepes 296 medium, before storage in 75 μL Ca^{2+} TL-Hepes 296 medium. Perform the fusion control after 30 min (*see* **Note 18**).
7. After fusion, the nuclear transfer complexes are electrically activated in the fusion chamber with Ca^{2+} Sor2 medium. Therefore, adapt the nuclear transfer complexes to the

activation medium by incubation for 1 min in 2 mL Ca^{2+} Tl-Hepes 296 medium, then in 2 mL Ca^{2+} Tl-Hepes 296 medium mixed with Ca^{2+} Sor2 medium in a ratio of 1:1 and at last, transfer them into 2 mL Ca^{2+} Sor2 medium. Place the nuclear transfer complexes between both electrodes of the fusion chamber in 100 μL Ca^{2+} Sor2 medium and activate them with an electrical pulse of 24 V for 45 μs. Incubate the activated nuclear transfer complexes for 1 min in 2 mL Ca^{2+} Tl-Hepes 296 medium mixed with Ca^{2+} Sor2 medium in a ratio of 1:1 and 2 mL Ca^{2+} Tl-Hepes 296 medium.

8. The electrical activation is followed by chemical activation. Therefore, wash the nuclear transfer complexes in 2 mL of PZM medium and two times in 495 μL PZM medium mixed with 5 μL Skriptaid (1:100 dilution), before incubation in 500 μL PZM medium with Skriptaid for 20 h at 39 °C and 5% CO_2 in a humidified chamber.
9. Finally, wash the nuclear transfer complexes in 3 mL of PZM medium and culture groups of 50 complexes in 500 μL PZM at 39 °C and 5% CO_2 until the next day.
10. For embryo transfer, place 90–100 one- and two-cell embryos in Ca^{2+} TL-Hepes 296 medium and aspirate them into straws using the following pattern: medium, air, medium with embryos, air, medium (*see* **Note 19**). Surgically transfer the embryos into the infundibulum of the oviduct of 7- to 9-month-old hormonally synchronized gilts (synchronization protocol equivalent to the microinjection).

4 Notes

1. To minimize potential off-target sites, the target sequence can be analyzed via BLAST (NCBI) (https://blast.ncbi.nlm.nih.gov/Blast.cgi), Cas-OFFinder (http://www.rgenome.net/cas-offinder/) or Off-Spotter (https://cm.jefferson.edu/Off-Spotter/).
2. For ligation, overnight incubation at 4 °C is possible.
3. Lysis time can be shortened to 2 h by increasing the incubation temperature to 55 °C.
4. Annealing temperature can be adapted to different primer pairs and if unspecific products occurred. Extension time can be adapted to polymerase (Taq polymerase amplifies 1,000 bp in 30 s).
5. Both RNP complexes can be stored at - 20 °C. Moreover, to eliminate solid particles, the RNP solution can be passed through a Millipore filter.

6. Temperature of ovaries should be between 28 °C and 32 °C when arriving at the laboratory.
7. A higher number of oocytes can be obtained, when the follicular fluid is aspirated while performing circular movements with the cannula.
8. Work should be carried out quickly at all steps. In particular, the FerTalp medium is very sensitive to pH changes.
9. Sperm motility should be rating at least between 30% and 40%.
10. Sexed sperm can be used for fertilization to predetermine the sex of the embryos.
11. Work should be carried out as quickly as possible at all steps.
12. Too high concentration of solid particles can block the injection pipette.
13. Be careful to break the injection pipette with small diameter.
14. Under optimal conditions, the oocytes are at the one-cell stage. If they are already at the two-cell stage, it is important to microinject into both cells. Successful microinjection can be visually controlled by "swelling" of the cytoplasm.
15. The air bubble in the straw allows a simple visible control of the correct application of the embryos into the uterus.
16. To improve the elution, DNA sample can be incubated at 37 °C for 1 h or at 4 °C overnight.
17. Exposure with UV light should be reduced to a minimum.
18. Fusion control should be on average 80–90%.
19. The air in straw allow a simple visible control of the correct application of the embryos into the fallopian tube.

Acknowledgments

The authors are grateful to the IVF and SCNT team, Dr. Andrea Lucas-Hahn, Dr. Monika Nowak-Imialek, Petra Hassel, Maren Ziegler, Roswitha Becker, and Antje Frenzel for their efforts in producing the SRY-KO pigs. We thank the staff from the pig facility for taking care of the pigs.

The pX330-U6-Chimeric_BB-CBh-hSpCas9 was a gift from Feng Zhang (Addgene plasmid # 42230; http://n2t.net/addgene:42230 ; RRID:Addgene_42230).

References

1. Kurtz S, Petersen B (2019) Pre-determination of sex in pigs by application of CRISPR/Cas system for genome editing. Theriogenology 137 (Proceedings of the IX International Conference of Boar Semen Preservation): 67–74
2. Babol J, Squires EJ (1995) Quality of meat from entire male pigs. Food Res Int 28(3): 201–212
3. Wesoly R, Weiler U (2012) Nutritional influences on skatole formation and skatole metabolism in the pig. Animals 2(2):221–242
4. Brooks R, Pearson A (1986) Steroid hormone pathways in the pig, with special emphasis on boar odor: a review. J Anim Sci 62(3):632–645
5. Bañón S, Andreu C, Laencina J, Ma-D G (2004) Fresh and eating pork quality from entire versus castrate heavy males. Food Qual Pref 15(3):293–300. https://doi.org/10.1016/S0950-3293(03)00069-7
6. Xue J (1997) Raising intact male pigs for meat: detecting and preventing boar taint. J Swine Health Prod 5(4):151–158
7. Xue J, Dial GD, Holton EE, Vickers Z, Squires EJ, Lou Y, Godbout D, Morel N (1996) Breed differences in boar taint: relationship between tissue levels boar taint compounds and sensory analysis of taint. J Anim Sci 74(9):2170–2177
8. Große-Brinkhaus C, Storck LC, Frieden L, Neuhoff C, Schellander K, Looft C, Tholen E (2015) Genome-wide association analyses for boar taint components and testicular traits revealed regions having pleiotropic effects. BMC Genet 16(1):36
9. Hansen LL, Mejer H, Thamsborg SM, Byrne DV, Roepstorff A, Karlsson AH, Hansen-Møller J, Jensen MT, Tuomola M (2006) Influence of chicory roots (*Cichorium intybus* L.) on boar taint in entire male and female pigs. Anim Sci 82(3):359–368
10. Lösel D, Claus R (2005) Dose-dependent effects of resistant potato starch in the diet on intestinal skatole formation and adipose tissue accumulation in the pig. J Vet Med Ser A 52(5):209–212
11. Bonneau M, Dufour R, Chouvet C, Roulet C, Meadus W, Squires E (1994) The effects of immunization against luteinizing hormone-releasing hormone on performance, sexual development, and levels of boar taint-related compounds in intact male pigs. J Anim Sci 72(1):14–20
12. Čandek-Potokar M, Škrlep M, Zamaratskaia G (2017) Immunocastration as alternative to surgical castration in pigs. In: Theriogenology. InTech
13. Aluwé M, Millet S, Bekaert K, Tuyttens F, Vanhaecke L, De Smet S, De Brabander D (2011) Influence of breed and slaughter weight on boar taint prevalence in entire male pigs. Animal 5(8):1283–1289
14. Aldal I, Andresen Ø, Egeli AK, Haugen J-E, Grødum A, Fjetland O, Eikaas JLH (2005) Levels of androstenone and skatole and the occurrence of boar taint in fat from young boars. Livestock Prod Sci 95(1-2):121–129
15. Valeeva NI, Backus GBC, Baltussen WHM (2009) Moving towards boar taint-free meat: an overview of alternatives to surgical castration from a chain perspective. Global agriculture from management, IFMA congres, Bloomington, 20–24 June 2009
16. Niemann H, Lucas-Hahn A, Petersen B (2019) The methodologies and application potential of genetically modified farm animals. Comp Biotechnol 4:466–480
17. Sekido R (2010) SRY: A transcriptional activator of mammalian testis determination. Int J Biochem Cell Biol 42(3):417–420
18. She Z-Y, Yang W-X (2014) Molecular mechanisms involved in mammalian primary sex determination. J Mol Endocrinol 53(1):R21–R37
19. Kashimada K, Koopman P (2010) Sry: the master switch in mammalian sex determination. Development 137(23):3921–3930. https://doi.org/10.1242/dev.048983
20. Kurtz S, Frenzel A, Lucas-Hahn A, Schlegelberger B, Göhring G, Niemann H, Mettenleiter TC, Petersen B (2021) Knockout of the HMG-Box Domain of the porcine SRY-gene causes sex reversal in gene-edited pigs. Proc Natl Acad Sci USA 118(2): e2008743118. https://doi.org/10.1073/PNAS.2008743118
21. Lucas-Hahn A, Petersen B, Nowak-Imialek M, Baulain U, Becker R, Eylers H-M, Hadeler K-G, Hassel P, Niemann H (2018) 122 A new maturation medium improves porcine embryo production in vitro. Reprod Fertil Dev 30(1):200–201. https://doi.org/10.1071/RDv30n1Ab122

Chapter 16

Generation of Double-Muscled Sheep and Goats by CRISPR/Cas9-Mediated Knockout of the Myostatin Gene

Peter Kalds, Martina Crispo, Chao Li, Laurent Tesson, Ignacio Anegón, Yulin Chen, Xiaolong Wang, and Alejo Menchaca

Abstract

The *myostatin* (*MSTN*) gene has shown to play a critical role in the regulation of skeletal muscle mass, and the translational inhibition of this gene has shown increased muscle mass, generating what is known as "double-muscling phenotype." Disruption of the *MSTN* gene expression using the CRISPR/Cas9 genome-editing system has shown improved muscle development and growth rates in livestock species, including sheep and goats. Here, we describe procedures for the generation of *MSTN* knockout sheep and goats using the microinjection approach of the CRISPR/Cas9 system, including the selection of targeting sgRNAs, the construction of CRISPR/Cas9 targeting vector, the in vitro examination of system efficiency, the in vivo targeting to generate *MSTN* knockout founders, the genomic and phenotypic characterization of the generated offspring, and the assessment of off-target effects in gene-edited founders through targeted validation of predicted off-target sites, as well as genome-wide off-target analysis by whole-genome sequencing. Editing the *MSTN* gene using the CRISPR/Cas9 system might be a rapid and promising alternative to promote meat production in livestock.

Key words Genome editing, CRISPR/Cas9, Knockout, Microinjection, *MSTN*, Meat production, Small ruminants, Sheep, Goats

1 Introduction

In the last few decades, unprecedented scientific and technological progress has been achieved to edit the genomes of many species up to the level of single nucleotide substitution with greater ease and accuracy [1–6]. Application of genome editing in animal breeding has enabled researchers to put their hands directly onto the genomic sequences and edit them to achieve the longstanding dreams of animal breeding in such a very short time with unprecedented precision. Domestic small ruminants, i.e., sheep (*Ovis aries*) and goats (*Capra hircus*), are ideal livestock species, and the application of genetic modification to manipulate the genomes of sheep and

Paul John Verma, Huseyin Sumer and Jun Liu (eds.), *Applications of Genome Modulation and Editing*,
Methods in Molecular Biology, vol. 2495, https://doi.org/10.1007/978-1-0716-2301-5_16,

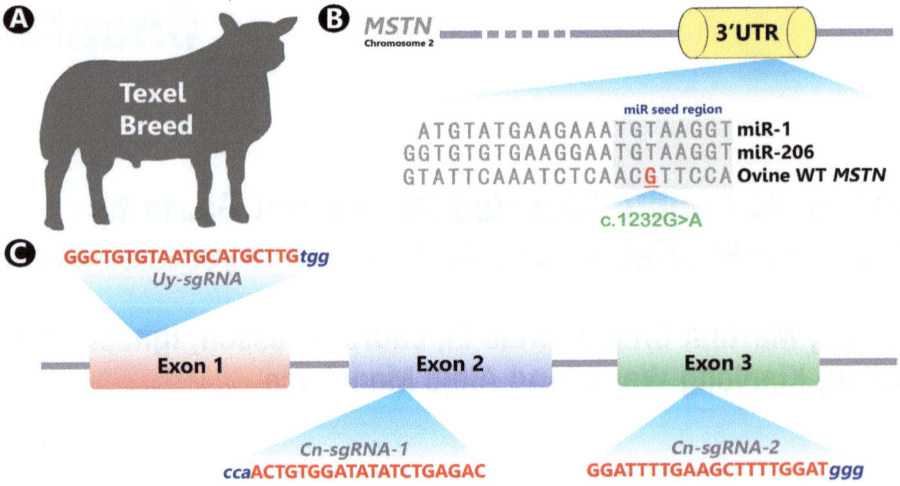

Fig. 1 The ovine *myostatin* (*MSTN*) gene, its natural causative mutation, and the positions of the Cas9/sgRNA-mediated ovine and caprine *MSTN* gene disruptions. (**a**) Texel sheep, the breed where the *MSTN* mutation that is responsible for the muscular hypertrophy phenotype was first discovered [14]. (**b**) The 3 prime untranslated region (3′UTR) of the ovine *MSTN* gene where the functional mutation is located (c.1232G>A; highlighted in red), including the alignment of the ovine wild-type (WT) *MSTN* sequence with sequences of the illegitimate microRNAs (miR-1 and miR-206; the highly expressed miRs in skeletal muscle). The G to A transition in the Texel sheep allows these miRs to bind to the *MSTN* gene, causing an expressional inhibition that generates the desirable double-muscling phenotype. The aligned sequences were adapted from Tellam et al. (2012) [13]. (**c**) The location of selected sgRNAs to target the ovine and caprine *MSTN* genes for the induction of CRISPR/Cas9-based gene loss-of-function insertions and deletions (indels) that mimic the inhibition influence of the *MSTN* c.1232G>A mutation

goats is of great importance to improve their utility in various fields [7–10]. The improvement of meat production is one of the ultimate goals of animal breeding and the demand for animal products is expected to be over 70% by 2050 due to the ever-increasing human population [11]. Myostatin (MSTN) is a member of the transforming growth factor-β (TGF-β) superfamily that showed to be a significant negative regulator of muscle differentiation and growth in various mammals [12, 13]. In sheep, a natural mutation has shown to cause translational inhibition of the *MSTN* gene that gives rise to the agriculturally desirable double-muscling phenotype [14] (Fig. 1a, b). Several *MSTN* knockout sheep and goat models have been generated using various genetic modification tools, including the clustered regularly interspaced short palindromic repeats/CRISPR-associated protein 9 (CRISPR/Cas9) system [9, 10]. These studies have reported the preferable increased muscle mass phenotype after targeting the *MSTN* gene, suggesting that genome editing is a significant approach to introduce economically desirable traits such as the promotion of growth rates and muscle development in sheep and goats. Based on the importance of genome editing as a promising approach, preferable phenotype of

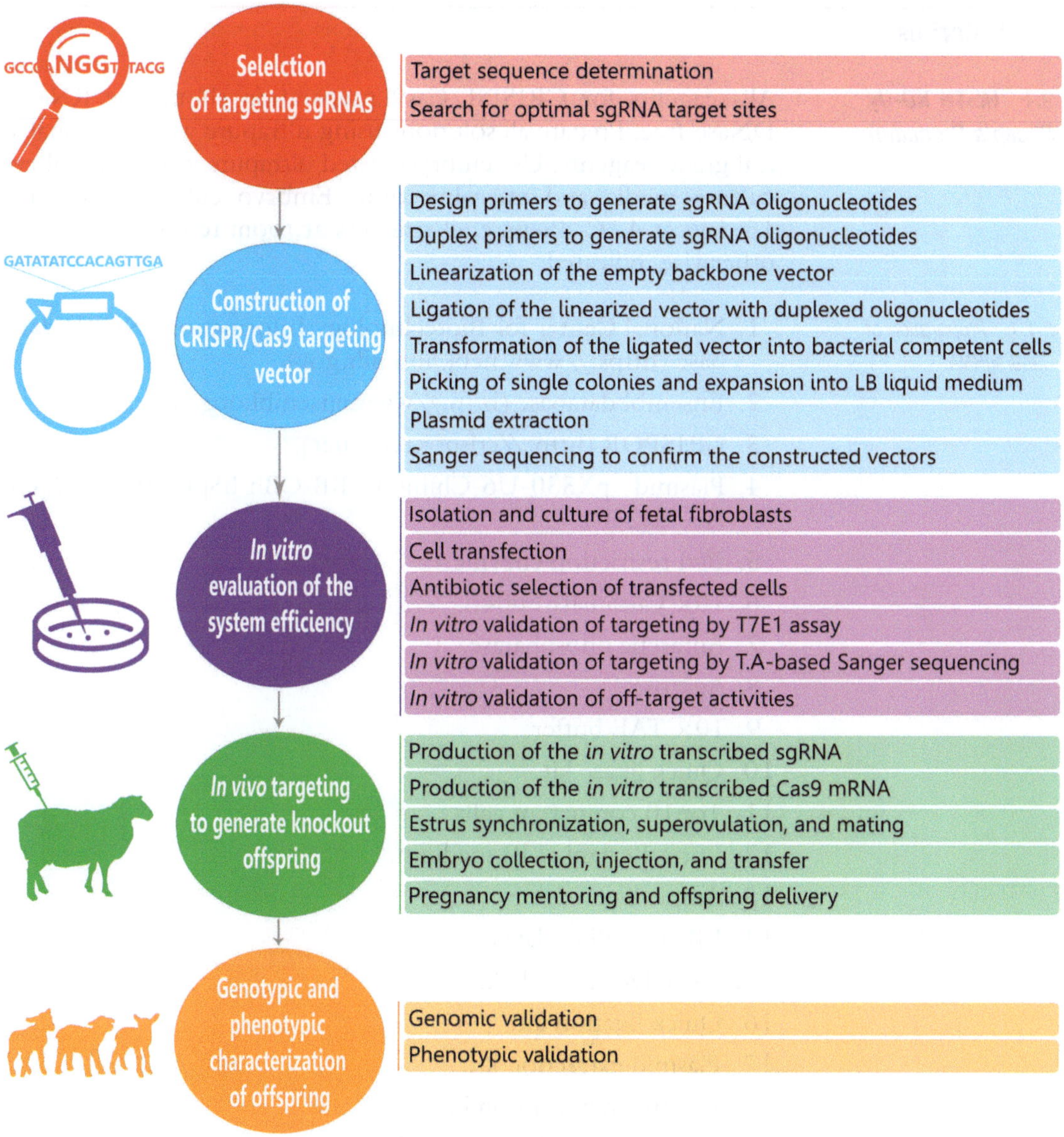

Fig. 2 The main steps to generate *myostatin* (*MSTN*) knockout sheep and goats using the CRISPR/Cas9 genome-editing system

the *MSTN* expressional inhibition, and small ruminants as an ideal livestock species, we provide the current procedures that describe the technical steps (Fig. 2) to generate *MSTN* knockout sheep and goats using the CRISPR/Cas9 system. The procedures described in this chapter were used to produce live *MSTN* gene-edited lambs/kids in Uruguay in 2015 (named as *MSTN* KO-Uy sheep [15]) as well as in China in 2015 and 2016 (named as *MSTN* KO-Cn sheep and goats [16, 17]).

2 Materials

2.1 MSTN *KO-Uy* *Sheep's Procedure*

All reagents for CRISPR/Cas9 preparation should be RNase/DNase-free. Prepare all solutions using ultrapure water and analytical grade reagents. Use embryo-tested ultrapure water for embryo culture media and microinjection. Embryo culture media last 15 days at 4 °C. Prepare all reagents at room temperature unless otherwise indicated.

2.1.1 Preparation of Cas9/sgRNA

1. National Center for Biotechnology Information (NCBI) database (https://www.ncbi.nlm.nih.gov).
2. Ensembl database (http://www.ensembl.org/index.html).
3. CRISPOR (http://crispor.tefor.net).
4. Plasmid pX330-U6-Chimeric_BB-CBh-hSpCas9 (Addgene No. 42230).
5. *Bbs*I restriction enzyme.
6. 10× CutSmart buffer.
7. Antarctic phosphatase.
8. Agarose.
9. 10× TAE buffer.
10. Quick ligase kit.
11. Top10 competent cells.
12. Luria-Bertani (LB) medium.
13. Ampicillin.
14. LB ampicillin plates.
15. T4 polynucleotide kinase.
16. Quick ligation kit.
17. Plasmid extraction kit.
18. In vitro transcription kit.
19. Megaclear kit (Life Technologies).

2.1.2 In Vitro Validation of the System Efficiency

1. A15 astroglial ovine cells.
2. Cell culture medium: Dulbecco's modified eagle medium (DMEM) supplemented with 10% fetal bovine serum (FBS), 10% glutamine, 2% sodium pyruvate, and 10% penicillin-streptomycin.
3. Lipofectamin LTX kit.
4. Nucleospin tissue.
5. CO_2 incubator.
6. Thermocycler.

7. Herculase II fusion polymerase.
8. T7 endonuclease I (T7E1).
9. Blastocysts DNA extraction kit.
10. Injection buffer: 10 mM Tris (pH 7.5) and 0.1 mM EDTA.
11. Synthetic oviduct fluid (SOF) medium supplemented with 5% BME essential amino acids, 2.5% MEM nonessential amino acids, and 0.4% bovine serum albumin (BSA).

2.1.3 In Vivo Targeting to Generate MSTN Knockout Founders

1. Searching medium: TCM199 containing 0.5% BSA, 25 mM HEPES, 50 IU/mL penicillin, and 50 μg/mL streptomycin.
2. Maturation medium: TCM199 containing 10% estrus goat/sheep serum, 10 μg/mL follicle-stimulating hormone (FSH), 10 μg/mL luteinizing hormone (LH), 100 μM cysteamine, 50 IU/mL penicillin, and 50 μg/mL streptomycin.
3. Fertilization medium: SOF medium supplemented with 2% estrus sheep serum, 10 μg/mL heparin, and 1 μg/mL hypotaurine.
4. Estrus sheep serum: Recovered from blood of females in estrus, heat-inactivated at 56 °C for 30 min, and frozen at −80 °C for further use.
5. Embryo culture medium: SOF medium containing 0.4% fat acid-free BSA, 5% essential, and 5% nonessential amino acids.
6. Embryo-tested mineral oil.
7. 35 mm, 60 mm, and 90 mm plastic Petri dishes.
8. 20 μL, 200 μL, and 1000 μL automatic pipettes.
9. CO_2 incubator.
10. Trigas chamber.
11. Stereomicroscope: 60× magnification.
12. Inverted microscope: 400× magnification.
13. Liquid nitrogen dewar.
14. RNA microinjection buffer: 10 mM Tris (pH 7.5) and 0.1 mM EDTA.
15. sgRNA and Cas9 mRNA.
16. 0.5 mL Eppendorf tubes.
17. Microinjection pulled capillary: Thin-walled borosilicate glass capillary with internal filament, 1 mm OD, and 0.78 mm ID.
18. Holding pipette: Bended, tip with 25 μm ID, and 100 μm OD.
19. Glass depression slide.
20. M2-buffered medium.
21. Microinjection station: Inverted microscope under the 200× magnification, two micromanipulators, and pressure injector.

22. Pipette puller.
23. Glass microforge.
24. Tomcat catheter: 1.0 mm in diameter.
25. Holding medium.
26. Shearing machine.
27. Stretcher.
28. Laparoscope.
29. Scalpel blade.
30. Atraumatic grasping forceps.
31. Paper clips.
32. Anesthesia: 5 mg/kg ketamine and 0.35 mg/kg diazepam in PBS.
33. Suture.
34. Antibiotic: 5 mg/kg ceftiofur.

2.1.4 Genotyping and Phenotyping of the Generated MSTN *Knockout Founders*

1. Ultrasound equipment (5 MHz probe) for pregnancy diagnosis and monitoring of fetal growth.
2. Cryovials.
3. Scalpel blade.
4. Monoclonal antibodies: Monoclonal mouse anti-myostatin and anti-GAPDH.
5. Secondary antibodies.
6. Horseradish peroxidase.
7. Scale and centimeter for phenotype determination at birth.

2.2 MSTN KO-Cn Sheep and Goats' Procedure

2.2.1 Preparation of Cas9/sgRNA

1. NCBI database (https://www.ncbi.nlm.nih.gov).
2. CHOPCHOP (https://chopchop.cbu.uib.no).
3. CRISPOR (http://crispor.tefor.net).
4. Thermocycler.
5. Warm water bath.
6. Autoclave.
7. Incubator.
8. Shaking incubator.
9. Pipettes.
10. Pipette tips.
11. Plasmid pGL3-U6-sgRNA-PGK-puromycin (Addgene No. 51133).
12. NEBuffer 2 (NEB).
13. CutSmart® Buffer (NEB).

14. *Bas*I restriction enzyme.
15. DNA purification kit.
16. 10× T4 DNA ligase buffer.
17. T4 DNA ligase.
18. *Escherichia coli* DH5α competent cells.
19. Ice container.
20. Tryptone.
21. Yeast extract.
22. NaCl (sodium chloride).
23. Agarose.
24. Ampicillin.
25. Kanamycin.
26. Glass flasks.
27. Petri plates.
28. Glass spreaders or glass beads.
29. Tongs.
30. 15 mL centrifuge tubes.
31. Plasmid extraction kit.
32. Micro-volume spectrophotometer.
33. Plasmid pUC57-sgRNA expression vector (Addgene No. 51132).
34. Plasmid pST1374-NLS-flag-linker-Cas9 (Addgene No. 44758).
35. High-fidelity DNA polymerase.
36. *Age*I restriction enzyme.
37. PCR purification kit.
38. RNase inactivation kit.
39. In vitro transcription kit.
40. Transcription clean-up kit.

2.2.2 In Vitro Validation of the System Efficiency

1. Dissection tools.
2. 0.25% trypsin and 0.02% EDTA solution.
3. Dimethyl sulfoxide (DMSO).
4. Cryovials.
5. Liquid nitrogen tank.
6. Lipofection kit.
7. Opti-MEM™ medium (ThermoFisher Scientific, Gibco™).
8. Puromycin dihydrochloride.

9. Blasticidin S hydrochloride.
10. Genomic DNA extraction kit.
11. T7E1.
12. High-fidelity DNA polymerase.
13. PCR purification kit.
14. TA cloning kit.
15. Cas-OFFinder (http://www.rgenome.net/cas-offinder).
16. Single cell whole genome amplification kit.

2.2.3 In Vivo Targeting to Generate MSTN Knockout Founders

1. EAZI-BREED CIDR (controlled internal drug release) Sheep and Goat Devices (EAZIBREED™).
2. Human menopausal gonadotrophin (hMG) or FSH.
3. Cloprostenol injection.
4. Luteinizing hormone-releasing hormone A_3 (LRH-A_3).
5. Pregnant mare serum gonadotropin (PMSG).
6. TCM199 medium.
7. Heated platform with the Olympus micromanipulation system ON3 (Olympus).
8. Eppendorf FemtoJet System (Eppendorf).
9. Quinn's advantage cleavage medium (QACM).

2.2.4 Genotyping and Phenotyping of the Generated MSTN *Knockout Founders*

1. Blood collection tube.
2. Blood and tissue DNA extraction kits.
3. Weighing scales.
4. Measuring tape and staff.

2.3 Ordering Following Oligos and Primers for the Study see Table 1

3 Methods

3.1 MSTN *KO-Uy* Sheep's Procedure

3.1.1 Design of Cas9/sgRNA

1. Target sequence determination: The *MSTN* gene sequence was obtained using the Ensembl database (http://www.ensembl.org/index.html). Gene ID: the ovine *MSTN* NCBI number is 443449.
2. Search for optimal sgRNA target sites: Search for efficient sgRNA target sites based on the main criterial sequence G $(N)_{20}$GG for the commonly used type-II *Streptococcus pyogenes* Cas9 (spCas9) protein using available online sgRNA design tools (Zhang Lab: Guide Design Resources https://zlab.bio/

guide-design-resources), for example, CRISPOR (http://crispor.tefor.net) [18] (*see* **Note 1**). Efficient sgRNA target sites with high-ranking score and low off-target matches should be selected for subsequent efficiency examination in vitro (i.e., at cellular and/or embryonic levels).

3.1.2 Construction of CRISPR/Cas9 Vector for In Vitro Targeting

1. Design primers to generate sgRNA oligonucleotides: After the selection of sgRNA target sites, design primers to generate the sgRNA double-strand oligonucleotides for the insertion into the sgRNA expression vector. The pX330-U6-Chimeric_BB-CBh-hSpCas9 vector is used to express sgRNA from the U6 promoter for the examination of system efficiency in vitro. To insert the selected sgRNA target site (Fig. 1c) into pX330-U6-Chimeric_BB-CBh-hSpCas9 vector, design a pair of single-stranded oligos with the addition of the required adapter sequences to both the Uy-sRNA-oligo-F and Uy-sRNA-oligo-R (Table 1).
2. Duplex primers to generate sgRNA oligonucleotides: After receiving the synthesized oligos, perform the following

Table 1
Oligo and primer sequences

Name	Sequences (5′–3′)[a]
Uy-sRNA-oligo-F	*CACC*GGGCTGTGTAATGCATGCTTG
Uy-sRNA-oligo-R	*AAAC*CAAGCATGCATTACACAGCCC
Uy-IVT-sgRNA-F	TTAATACGACTCACTATAGGCTGTGTAATGCATGCTTG
Uy-IVT-sgRNA-R	TTTAAAAGCACCGACTCGGTGCC
Uy-T7E1-F	TCACTGGTGTGGCAAGTTGT
Uy-T7E1-R	AAAAGCTCTTTGCCCTCCTC
Cn-sRNA-Ex2-oligo-F	*ACCG*TCTCAGATATATCCACAGT
Cn-sRNA-Ex2-oligo-R	*AAAC*ACTGTGGATATATCTGAGA
Cn-sRNA-Ex3-oligo-F	*ACCG*ATTTTGAAGCTTTTGGAT
Cn-sRNA-Ex3-oligo-R	*AAAC*ATCCAAAAGCTTCAAAAT
Cn-IVT-sgRNA-F	TCTCGCGCGTTTCGGTGATGACGG
Cn-IVT-sgRNA-R	AAAAAAAGCACCGACTCGGTGCCACTTTTTC
hU6 primer	GACTATCATATGCTTACCGT
RVprimer3	CTAGCAAAATAGGCTGTCCC
M13F(-20) primer	GTAAAACGACGGCCAGT
M13F(-47) primer	CGCCAGGGTTTTCCCAGTCACGAC

[a]Adapter sequences are italicized and underlined

reaction in a thermocycler to duplex "anneal" each pair of each target site. To prepare a 20 μL reaction, use 5 μL of 100 μM forward oligo, 5 μL of 100 μM reverse oligo, and 10 μL TEN 10:1:50 buffer. Run the following annealing program: 95 °C for 5 min, followed by 95–85 °C at −2 °C/s, 85–25 °C at −0.1 °C/s, and then hold at 4 °C.

3. Linearization of the empty backbone vector: Subject the pX330-U6-Chimeric_BB-CBh-hSpCas9 vector to digestion by *Bbs*I restriction enzyme. Add 3 μg of the empty backbone vector, 5 μL CutSmart® Buffer, 1.5 μL *Bbs*I-HF, and sterile water to 50 μL. Incubate at 37 °C for 1 h. Perform gel electrophoresis to check the result of vector digestion. Purify the digestion products using a purification kit.
4. Ligation of the linearized vector with the duplexed sgRNA oligonucleotides: Add 10 μL 2× T4 DNA Quick ligase buffer, 1 μL T4 DNA Quick ligase, 50 ng linearized vector, 1 μL duplexed sgRNA oligonucleotides, and sterile water to 20 μL. Incubate 30 min at room temperature.
5. Transformation of the ligated vector into bacterial competent cells: Mix 1 μL ligated vector with 30–50 μL of *Escherichia coli* Top10 competent cells. Incubate the competent cells/ligated vector mixture into ice container for 30 min. Heat shock the transformation tube at 42 °C for 30 s. Put the tubes back into the ice container for 2 min. Spread the mix onto a prewarmed LB solid medium using sterile glass spreaders or glass beads (*see* **Note 2**). Incubate plates at 37 °C overnight.
6. Picking of single colonies and expansion into LB liquid medium: Pick up monoclonal colonies, about four colonies from each inoculated plate, and separately culture them into 15 mL centrifuge tubes containing 5 mL LB liquid medium (*see* **Note 2**). Incubate into shaking incubator (37 °C/ 180–220 rpm) for about 16–18 h.
7. Plasmid extraction: Extract constructed vectors from transformed competent cells using a plasmid extraction kit following the manufacturer's instructions. Use a micro-volume spectrophotometer for nucleic acid quantification of extracted vectors. Check positive colonies by restriction enzyme digestion with *Eco*RI and *Bbs*I.
8. Sanger sequencing to confirm the constructed vectors: Confirm the correctly constructed vectors by Sanger sequencing using a commonly used hU6 primer (Table 1) for sequencing.

3.1.3 Production of sgRNA/Cas9 mRNA for Embryo Microinjection

1. Production of the in vitro transcribed sgRNA: Construct the in vitro transcription matrix of the sgRNA by inserting the T7 promoter to the sgRNA duplexed oligonucleotides previously cloned into the pX330-U6-Chimeric_BB-CBh-hSpCas9

expression vector by PCR. Amplify the sgRNA-containing region using a high-fidelity Herculase II fusion DNA polymerase and primers: Uy-IVT-sgRNA-F and Uy-IVT-sgRNA-R (Table 1). Use the following PCR program: 94 °C for 2 min, 35 cycles of (98 °C for 10 s, 60 °C for 10 s, and 72 °C for 30 s), 72 °C for 3 min, and hold at 4 °C. Run the PCR products on a gel electrophoresis and recover the desired product using a gel extraction kit. Transcribe sgRNA using an MEGAshortscript™ T7 Transcription Kit following the manufacturer's instructions. Purify transcription products using a MEGAclear™ Transcription Clean-Up Kit following the manufacturer's instructions. Load the purified products into 1% agarose gel to check the quality of generated sgRNA.

2. Production of the in vitro transcribed Cas9 mRNA: Use the JDS246 vector to generate Cas9 mRNA. Amplify Cas9 vector and then extract vector using a plasmid extraction kit. Linearize the extracted vector by *Pme*I resection enzyme and using the following solution, 10 μg vector, 10 μL CutSmart® Buffer, 5 μL *Pme*I, and sterile water to 100 μL. Incubate the reaction at 37 °C for 1 h. Run gel electrophoresis to confirm vector linearization and to extract the desired digestion products using a gel extraction kit. Transcribe Cas9 using an mMESSAGE mMACHINE™ T7 ULTRA Transcription Kit, and DNase I treatment. Purify transcription products using a MEGAclear™ Transcription Clean-Up Kit following the manufacturer's instructions. Load the purified products into 1% agarose gel to check the quality of generated Cas9 mRNA.

3.1.4 In Vitro Validation of the System Efficiency in Cells and Embryos

1. Cell transfection: Use lipofection kit to deliver CRISPR reagents into A15 astroglial sheep cells according to the instructions. The following conditions are for the transfection using 24-well plates. In day 0, seed cells to be 70–90% confluent at transfection (0.5–2 × 10^5). In day 1, dilute 5 μL Lipofectamine™ LTX Reagent in 50 μL Opti-MEM™ medium and mix well to generate Diluted Lipofectamine™ LXT Reagent. Prepare a master mix of DNA by diluting 2 μg DNA in 250 μL Opti-MEM™ medium in each tube, then add 5 μL PLUS™ Reagent and mix well to generate Diluted DNA with PLUS™ Reagent. Add 50 μL Diluted DNA with PLUS™ Reagent to each 50 μL Diluted Lipofectamine™ LTX Reagent to generate DNA-lipid complex. Incubate for 5 min at room temperature then add DNA-lipid complex to cells. Incubate cells at 37 °C for 2–4 days.
2. Genomic DNA extraction: Extract genomic DNA of the transfected cells and Day-6 embryos using the standard protocol of phenol-chloroform extraction/alcohol-sodium acetate

precipitation or using a genomic DNA extraction kit following the manufacturer's instructions. After extraction, use a micro-volume spectrophotometer to quantify the extracted genomic DNA.

3. In vitro validation of targeting activities by T7E1 assay: Preform T7E1 cleavage assay to confirm the occurrence of editing activities using primers Uy-T7E1-F and Uy-T7E1-R (Table 1). Amplify targeted fragments from the extracted DNA using Herculase II Fusion DNA polymerase. PCR solution contains 5 μL 5× Herculase buffer, 2.5 μL of 2.5 mM dNTPs, 1 μL of 1.0 U/μL Herculase II fusion DNA polymerase, 0.625 μL of 10 μM forward primer, 0.625 μL of 10 μM reverse primer, 10–50 ng template DNA, and sterile water to 25 μL. Run PCR program: 94 °C for 2 min, 35 cycles of (98 °C for 10 s, 60 °C for 10 s, and 72 °C for 30 s), 72 °C for 3 min, and then hold at 4 °C. Check the generated PCR products by gel electrophoresis. Denature and reanneal the purified PCR products in NEB-uffer 2 using a thermocycler according to the following protocol; 95 °C for 5 min, 95–85 °C at −2 °C/s, 85–25 °C at −0.1 °C/s, and then hold at 4 °C. Digest the hybridized PCR products with T7E1 for 30 min at 37 °C. Load the T7E1-digested products into 2–2.5% agarose gel or on microfluidic electrophoresis device (LabChip GXII, Caliper) [19]. The ratio of the cleaved products (255 bp and 379 bp) to the uncleaved products (634 bp) was used to calculate non-homologous end joining (NHEJ) frequency using Image J software. NHEJ frequency was calculated as follows: % gene modification = 100 × (1-fraction cleaved)^1/2.
4. In vitro validation of off-target activities: The sgRNA selection tools can provide a list of potential off-target sites. Design primers to amplify potential off-target sites. Use the extracted genomic DNA to check the potential off-targets primarily using PCR-based Sanger sequencing and further validation strategies based on T7E1 assay, T.A cloning-based Sanger sequencing, and targeted deep sequencing.

3.1.5 Embryo Production for Microinjection

Zygotes for microinjection are produced using in vitro embryo technology (IVEP): immature cumulus-oocyte complexes (COCs) are collected from the ovaries and subjected to in vitro maturation (IVM). The matured oocytes are fertilized in vitro (IVF) and cultured until embryo transfer. Methods used in our laboratory are further described elsewhere [20]. Briefly, the procedure is performed as follows:

1. COCs are retrieved from ovaries collected in slaughterhouse, through aspiration from 2 to 8 mm ovarian follicles by using an appropriate syringe and needle. About 250–300 ovaries can be

processed in a single day and 300–350 good quality COCs can be retrieved.

2. Load retrieved COCs into a 90 mm Petri dish and add prewarmed searching medium to fill the Petri dish.
3. Pick up COCs using a glass capillary pipette under a stereomicroscope (40×) and transfer to a 35 mm Petri dish containing search medium.
4. Select Grade 1 and 2 COCs (*see* **Note 3**) and wash in drops of search medium.
5. For IVM, place 20–25 selected COCs into 100 μL drops of maturation medium (1 oocyte/2–5 μL of medium) under mineral oil in 35 mm Petri dishes and incubate for 24–27 h at 39 °C in a humidified atmosphere containing 5% CO_2.
6. Sperm selection and capacitation (*see* **Note 4**): Trespass the semen sample from the straw in 1 mL of capacitation/fertilization medium into a 15 mL Falcon conical tube and incubate in 45° angle position for 15 min at 39 °C in a humidified atmosphere with 5% CO_2. After incubation, trespass 840 μL of the supernatant medium into a prewarmed 1.5 mL plastic tube.
7. After sperm selection, evaluate the percentage of viable sperm and individual progressive sperm motility under inverted microscope at 400× magnification. Determine sperm concentration in a Neubauer chamber and calculate the insemination dose.
8. For IVF, co-incubate matured COCs with 1×10^6 sperm into 100 μL of fertilization medium drops, covered by mineral oil in 35 mm Petri dishes, for 16–18 h at 39 °C in humidified atmosphere containing 5% CO_2 (*see* **Note 5**).
9. After IVF, the presumptive zygotes are ready to be subjected to microinjection.

3.1.6 Zygote Microinjection

Eighteen hours after fertilization, denude presumptive zygotes by gently pipetting COCs with a 1000 μL micropipette in a 500 μL drop of buffered SOF medium until all cumulus cells have been removed.

1. Preparation of capillary glass: Pull injection pipettes. The tip of pulled pipettes may be bended at an angle of 10–15° with a microforge. Prepare at least 20 pulled pipettes for a typical microinjection day (*see* **Note 6**). Glass holding pipettes with 15° angle can be purchased from a commercial vendor or be homemade.
2. Microinjection mix preparation: Mix 5 ng/μL of sgRNA and 20 ng/μL of Cas9 mRNA in RNA microinjection buffer.

Prepare at list 10 μL of mix for a standard microinjection day (up to 500 zygotes) (*see* **Notes 7** and **8**).

3. Microinjection setup: Switch on the microinjection pressure device to stabilize pressure at list 1 h before starting the first microinjection session. Set the injection pressure at 500–1000 hPa and the compensation pressure at 30–50 hPa (*see* **Note 9**). Load 20 μL of M2 medium on the depression of a glass slide injection chamber, cover it with mineral oil, and place it on the microscope stage. Insert the holding pipette into the holder of the left-hand-side micromanipulator. Low the pipette into the M2 drop through the oil and adjust it so that the tip is parallel to the bottom of the chamber. Backfill the injection pipette by dipping the back into the 0.5 mL tube containing the CRISPR/Cas9 solution and let the mix enter the tip of the pipette by capillary force. Assemble the pipette into the holder of the right-hand-side micromanipulator. Low it into the microinjection drop and adjust the pipette in the same focus as the holding pipette, parallel to the bottom of the injection chamber. Carefully tap the tip of the injection pipette against the tip of the holding pipette to break it. Test the microinjection flow by applying pressure.
4. Microinjection session: Transfer a group of 100–150 denuded zygotes to the upper region of the injection chamber. Attach one zygote to the tip of the holding pipette by applying negative pressure with the air device attached to the holder of the left-hand micromanipulator. Adjust the zona pellucida (ZP) and the tip of the injection pipette on the same level. Insert the injection pipette through the ZP and the cell membrane into the cytoplasm and apply injection pressure through the injector. When the cytoplasm swells, quickly pull the pipette out of the zygote's cytoplasm. Place the injected zygote on the lower region of the M2 drop, far away from the non-injected zygotes. Inject the 100–150 zygotes one by one within 30 min until the whole group is ready (*see* **Notes 10** and **11**). Transfer injected zygotes into a fresh drop of warm culture medium for washing and further culture. Load a new group of embryos to the chamber and proceed to inject them with a freshly loaded injection pipette. Incubate injected zygotes in 100 μL of culture medium drops in a tri-gas chamber with 5% CO_2, 5% O_2, and 90% N_2 in humidified atmosphere at 39 °C (*see* **Note 12**) until DNA extraction for genotyping and sequencing [15], cryopreservation [21], or embryo transfer to recipients [22].

3.1.7 Embryo Transfer to Generate MSTN *KO-Uy Sheep Founders*

1. Synchronize recipient's estrous cycle with the stage of the embryos to be transferred (*see* **Notes 13** and **14**). The day of ovulation of the recipients should be synchronized with the day of IVF (*see* **Notes 15 and 16**).

2. Place the recipient in a stretcher. Shear and clip cranial to the udder, wash, and sanitize the skin.
3. Administer antibiotics and anesthesia prior to the transfer procedure.
4. Make a ~3 cm incision in the midventral region just cranial to the udder and introduce atraumatic grasping forceps into the abdominal cavity.
5. Examine both ovaries with the laparoscope to find at least one *corpus hemorrhagicum*. No large follicles should be present. Females bearing either a small or pale corpus luteum should be rejected.
6. Exteriorize the horn to be transferred through the incision in the right paramedian region. Hold the horn between the thumb and index finger.
7. Perform a puncture through the uterine wall using a sterilized small diameter instrument (e.g., a paper clip or blunt 20 g needle).
8. Load the embryos into a Tomcat catheter (i.e., 1.0 mm in diameter) with a small volume of holding medium and 1 or 2 air bubbles positioned on both sides of the embryos (i.e., air-medium-air-embryo-air-medium-air).
9. Insert the tip of the Tomcat catheter at least 1–2 cm into the uterine horn and direct it cranially into the horn. Inject the contents of the catheter into the lumen, ensuring no bubbling reflux of fluid occurs. Push the oviduct and uterine horn back into the abdomen.
10. Suture body cavity by picking up the peritoneum and the abdominal wall by the *linea alba* as one layer. Skin incision may be closed with a non-absorbable suture material using a continuous suture pattern or metal clips.

3.1.8 Pregnancy, Birth, and Offspring Assessment

1. Check pregnancy and fetal development using B-mode trans-abdominal ultrasonography with a 5 Mhz probe at Day 30. Check single, twin, or triplet pregnancy and register. At Day 105, monitor thoracic diameter, biparietal diameter, occipitonasal length, and heart rate.
2. At birth, register length of gestation, gender, rectal temperature, heart and respiratory rates, body weight, thoracic perimeter, biparietal diameter, crown-rump and occipitonasal length, height at withers, height at hips, width at hips and width at chest, and any anomalous observation.
3. Register offspring survival during the first month of life. Register body weight and morphometric variables at Day 15, 30, and 60.

3.1.9 Genotyping of the Generated MSTN KO-Uy Founders

When the lambs reach 1 week, recover tissue samples from skin and muscle with a scalpel, and freeze at −20 °C until genotyping. Withdraw blood from the jugular vein, and store until DNA extraction. Fix samples from deltoid and biceps femoris muscles in 4% formaldehyde solution until phenotype analysis.

1. Extract genomic DNA using blood and tissue DNA extraction kit as described in Subheading 3.1.4.
2. Perform PCR amplification of *MSTN* target site as described in Subheading 3.1.4.
3. Validate the on-target activities using PCR-based Sanger sequencing, T7E1 assay, T.A cloning-based Sanger sequencing as described above in Subheading 3.1.4.
4. To validate off-target activities, amplify potential off-target sites by PCR and follow the same steps of genomic validation of on-target activities as described in Subheading 3.1.4.

3.1.10 Phenotyping of the Generated MSTN KO-Uy Founders

1. To analyze the *MSTN* expression in muscle fiber, extract total proteins to perform Western blotting. Run equal amounts of protein on a 12% (v/v) gel electrophoresis and electrophoretically transfer to a PVDF membrane. Use monoclonal mouse anti-myostatin and anti-GAPDH antibodies for the Western blotting. Incubate washed membranes with 1:50,000 dilution of secondary antibody linked to horseradish peroxidase (HPR). Detect HPR activity using Western blot chemiluminescence.
2. For histology, include fixed samples in paraffin, section 1 mm slides, and stain with hematoxylin-eosin to study muscle morphology. Measure muscle fibers (min 250 per sample) with respect to their minimum Feret (MinFeret) diameter. Calculate mean area. A representative example of an *MSTN* KO lamb generated using this procedure is shown in Fig. 3d.

3.2 MSTN *KO-Cn* Sheep and Goats' Procedure

3.2.1 Design of Cas9/sgRNA

1. Target sequence determination: After the determination of the target gene, in the case of this protocol is the *MSTN* gene, search and obtain gene sequence using the NCBI database (https://www.ncbi.nlm.nih.gov). Gene ID: the ovine *MSTN* is 443449 and the caprine *MSTN* is 100860887.
2. Search for optimal sgRNA target sites: Search for efficient sgRNA target sites based on the main criterial sequence G $(N)_{20}GG$ for the commonly used type-II spCas9 protein using available online sgRNA design tools (Zhang Lab: Guide Design Resources https://zlab.bio/guide-design-resources), for example, CHOPCHOP (https://chopchop.cbu.uib.no) [23] or CRISPOR (http://crispor.tefor.net) [18] (*see* **Note 1**). Efficient sgRNA target sites with high-ranking score and low off-target matches should be selected for subsequent efficiency examination in vitro.

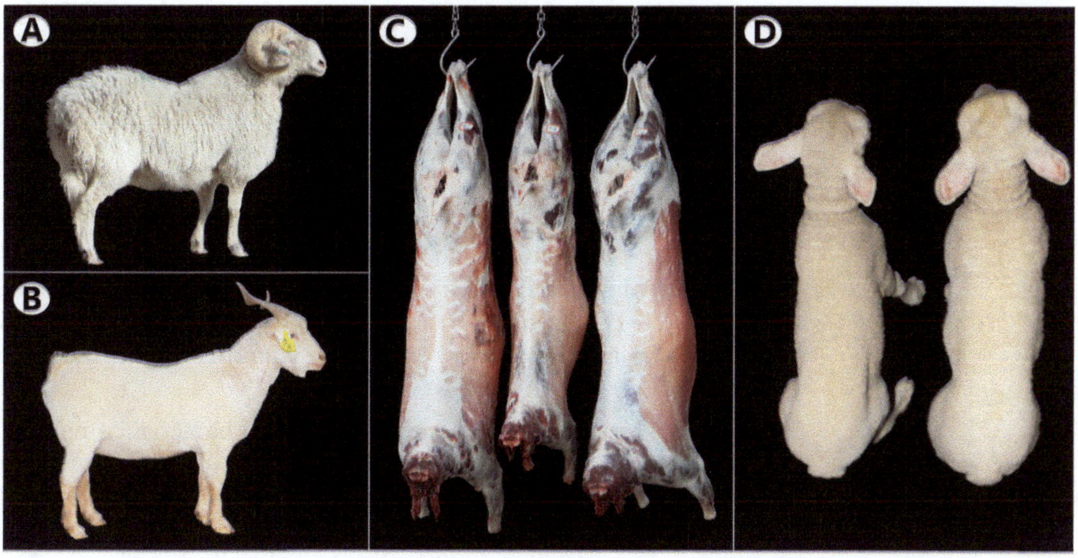

Fig. 3 Representative examples of the CRISPR/Cas9-generated *myostatin* (*MSTN*) knockout small ruminants. (**a**) A view of the *MSTN* knockout Chinese Tan sheep (*MSTN* KO-Cn sheep). (**b**) A lateral view of the *MSTN* knockout Chinese Shaanbei white Cashmere goat (*MSTN* KO-Cn goat). (**c**) Representative comparison between the carcasses of the *MSTN* knockout (right and left) and wild-type (middle) individuals (*MSTN* KO-Cn goats). (**d**) A top view of the *MSTN* Knockout Merino sheep (right) and its wild-type counterpart (left) from Uruguay (*MSTN* KO-Uy sheep)

3.2.2 Construction of CRISPR/Cas9 Vector for In Vitro Targeting

1. Design primers to generate sgRNA oligonucleotides: After the selection of sgRNA target sites, design primers to generate the sgRNA double-strand oligonucleotides for the insertion into the sgRNA expression vector. The pGL3-U6-sgRNA-PGK-puromycin vector is used to express sgRNA from the U6 promoter for the examination of system efficiency in vitro. To insert the selected sgRNA (Fig. 1c) into the pGL3-U6-sgRNA-PGK-puromycin vector, design a pair of single-stranded oligos with the addition of the required adapter sequences for both the forward oligo (5′-ACCG-3′) and the reverse oligo (5′-AAAC-3′). Two pairs of oligos Cn-sRNA-Ex2-oligo-F/Cn-sRNA-Ex2-oligo-R and Cn-sRNA-Ex3-oligo-F/Cn-sRNA-Ex3-oligo-R (Table 1) were used for targeting ovine [17] and caprine [16] *MSTN* in our previous studies.
2. Duplex primers to generate sgRNA oligonucleotides: To prepare a 10 μL annealing reaction solution, add 4.5 μL of 100 μM forward oligo, 4.5 μL of 100 μM reverse oligo, and 1 μL NEBuffer 2. Run the following annealing program: 95 °C for 5 min, followed by 95–85 °C at −2 °C/s, 85–25 °C at −0.1 °C/s, and then hold at 4 °C.

3. Linearization of the backbone vector: Subject the pGL3-U6-sgRNA-PGK-puromycin vector to digestion using *BasI* restriction enzyme. Add 3 μg of the backbone vector, 5 μL CutSmart® Buffer (NEB), 1.5 μL *BasI*-HF (NEB), and sterile water to 50 μL. Incubate at 37 °C for a sufficient time (e.g., overnight). Perform gel electrophoresis to check the result of vector digestion. Purify the digestion products using a purification kit.
4. Ligation of the linearized backbone vector with the duplexed sgRNA oligonucleotides: Add 1 μL 10× T4 DNA ligase buffer, 1 μL T4 DNA ligase, 50 ng linearized vector, 1 μL duplexed sgRNA oligonucleotides, and sterile water to 10 μL. Incubate at 16 °C overnight.
5. Transformation of the ligated vector into bacterial competent cells: Mix 5 μL ligated vector with 50 μL of *Escherichia coli* DH5α competent cells. Incubate the competent cells/ligated vector mixture on ice for 30 min. Heat shock the transformation tube at 42 °C for 90 s. Put the tubes back on ice for 2–3 min. Spread the mix onto a prewarmed LB-agar plate using sterile glass spreaders or glass beads (*see* **Note 2**). Incubate plates at 37 °C overnight.
6. Picking of single colonies and expansion into LB liquid medium: Pick up monoclonal colonies (about four colonies from each inoculated plate) and separately culture them into 15 mL centrifuge tubes containing 5 mL LB medium (*see* **Note 2**). Incubate in a shaking incubator at 37 °C, 180–220 rpm for about 16–18 h.
7. Plasmid extraction: Extract constructed vectors from transformed competent cells using a plasmid extraction kit following the manufacturer's instructions. Use a micro-volume spectrophotometer for nucleic acid quantification of extracted vectors.
8. Sanger sequencing to confirm the constructed vectors: Confirm the correctly constructed vectors by Sanger sequencing using RVprimer3 (Table 1).

3.2.3 Production of sgRNA/Cas9 mRNA for Microinjection

1. Production of the in vitro transcribed sgRNA: Construct the in vitro transcription vector of the sgRNA by inserting the sgRNA duplexed oligonucleotides with the required adapter sequences (5′-TAGG-3′ for the forward oligo and 5′-AAAC-3′ for the reverse oligo) into the pUC57-sgRNA expression vector as the backbone for in vitro transcription of sgRNA driven by T7 promoter as described in Subheading 3.2.2. Validate the constructed vector by Sanger sequencing using M13F(-20) primer (Table 1). After the confirmation of the correctly constructed vector, perform the following PCR to amplify the sgRNA-containing region using a high-fidelity KOD DNA Polymerase and the following primers: Cn-IVT-sgRNA-F and

Cn-IVT-sgRNA-R (Table 1). PCR solution contains 5 μL 10× PCR buffer for KOD-Plus-Neo, 5 μL of 2 mM dNTPs, 3 μL of 25 mM $MgSO_4$, 1 μL of 1.0 U/μL KOD-Plus-Neo DNA polymerase, 1 μL of 10 μm forward primer Cn-IVT-sgRNA-F, 1 μL of 10 μm reverse primer Cn-IVT-sgRNA-R, 100–200 ng template DNA, and sterile water to 50 μL. PCR program: 94 °C for 2 min, 34 cycles of (98 °C for 10 s, 58 °C for 30 s, and 68 °C for 20 s), 68 °C for 5 min, and hold at 4 °C. Check the PCR products using gel electrophoresis and purify the PCR products using a PCR purification kit. After inactivating RNases using an RNase inactivation kit, transcribe sgRNA using MEGAshortscript™ T7 Transcription Kit following the manufacturer's instructions. After DNase treatment, purify sgRNA IVT products using MEGAclear™ Transcription Clean-Up Kit following the manufacturer's instructions. Load the purified products into 1% agarose gel to check the quality of generated sgRNA.

2. Production of the in vitro transcribed Cas9 mRNA: Use the pST1374-NLS-flag-linker-Cas9 vector as DNA template to perform IVT to generate Cas9 mRNA. Linearize the vector using *Age*I resection enzyme and in the following solution: 10 μg of the DNA template vector, 10 μL CutSmart® Buffer, 5 μL *Age*I-HF, and sterile water to 100 μL, incubate at 37 °C overnight. Run a gel electrophoresis and purify the linearized vector using a purification kit from the gel. After inactivating RNases using an RNase inactivation kit, transcribe Cas9 mRNA using mMESSAGE mMACHINE™ T7 ULTRA Transcription Kit. After DNase treatment, add poly-A tail to the IVT products using reagents provided in the mMESSAGE mMACHINE™ T7 ULTRA Kit. Purify the transcription products using RNeasy Mini Kit following the manufacturer's instructions. Load the purified products into 1% agarose gel to check the quality of the generated Cas9 mRNA.

3.2.4 In Vitro Validation of the System Efficiency in Cells and Embryos

1. Isolation and culture of fetal fibroblasts: Obtain viable fetuses (about 45 days old) from a pregnant ewe/doe. Transport fetuses to the laboratory and remove the heads, internal organs (innards), and legs. Cut the remaining tissues into small pieces (several mm in size). Culture the minced tissue in Petri dishes with culture medium at 37 °C in a humidified atmosphere of 5% CO_2. When the expanding cells become confluent, remove the explants and trypsinize the cells using 0.25% trypsin and 0.02% EDTA solution for propagations. At passages 3 to 5, freeze the fetal fibroblasts in DMEM freezing medium containing 10% FBS and 10% DMSO and store in liquid nitrogen until use (*see* **Note 17**).
2. Cell transfection: Use, for example, Lipofectamine™ 3000 kit to deliver CRISPR reagents into cells according to the

instructions. The following conditions are for the transfection using 6-well plates. At day 0: seed 0.25-1 × 10^6 cells. At day 1: the cells are at 70–90% confluency, dilute 3.75 μL and/or 7.5 μL Lipofectamine™ 3000 in 125 μL Opti-MEM™ medium and mix well to generate Diluted Lipofectamine 3000; dilute 5 μg DNA (*see* **Note 18**) in 125 μL Opti-MEM™ medium, then add 10 μL P3000 provided in the kit and mix well to generate Diluted DNA with P3000. Add 125 μL Diluted DNA with P3000 to 125 μL Diluted Lipofectamine 3000 to generate 250 μL DNA-lipid complex. Incubate the complex at room temperature for 10–15 min then add it to the cells. Culture the cells in the culture medium at 37 °C in a humidified atmosphere of 5% CO_2 for 2–4 days.

3. Antibiotic selection of the transfected cells: For antibiotic selection, use puromycin dihydrochloride to select sgRNA-expressing cells and blasticidin S hydrochloride to select Cas9-expressing cells. Ideally, it is better to first determine the optimal antibiotic concentrations required for screening the used fibroblast cells. This can be performed by examining gradual concentrations. An example of previously used concentrations includes adding 1 μg/mL puromycin dihydrochloride in sheep and goat fibroblasts cultures, and 5 μg/mL or 12 μg/mL blasticidin S hydrochloride in sheep or goat fibroblasts cultures, respectively, 48 h after transfection. Change the medium to antibiotic-free medium about 36–48 h after selection. Expand the survival cells into fresh medium until >90% confluency and then extract genomic DNA from the cells.
4. Genomic DNA extraction: Extract genomic DNA using the standard protocol of phenol-chloroform extraction/alcohol-sodium acetate precipitation or genomic DNA extraction kits. Quantify the extracted genomic DNA using a micro-volume spectrophotometer.
5. In vitro validation of targeting activities using T7E1 assay: Amplify targeted fragments from the extracted DNA using a high-fidelity KOD DNA polymerase and the designed T7E1 primers. The PCR reaction setup is described in Subheading 3.2.3 and PCR program is: 94 °C for 2 min, 34 cycles of (98 °C for 10 s, AT*°C for 30 s, and 68 °C for ET*), 68 °C for 5 min, and then hold at 4 °C forever (*see* **Note 19**). Purify the PCR products using a purification kit according to the manufacturer's instructions. Denature and reanneal the purified PCR products in NEBuffer 2 using a thermocycler according to the following program: 95 °C for 5 min, 5 cycles of (95–85 °C at −2 °C/cyc for 5 s), 600 cycles of (85–25°C at −0.1 °C/cyc for 5 s), and then hold at 4 °C. Digest the hybridized PCR products with T7E1 at 37 °C for 30 min. Load the T7E1-digested products into 2–2.5% agarose gel for gel electrophoresis.

6. In vitro validation of targeting activities using T.A cloning-based Sanger sequencing: Subclone the PCR products into T vectors using a TA cloning kit according to the manufacturer's instructions. Transform the generated vectors into DH5α competent cells as mentioned above (Subheading 3.2.2). Randomly pick up 16 colonies from each sample and expand into LB liquid medium. Extract plasmids using Mini-prep kits and subject to Sanger sequencing using M13F(-47) primer (Table 1) (*see* **Note 20**).
7. In vitro validation of potential off-target activities: The sgRNA selection tools can provide a list of potential off-targets of the selected sgRNAs; however, other software tools can be used to search for potential off-targets, for example, Cas-OFFinder (http://www.rgenome.net/cas-offinder) [24]. The strategies are described in Subheading 3.1.4 (*see* **Note 21**).
8. The preparations to obtain, handle, and treat embryos are described in the following sections. After 48 h of embryo microinjection, genomic DNA is amplified and extracted from embryos using a single cell whole genome amplification kit (e.g., Discover-sc Single Cell Kit). In vitro validation of targeting activities and potential off-target activities are evaluated using the embryo genomic DNA as described above.

3.2.5 *In Vivo Targeting to Generate* MSTN *Knockout Founders*

Zygotes for microinjection are obtained from live ewes and does after superovulation, mating, and surgical collection from fallopian tubes. The collected zygotes are subject to microinjection and embryo transfer for offspring delivery.

1. Select healthy ewes/does (about 2–5 years old) with normal/regular estrous cycles as donors for zygote collection.
2. Treat donor ewes/does with EAZI-BREED CIDR Sheep and Goat Devices (EAZI-BREED containing 300 mg of progesterone) *via* insertion into the vagina for 12 days in sheep and 14 days in goats.
3. Perform superovulation 60 h prior to CIDR device removal and another 12 h post CIDR device removal, a total of 72 h, using intramuscular injection of hMG (Ningbo Second Hormone Factory, China). Administer a total of 700 units of hMG in 7 injections of 150, 125, 125, 100, 75, 75, and 50 units or 300 units of FSH in 7 injections of 75, 50, 50, 37.5, 37.5, 25, and 25 units at 12 h intervals.
4. Treat donor ewes/does with 0.1 mg cloprostenol injection at the time of CIDR device removal and monitor the estrous status of the donors 12–28 h after the CIDR removal for mating.

5. Inject 12.5 μg LRH-A_3 at the first mating, then repeat mating at 8 h intervals (*see* **Note 22**).
6. Collect one-cell stage zygotes by surgical oviduct flushing from oviducts 10–16 h (average 14 h) after the last mating and immediately transfer to TCM199 medium (*see* **Note 23**).
7. Conduct microinjection in the manipulation medium TCM199 on a heated platform with the Olympus micromanipulation system ON3 (Olympus). Co-inject 100 ng/μL Cas9 mRNA and 50 ng/μL for each sgRNA into the cytoplasm of collected zygotes using the Eppendorf FemtoJet System (Eppendorf). Set injection pressure, compensatory pressure, and time parameter as 45 kPa, 7 kPa, and 0.1 s, respectively.
8. Culture the injected embryos in QACM at 37 °C, 5% CO_2, and saturated humidity conditions for 24 h.
9. Determine the surrogates according to their natural estrus cycles (*see* **Note 24**).
10. Transfer about 2–3 divisive embryos into the ampullary-isthmic junction of the oviducts.
11. Confirm pregnancy by observing estrous behaviors of surrogates at each ovulation cycle.
12. Monitor the birth of lambs/kids after about 150 days of pregnancy and provide the necessary care management to newborns.

3.2.6 *Genotyping and Phenotyping of the Generated* MSTN *knockout Founders*

1. Collect blood samples from jugular venous or peripheral venous, as well as tissue samples.
2. Extract genomic DNA using blood and tissue DNA extraction kits.
3. Perform PCR amplification of target sites. Validate the on-target activities using PCR-based Sanger sequencing, T7E1 assay, or T.A cloning-based Sanger sequencing as described above (Subheading 3.2.4) or using targeted deep sequencing (*see* **Note 21**).
4. To validate off-target activities, amplify potential off-target sites by PCR and follow the same steps of genomic validation of off-target activities as mentioned above (Subheading 3.2.4). Additionally, whole-genome sequencing (WGS) is performed for further validation of off-target activities (*see* **Note 25**).
5. Compare the body weight (BW) between mutant and wild-type founders. Weight the newborns at birth and each month after birth and calculate the average daily gain (ADG).
6. Further analyses can be performed to phenotypically confirm the editing events. These include muscle fiber histology by hematoxylin-eosin staining and transmission electron

microscopy analysis, as well as quantitative RT-PCR (qRT-PCR) and Western blotting to assess the expression level of the disrupted *MSTN* gene (*see* **Note 26**). Representative examples of the *MSTN* KO sheep and goats generated using this procedure are shown in Fig. 3a–c.

4 Notes

1. Various sgRNA selection tools can be used for the search of optimal sgRNA target sites [25–28]. Targeting the first exons of the gene of interest is preferable to ensure efficient gene function disruption. When using sgRNA transcription driven by hU6 promoter and in the absence of guanine (G) nucleotide at the beginning of 5′ end of the sgRNA, additional "G" is favored to be added to the sgRNA oligonucleotide during the sgRNA design. This additional "G" is helpful for sgRNA transcription [2, 3].
2. To prepare LB solid medium (1 L) combine the following reagents with 1 L water volume into a glass flask; 10 g tryptone, 5 g yeast extract, 10 g NaCl, and 15 g agar. After autoclave (120 °C/1 h) and cooling down the medium, add 100 μL ampicillin to each 100 mL (100 ng/mL) medium to prepare plates for U6 promoter-driven sgRNA vectors or 50 μL (50 ng/ml) kanamycin to each 100 mL medium to prepare plates for T7 promoter-driven sgRNA vectors. Pour the medium into Petri plates. When the medium becomes hardened, invert the plates and store at 4 °C until use. Note that exclude the addition of agar for the preparation of the LB liquid medium.
3. Oocyte quality and competence is a key factor for the IVEP success and the method for COC evaluation is described at the International Embryo Technology Manual [29].
4. In vitro sperm capacitation is achieved by adding capacitating agents during sperm washing and incubation, and also during sperm co-incubation in IVF. Chemical capacitating agents include heparin, hypotaurine, ionomycin, or a mixture of penicillamine, hypotaurine, and epinephrine.
5. Variability in sperm fertility occurs between bucks or rams, and also between ejaculates from the same male. Sperm evaluation by motility and morphology should be conducted according to the standardized procedure. In practice, because there is no single in vitro measurement that reflects the real fertility of the sperm, it is recommended to test each batch in one IVEP session by comparing it with other sperm batch of known fertility previously used.

6. A good injection pipette should have a tip with a diameter less than 1 μm. Pull the pipettes the same day to avoid dust accumulation.
7. Use RNase-free materials for the CRISPR mix preparation and manipulate the mix in ice during the whole microinjection session using powder-free disposable gloves.
8. Aliquots of Cas9 RNA microinjection mix ready to inject can be frozen at −80 °C for several months.
9. Adjust the injection pressure according to the flow of microinjection mix when injecting an embryo.
10. If the tip of the injection pipette becomes too large, discard the pipette and use a new one. Wide openings can damage the zygote and produce high lysis rates.
11. If the cytoplasm swells visibly, it has been successfully injected (approx. 5–10 pl). If the cytoplasm does not swell, the pipette is clogged or has not enter the plasma membrane. Unclog the pipette and reinject the zygote.
12. Culture medium must be equilibrated in the incubator at 39 °C and 5% CO_2 for at least 2 h prior to use. Replace 80% of the culture medium per drop on Day 3 and Day 6 [30].
13. Cleavage rate is recorded on Day 2 after fertilization (2–4 cell embryos/total oocytes). Development rate of morulae and blastocysts stages are recorded on Day 6. As reference, the expected blastocyst rate applying this procedure is around 30–40%.
14. Embryos can be transferred at the blastocyst stage (i.e., 6–7 days after IVF) into the uterus (late embryo transfer), or at 2- to 4-cell stage (i.e., within 2 days after IVF) into the oviduct (early embryo transfer). Often 3–5 embryos are placed per recipient that usually results in singleton or twin pregnancies.
15. Fixed-time embryo transfer is performed in a single day in those females that showed estrus behavior within 48 h after device removal, which requires the use of paint-marked vasectomized males from 24 to 48 h (or alternatively, estrus detection).
16. Ideally, recipients should be multiparous, preferably with a recent history of having successfully delivered and raised offspring. Select healthy females, in good body condition score, with good teeth and feet, and pay particular attention to udder condition (e.g., mastitis and damaged teats during shearing). Since for IVEP usually two or more embryos are transferred per recipient, body size should consider the possibility of pregnancies bearing twins.

17. Various cell types have been used to test the effect of targeting *MSTN* using CRISPR/Cas9 in sheep and goats. These include fetal fibroblast cells [16, 17, 31, 32], A15 astroglial cells [15], and skeletal muscle satellite cells [33], with a wide use of fetal fibroblasts, which are easy to culture and maintain in vitro.
18. During the preparation of DNA for transfection, add the constructed and validated sgRNA-expressing vector alongside with the Cas9-expression vector at ratio 2-3:1 Cas9:sgRNA(s).
19. AT* indicates Annealing Time, while ET* indicates Extension Temperature. These are changeable based on the GC/TA content of the designed primers and the length of PCR products, respectively. Calculate these values based on the conditions of selected primers and the expected size of PCR products.
20. PCR-based Sanger sequencing, T7El assay, and T.A cloning-based Sanger sequencing are used to detect editing events. Additionally, other strategies can also be employed for molecular validation such as the PCR-RFLP (restriction fragment length polymorphism) test in the presence of restriction enzyme target sites [32, 34–36] and targeted deep sequencing (*see* **Note 21**).
21. Targeted deep sequencing is used for the confirmation of both on- and off-target occurrence. The procedure of targeted deep sequencing briefly includes (i) amplification of on- and off-target regions from genomic DNA using a high-fidelity DNA polymerase, for example, PrimeSTAR HS (Takara), (ii) fragmentation of the amplified PCR products with a Covaris S220 ultrasonicator, (iii) preparation of the fragmented PCR products using the TruSeq CHIP Sample Preparation Kit (Illumina), (iv) quantification with a Qubit High-Sensitivity DNA Kit, and (v) pooling and sequencing of the PCR products with different tags using Illumina platform and following standard protocols. CRISPResso2 (https://crispresso.pinellolab.partners.org) is used for data analysis [37].
22. The same male should repeatedly mate the same donor female each mating time, except special cases.
23. Animals participated in surgical embryo collection and transfer should be fasted 24 h prior to admission for surgery.
24. To prepare recipients, select healthy ewes/does (about 2–5 years old) with normal/regular estrous cycles. Treat recipient ewes/does with progesterone-containing sponge suppository *via* insertion into the vagina for (12 days in sheep and 14 days in goats). Inject 150 units hMG or 400 units PMSG 24–36 h before the sponge suppository removal. Treat with 0.1 mg cloprostenol injection at the time of sponge suppository

removal and monitor the estrous status of the donors 12–24 h after the CIDR removal. Inject 12.5 μg LRH-A_3 to recipients that show estrus behavioral signals.

25. WGS is used for sufficient validation of off-target activity. The procedure of WGS briefly includes; (i) extraction of the genomic DNA from blood samples using, for example, Qiagen DNeasy Blood and Tissue Kit (Qiagen), (ii) fragmentation of extracted genomic DNA (~300 bp) by ultrasonication using a Covaris S2 system, (iii) construction of WGS library using 1 μg of the fragmented genomic DNA and Illumina TruSeq DNA Library Preparation Kit (Novogene), (iv) mapping the qualified reads to the reference genome (sheep [38] and goats [39]) using Burrows–Wheeler Aligner (BWA; v.0.7.13) tools (http://bio-bwa.sourceforge.net/) [40], (v) local realignment and base quality re-calibration using the Genome Analysis Toolkit (GATK) [41], (vi) calling the SNPs and small insertion and deletions (indels; <50 bp) using both GATK [41] and Samtools (http://samtools.sourceforge.net/) [42], (vii) filtration of the called SNPs according to the following criteria: (a) choosing the overlapped SNPs that are identified by GATK [41] and Samtools [41]; (b) filtering the SNPs that exist in SNP databases (e.g., http://animal.nwsuaf.edu.cn); (c) filtering common SNPs that exist in the control groups, for indels, the same filtration steps as SNPs are followed, and (viii) comparison of the positions of the candidate SNPs and indels obtained by the final filtration, whether they are located at the predicted off-target sites. SNPs and indels that coincide the predicted off-target sites are considered as potential off-targets.

26. Further validations of the *MSTN* editing events (including on- and off-target validations) and the accompanying editing effects in both mutants and wild-type counterparts have been studied in both sheep and goat models. These include (i) laboratory analyses of key chemistry parameters using blood samples; (ii) monitoring of body weight and body measurement traits (including body height, body length, chest circumference, chest depth, and chest width); (iii) confirmation of the occurrence of editing events in various tissues (heart, liver, spleen, lung, kidney, skin, muscle, testis, and ovary) from the three germ layers; (iv) presence of mosaicism; (v) germline transmission; (vi) changes in *MSTN* expressional level; (vii) muscle fiber histology; (viii) transcriptomic analyses [16, 17, 43, 44]; and (xi) the occurrence of off-target activities based on family trio-based deep sequencing [45, 46].

Acknowledgments

The technique described in *MSTN* KO-Uy was partially supported by Agencia Nacional de Investigación e Innovación (ANII), PEDECIBA, Institut Pasteur de Montevideo (IPMontevideo), and Instituto de Reproducción Animal de Uruguay (IRAUy). The authors of the *MSTN* KO-Cn sheep and goats' section are grateful to grants from the Natural Science Foundation of China (31972526 and 31772571) and local grants (2018KJXX-009, 2020KW-028, and NXTS-2020-001), as well as thankful to all of the colleagues in Prof. Yulin Chen's laboratory for their valuable discussions and suggestions.

References

1. Jinek M, Chylinski K, Fonfara I et al (2012) A programmable dual-RNA-guided DNA endonuclease in adaptive bacterial immunity. Science 337:816–821. https://doi.org/10.1126/science.1225829
2. Cong L, Ran FA, Cox D et al (2013) Multiplex genome engineering using CRISPR/Cas systems. Science 339:819–823. https://doi.org/10.1126/science.1231143
3. Mali P, Yang L, Esvelt KM et al (2013) RNA-guided human genome engineering via Cas9. Science 339:823–826. https://doi.org/10.1126/science.1232033
4. Komor AC, Kim YB, Packer MS et al (2016) Programmable editing of a target base in genomic DNA without double-stranded DNA cleavage. Nature 533:420–424. https://doi.org/10.1038/nature17946
5. Gaudelli NM, Komor AC, Rees HA et al (2017) Programmable base editing of A•T to G•C in genomic DNA without DNA cleavage. Nature 551:464–471. https://doi.org/10.1038/nature24644
6. Anzalone AV, Randolph PB, Davis JR et al (2019) Search-and-replace genome editing without double-strand breaks or donor DNA. Nature 576(7785):149–157. https://doi.org/10.1038/s41586-019-1711-4
7. Menchaca A, Anegon I, Whitelaw CBA et al (2016) New insights and current tools for genetically engineered (GE) sheep and goats. Theriogenology 86:160–169. https://doi.org/10.1016/j.theriogenology.2016.04.028
8. Menchaca A, dos Santos-Neto PC, Mulet AP, Crispo M (2020) CRISPR in livestock: from editing to printing. Theriogenology 150:247–254. https://doi.org/10.1016/j.theriogenology.2020.01.063
9. Kalds P, Zhou S, Cai B et al (2019) Sheep and goat genome engineering: from random transgenesis to the CRISPR era. Front Genet 10:750. https://doi.org/10.3389/fgene.2019.00750
10. Kalds P, Gao Y, Zhou S et al (2020) Redesigning small ruminant genomes with CRISPR toolkit: Overview and perspectives. Theriogenology 147:25–33. https://doi.org/10.1016/j.theriogenology.2020.02.015
11. Herrero M, Wirsenius S, Henderson B et al (2015) Livestock and the environment: What have we learned in the past decade? Annu Rev Environ Resour 40:177–202. https://doi.org/10.1146/annurev-environ-031113-093503
12. Aiello D, Patel K, Lasagna E (2018) The *myostatin* gene: an overview of mechanisms of action and its relevance to livestock animals. Anim Genet 49:505–519. https://doi.org/10.1111/age.12696
13. Tellam RL, Cockett NE, Vuocolo T, Bidwell CA (2012) Genes contributing to genetic variation of muscling in sheep. Front Genet 3:164. https://doi.org/10.3389/fgene.2012.00164
14. Clop A, Marcq F, Takeda H et al (2006) A mutation creating a potential illegitimate microRNA target site in the *myostatin* gene affects muscularity in sheep. Nat Genet 38:813–818. https://doi.org/10.1038/ng1810
15. Crispo M, Mulet AP, Tesson L et al (2015) Efficient generation of *myostatin* knock-out sheep using CRISPR/Cas9 technology and microinjection into zygotes. PLoS One 10:e0136690. https://doi.org/10.1371/journal.pone.0136690
16. Wang X, Yu H, Lei A et al (2015) Generation of gene-modified goats targeting *MSTN* and

FGF5 via zygote injection of CRISPR/Cas9 system. Sci Rep 5:13878. https://doi.org/10.1038/srep13878

17. Wang X, Niu Y, Zhou J et al (2016) Multiplex gene editing via CRISPR/Cas9 exhibits desirable muscle hypertrophy without detectable off-target effects in sheep. Sci Rep 6:32271. https://doi.org/10.1038/srep32271
18. Concordet J-P, Haeussler M (2018) CRISPOR: intuitive guide selection for CRISPR/Cas9 genome editing experiments and screens. Nucleic Acids Res 46:W242–W245. https://doi.org/10.1093/nar/gky354
19. Chenouard V, Brusselle L, Heslan J-M et al (2016) A rapid and cost-effective method for genotyping genome-edited animals: a heteroduplex mobility assay using microfluidic capillary electrophoresis. J Genet Genomics 43: 341–348. https://doi.org/10.1016/J.JGG.2016.04.005
20. Menchaca A, Barrera N, dos Santos Neto PC et al (2016) Advances and limitations of in vitro embryo production in sheep and goats. Anim Reprod 13:273–278. https://doi.org/10.21451/1984-3143-AR871
21. Dos Santos-Neto PC, Cuadro F, Barrera N et al (2017) Embryo survival and birth rate after minimum volume vitrification or slow freezing of in vivo and in vitro produced ovine embryos. Cryobiology 78:8–14. https://doi.org/10.1016/j.cryobiol.2017.08.002
22. Menchaca A, Vilariño M, Crispo M et al (2010) New approaches to superovulation and embryo transfer in small ruminants. Reprod Fertil Dev 22:113. https://doi.org/10.1071/RD09222
23. Labun K, Montague TG, Krause M et al (2019) CHOPCHOP v3: expanding the CRISPR web toolbox beyond genome editing. Nucleic Acids Res 47:W171–W174. https://doi.org/10.1093/nar/gkz365
24. Bae S, Park J, Kim J-S (2014) Cas-OFFinder: a fast and versatile algorithm that searches for potential off-target sites of Cas9 RNA-guided endonucleases. Bioinformatics 30:1473–1475. https://doi.org/10.1093/bioinformatics/btu048
25. Cui Y, Xu J, Cheng M et al (2018) Review of CRISPR/Cas9 sgRNA design tools. Interdiscip Sci Comput Life Sci 10:455–465. https://doi.org/10.1007/s12539-018-0298-z
26. Pallarès Masmitjà M, Knödlseder N, Güell M (2019) CRISPR-gRNA design. In: CRISPR gene editing: methods and protocols, methods in molecular biology, vol 1961. Humana Press, New York, NY, pp 3–11. https://doi.org/10.1007/978-1-4939-9170-9_1
27. Torres-Perez R, Garcia-Martin JA, Montoliu L et al (2019) WeReview: CRISPR tools-live repository of computational tools for assisting CRISPR/Cas experiments. Bioeng (Basel, Switzerland) 6:63. https://doi.org/10.3390/bioengineering6030063
28. Liu G, Zhang Y, Zhang T (2020) Computational approaches for effective CRISPR guide RNA design and evaluation. Comput Struct Biotechnol J 18:35–44. https://doi.org/10.1016/J.CSBJ.2019.11.006
29. Menchaca A, Souza-Neves M (2020) Chapter 4b. SOP for small ruminant embryo technology. Part 2: In vitro embryo production. In: Manual of the International Embryo Technology Society, 5th edn, pp 4b1–4b9
30. Vilariño M, Crispo M, dos Santos-Neto P et al (2012) The effect of culture medium changes on in vitro production of sheep embryos. Reprod Domest Anim 47:1806
31. Han H, Ma Y, Wang T et al (2014) One-step generation of *myostatin* gene knockout sheep via the CRISPR/Cas9 system. Front Agric Sci Eng 1:2–5. https://doi.org/10.15302/J-FASE-2014007
32. Ni W, Qiao J, Hu S et al (2014) Efficient gene knockout in goats using CRISPR/Cas9 system. PLoS One 9:e106718. https://doi.org/10.1371/journal.pone.0106718
33. Zhang Y, Wang Y, Yulin B et al (2018) CRISPR/Cas9-mediated sheep *MSTN* gene knockout and promote sSMSCs differentiation. J Cell Biochem 120:1794–1806. https://doi.org/10.1002/jcb.27474
34. Fan Z, Perisse IV, Cotton CU et al (2018) A sheep model of cystic fibrosis generated by CRISPR/Cas9 disruption of the *CFTR* gene. JCI Insight 3:e123529. https://doi.org/10.1172/jci.insight.123529
35. McLean Z, Appleby SJ, Wei J et al (2020) Testes of *DAZL* null neonatal sheep lack prospermatogonia but maintain normal somatic cell morphology and marker expression. Mol Reprod Dev:23443. https://doi.org/10.1002/mrd.23443
36. Menchaca A, Dos Santos-Neto PC, Souza-Neves M et al (2020) *Otoferlin* gene editing in sheep via CRISPR-assisted ssODN-mediated homology directed repair. Sci Rep 10:5995. https://doi.org/10.1038/s41598-020-62879-y
37. Clement K, Rees H, Canver MC et al (2019) CRISPResso2 provides accurate and rapid genome editing sequence analysis. Nature

Biotechnology 37(3):224–226. https://doi.org/10.1038/s41587-019-0032-3

38. Jiang Y, Xie M, Chen W et al (2014) The sheep genome illuminates biology of the rumen and lipid metabolism. Science 344:1168–1173. https://doi.org/10.1126/science.1252806
39. Bickhart DM, Rosen BD, Koren S et al (2017) Single-molecule sequencing and chromatin conformation capture enable de novo reference assembly of the domestic goat genome. Nat Genet 49:643–650. https://doi.org/10.1038/ng.3802
40. Li H, Durbin R (2010) Fast and accurate long-read alignment with Burrows–Wheeler transform. Bioinformatics 26:589–595. https://doi.org/10.1093/bioinformatics/btp698
41. McKenna A, Hanna M, Banks E et al (2010) The Genome Analysis Toolkit: a MapReduce framework for analyzing next-generation DNA sequencing data. Genome Res 20:1297–1303. https://doi.org/10.1101/gr.107524.110
42. Li H, Handsaker B, Wysoker A et al (2009) The sequence alignment/map format and SAMtools. Bioinformatics 25:2078–2079. https://doi.org/10.1093/bioinformatics/btp352
43. Wang X, Niu Y, Zhou J et al (2018) CRISPR/Cas9-mediated *MSTN* disruption and heritable mutagenesis in goats causes increased body mass. Anim Genet 49:43–51. https://doi.org/10.1111/age.12626
44. Wang L, Cai B, Zhou S et al (2017) RNA-seq reveals transcriptome changes in goats following *myostatin* gene knockout. PLoS One 12:e0187966. https://doi.org/10.1371/journal.pone.0187966
45. Wang X, Liu J, Niu Y et al (2018) Low incidence of SNVs and indels in trio genomes of Cas9-mediated multiplex edited sheep. BMC Genomics 19:397. https://doi.org/10.1186/s12864-018-4712-z
46. Li C, Zhou S, Li Y et al (2018) Trio-based deep sequencing reveals a low incidence of off-target mutations in the offspring of genetically edited goats. Front Genet 9:499. https://doi.org/10.3389/fgene.2018.00449

Part VI

Concluding Remarks

Chapter 17

Regulatory and Policy Considerations Around Genome Editing in Agriculture

Steffi Friedrichs, Karinne Ludlow, and Peter Kearns

Abstract

The increasingly widespread use of genome editing brought with it a fierce debate about the most adequate regulation of this latest innovation in modern biotechnology and the products resulting from it. In almost all cases, this debate has become a repetition or continuation of the deliberations concerning genetically modified organisms (GMOs) of the 1990s and early 2000s.

This chapter aims to untangle the historically influenced and often biased arguments of the debates by addressing the complex question of the correct interpretation of relevant underlying law and its applicability. In doing so, the chapter considers 25 countries and regions that have published results or ongoing investigations and discussions pertaining to the governance of genome editing in their jurisdictions: 16 have published policies or signed statements that exempt gene edited plants from GMO-regulations, as long as no foreign DNA or transgene remained in the final product. Such exemptions are based on the widely supported acceptance that the products of the underlying genome editing processes resemble those of "conventional breeding" techniques. These policies and statements often refer to the important role that modern precision biotechnologies, of which genome editing is one, play in addressing some of the world's overarching challenges, such as the loss of biodiversity, pest and disease control, and climate change; they are furthermore shown to exhibit an adherence to the four universal principles of good regulation: (a) proportionality, (b) non-discrimination, (c) predictability, and (d) enforceability. And while it is the right of jurisdictions to develop their own regulations independent from that of their neighbors, it is specifically the principle of "enforceability" that may become the ultimate litmus test of those regulations that do not grant exemptions from GMO-regulations.

Key words New (plant) breeding techniques, Genome editing, Genetic modification, Genetically modified organisms, Site-directed nuclease, Regulation, Legislation, Policy

1 Introduction: Genome Editing, New Breeding Techniques, and Genetic Modification

Genome editing is celebrated as the latest and most precise and impactful innovation in the toolbox of plant breeders. However, genome editing has also brought with it a regulatory conundrum: whether all plants resulting from genome editing are classified and

Paul John Verma, Huseyin Sumer and Jun Liu (eds.), *Applications of Genome Modulation and Editing*,
Methods in Molecular Biology, vol. 2495, https://doi.org/10.1007/978-1-0716-2301-5_17,

thus regulated as genetically modified organisms (GMOs) and whether they should be.

What may at first seem like semantics, turns out–under detailed legal analysis–to be a complex question of the correct interpretation of relevant underlying law and its applicability. Few modern-day laws pertaining to technological processes or products demonstrate better just how convoluted legal interpretation can be, and how difficult, yet important, it is to get it right.

Much of the current discussions pertaining to the regulation or deregulation of gene edited plants hinges on the identification of a "trigger," whose presence would cause a specific action/decision with regard to regulatory provisions and requirements. In the following section, we will describe and discuss the regulatory provisions in different jurisdictions, aiming to highlight the underlying "triggers" in each case.[1]

In this chapter, we provide an overview of national rules and regulations pertaining to genome editing techniques and the products resulting from them. In doing so, we highlight relevant cornerstones of the underlying regulatory frameworks, with a special emphasis on identifying the triggers of existing GMO regulations that genome editing governance models build on. Table 1 summarizes GMO regulations, and the current status of (known) regulatory arrangements pertaining to genome editing, in a sample of countries, as well as official statements or notifications by (coalitions of) countries to international bodies, such as the World Trade Organization (WTO). Specific regulatory arrangements and their historic developments are discussed in more detail in the section "Policy Considerations regarding Genome Editing" below for selected countries listed in Table 1. Variations in regulatory responses matter, as these may give rise to—for example—trade obstacles and other disruptions.

This chapter focuses on plant breeding and crop agriculture; in a few cases, however, where the relevant regulation includes animals, or draws on language concerning both plants and animals, references to regulations pertaining to the latter have also been included.

The chapter begins by considering relevant definitions—the building blocks of regulation that have caused great difficulty in the responses to genome edited plants; it then summarizes the regulations that govern the cultivation and/or import of genome edited crops in several jurisdictions around the globe, highlighting regulatory backgrounds and specifics of the preceding debates and policy development processes, where relevant.

The penultimate section discusses the adequacy of the specific regulations passed on genome edited plants with regard to the

[1] Terminologies, such as "product-based" or "process-based," that are often utilized during the analyses and discussions of regulatory frameworks/arrangements, are much less accurate then the reference to a "trigger," and may even mean different things in different contexts.

Table 1
Simplified overview of the (currently known) policies in different countries and regions; specific regulatory arrangements are discussed in the section "Policy Considerations regarding Genome Editing" below (SDN = site-directed nuclease)

	GMOs allowed?					Joint Statement by selected South American Countries [a] (Sept. 2018)	Statement by a Coalition of 14 Countries and Regions [b] (Nov. 2019)
	Cultiva-tion	Import	Exemption of genome editing from GMO regulations?				
			SDN1	SDN2	SDN3		
Argentina	YES	YES	YES	YES[c]	NO	X	X
Econom. Community of West Africa							X
Australia	YES	YES	YES	YES[c]	NO		X
Brazil	YES	YES	YES	YES[c]	NO	X	X
Canada[d]	YES	YES	YES	YES	YES		X
Chile	YES	YES	YES	YES[c]	NO	X	
China	YES	NO[e]					
Colombia	YES	YES	YES	YES[c]	NO		X
Dominican Republic							X
Ecuador	NO	YES	YES	YES[c]			
European Union	YES[f]	YES[f]	NO	NO	NO		
Guatemala							X
Honduras							X
India[g]	YES	YES	NO	NO	NO		
Israel	NO	YES	YES	YES[c]	NO		
Japan			YES	YES[c]	NO		
Jordan							X
New Zealand	YES[f]	YES[f]	NO	NO	NO		
Nigeria	YES	YES	YES	NO	NO		
Paraguay	YES	YES	YES	YES[c]	NO	X	X
Russia[h]	NO	NO	n/a	n/a	n/a		
United States	YES	YES	YES	YES[c]	NO		X
Uruguay	YES	YES	n/a	n/a	n/a	X	X
Norway	NO	YES	NO	NO	NO		
Vietnam							X

Note: Blank fields denote a lack of reliable information

[a]Genetic Editing Techniques, Southern Agricultural Council (CAS) XXXV Regular Meeting, 20th September 2018. (accessed: 1st March 2021)

[b]WTO, G/SPS/GEN/1658/Rev3, 1st November 2018, Committee on Sanitary and Phytosanitary Measures, International Statement on Agricultural Applications of Precision Biotechnology (accessed: 1st March 2021)

[c]If no foreign DNA / transgene remains in the product

[d]Canada requires a *pre-market safety assessment* for agriculture biotechnology products, including products produced through genome editing, only if they were novel (i.e. express a new characteristic) and could therefore pose a *new risk*

[e]China strictly limits the import of GMOs [1]

[f]Cultivation and import are not prohibited; obtaining permission is technically possible, but very difficult to achieve

[g]In January 2020, India published a *Draft Document on Genome Edited Organisms: Regulatory Framework and Guidelines for Risk Assessment*, which *'[. . .] covers genome editing, process and product,'* and which is explicitly applicable to both genome-edited plants, animals, and human cells

[h]Russia announced a large funding programme, under which 30 new varieties of genome edited crops and animals would be created, but very little is known about the underlying policies

attributes of (a) proportionality, (b) non-discrimination, (c) predictability, and (d) enforceability, providing examples of the manifestation and impact indicators of some of the attributes.

The final section draws some conclusions regarding those regulations pertaining to genome editing in plants that have been passed and those that are still being debated. The section points to the drivers and forces that continue to influence the policy development process, and end in highlighting some of the remaining challenges.

1.1 Definitions

Countries can define fundamental terms for the purposes of regulation as they prefer, unless restricted by their membership of an international agreement, such as multilateral agreements (e.g., those of the World Trade Organization), or free trade agreements. One widely adopted definition of GMO is that provided in the Cartagena Protocol on Biosafety (CPB), where GMOs are called living modified organisms (LMOs) (see Box 1).[2,3] This definition though, is not universal.

Box 1: Excerpt from the Cartagena Protocol on Biosafety (CPB)—Use of Terms[4]

Article 3

Use of Terms

[...]

(g) *"Living modified organism" means any living organism that possesses a novel combination of genetic material obtained through the use of modern biotechnology*

[...]

(i) *"Modern biotechnology" means the application of:*

(a) *In vitro nucleic acid techniques, including recombinant deoxyribonucleic acid (DNA) and direct injection of nucleic acid into cells or organelles*

(b) *Fusion of cells beyond the taxonomic family*

that overcome natural physiological reproductive or recombination barriers and that are not techniques used in traditional breeding and selection [...]

---

[2] Website: Cartagena Protocol on Biosafety (accessed: 10th March 2021).

[3] According to Food and Agriculture Organization of the United Nations (FAO), the *"so-called Cartagena Protocol on Biosafety to the 1992 Convention on Biological Diversity (2000)* (Secretariat of the Convention on Biological Diversity, 2000 [2]) *[...] does not refer to genetically modified organisms but rather, for reasons that are not explicit, to "living modified organisms" but it is clear that the two terms should be regarded as synonymous."*

[4] Russia announced a large funding programme, under which 30 new varieties of genome edited crops and animals would be created, but very little is known about the underlying policies.

Until recently, the delineation between GMO and non-GMO had been relatively simple, in that selection- and cross-breeding techniques yielded non-GMOs, because they were treated as "traditional" techniques; in some jurisdictions, such plants are GMOs but they are excluded from regulation because they are considered "conventionally bred." This can either be on the basis that the technique used is excluded from the definition of the science of genetic modification or because the resulting plants are excluded from being GMOs or both. Other techniques, such as transgenesis, are considered to be modern biotechnology and the products of such techniques are therefore GMOs (see Box 2, Fig. 1) [4, 5].

Box 2: Plant Breeding Techniques (After Biology Fortified[5] and [3])

Plant Breeding Techniques

- "Cross-breeding" between two sexually compatible species is the only form of "traditional" breeding technique (used for both plants and farm animals).
- "Mutation breeding" refers to techniques developed and refined since the 1930s, in which mutagens, such as radioactivity or chemicals are used to trigger the creation of random mutations in plants, aiming to obtain specific desirable phenotypes among the many resulting mutants.
- "Polyploidy breeding" is another selection- and cross-breeding technique that utilizes the occurrence of polyploidy (i.e., presence (and artificially induced occurrence) of more than two paired sets of chromosomes in a cell) in plants, in order to either induce or overcome sterility (e.g., rendering fruits seedless, or propagating otherwise sterile hybrids, respectively.
- "Protoplast fusion" is a technique that combines two nuclear genomes in a living cell by removal of their cell walls, in order to transfer traits between species and thus create hybrids that cannot be obtained through sexual crossing.
- "Transgenetic breeding" describes a process of modern biotechnology, during which a gene of a known desired phenotype of one species is inserted into the gene of another species.
- "Genome editing" describes a range of techniques, in which specialized enzymes that have been modified can insert, replace, or remove DNA from a genome with a high degree of specificity.

[5] Website: Biology Fortified (accessed: 29th April 2021).

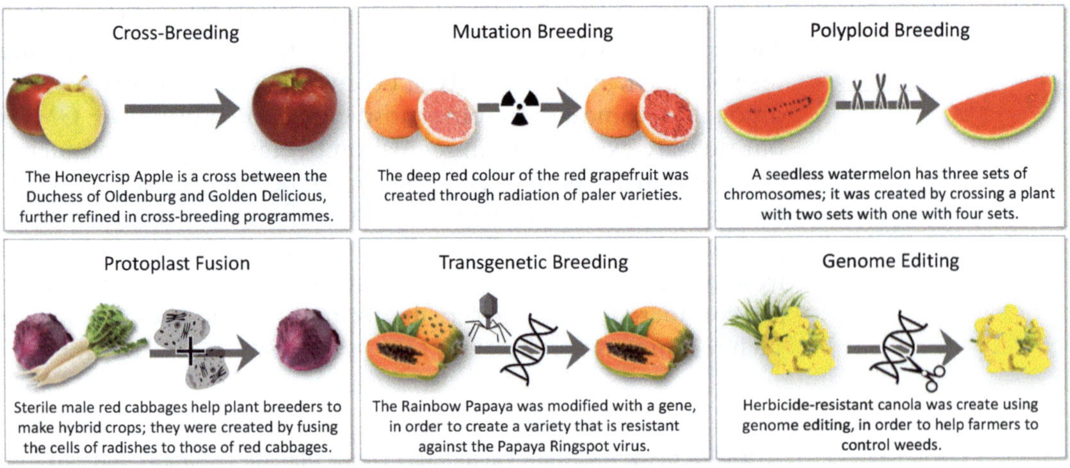

Fig. 1 Plant breeding techniques (incl. genome editing) (source Biology Fortified[11] and [3])

Genome editing is the latest addition to the so-called new breeding techniques (NBTs) (AKA new plant breeding techniques (NPBTs))—a group of recent innovations in plant breeding that use molecular biology tools, in order to achieve a targeted change of trait in a plant by inducing specific changes into the plant DNA; these NBT-induced DNA modifications can vary in scale from altering a single base to inserting or removing one or more genes [5].[6,7] Box 3 provides an overview of the history and types of genome editing techniques.

Box 3: Brief History and Different Types of Genome Editing (After [6, 8])

The concept of editing a genome with the help of a programmable nuclease had been known and applied in diverse fields of agricultural and medical biotechnology since the early 2000s. Until the discovery of the CRISPR (Clustered Regularly Interspaced Short Palindromic Repeats)/Cas9 systems in 2012, however, the proliferation of the technique had been hampered by the slow, expensive, and difficult synthesis of the hitherto known gene editors; the CRISPR/Cas tool, by contrast, is guided by a piece of RNA, which can be easily and rapidly assembled (Table 2).

(continued)

[6] Website: New (Plant) Breeding Techniques (accessed: 10th March 2021).

[7] NBTs include: cisgenesis, intragenesis (technologies using transformation with genetic material restricted to the species' own gene-pool), emerging techniques to induce controlled mutagenesis or insertion (ODM, Zinc Finger Nuclease technologies 1–3), other applications such as grafting on GM rootstocks or reverse breeding, and the three different site-directed nuclease (SDN) types of genome editing: SND1, SDN2, and SDN3 [6].

Box 3 (continued)

Table 2
Properties of genome editing tools (after [7])

The four families of gene editors	Meganuclease (1985)	Zinc finger nucleases (2003)	TALENs (2010)	CRISPR/Cas (2012)
Number of proteins	1	2	2	1 + 1 RNA
Production	Difficult	Not very easy	Easy	Very easy
Cost of production	50,000 Euros	5000 Euros	1000 Euros	10 Euros
Time needed for an experiment	months	months	weeks	days

Three types of genome editing need to be distinguished, especially with regard to potential discrepancies in the regulatory approaches to the techniques and/or their products (Table 3).

Table 3
The three main types of genome editing

Genome editing type	Description
SDN1[a]	Involves the unguided repair of a targeted double-strand break (DSB) by the mechanism called nonhomologous end joining. The spontaneous repair of this break can lead to a mutation causing gene silencing, gene knock-out or a change in the activity of a gene. Efficient method, with many applications already.
SDN2[a]	Involves a template-guided repair of a targeted DSB using a sequence donor, typically short single-stranded DNA. The donor carries one or several small mutations flanked by two sequences matching both ends of the DSB and is thus recognized as a repair template, allowing the introduction of the mutation(s) at the target site. The efficiency of the technique is lower than SDN1, but strongly varies according to the species, donor design, the time and method of delivery, and other conditions.
SDN3[a]	Involves a template-guided repair of a targeted DSB using a sequence donor, typically double-stranded DNA containing an entire gene or an even longer genetic element(s). Both ends of the donor are homologous to the DSB ends (usually more than 800 bp each), which therefore recognize the donor as a repair template, allowing the introduction of the gene or genetic element(s) at the target site. Efficiency is lower than SDN1, but strongly varies according to the species, donor design, the time and method of delivery, and other conditions.

[a]SDN: Site-Directed Nuclease

While there is no unique, or legally enforced definition of NBTs, or, indeed, genome editing, the term was coined, in order to differentiate the innovative techniques discovered in the early 2000s, and that exhibit greater accuracy and selectivity in DNA manipulation than the biotechnology-assisted breeding techniques that had been utilized since the 1930s.

The recent-day confusion arises from the fact that the deliberate application of molecular biology tools in NBTs, such as genome editing, resembles the methods of "modern biotechnology" processes, while the genetic changes introduced by (at least some) NBTs are akin to those obtained through "traditional" or cross-breeding techniques or mutagenesis.[8]

1.2 Policy Considerations Regarding Genome Editing

1.2.1 South America

Argentina: The Frontrunner

Argentina was the first country to consider and provide clarification on whether the application of NBTs should be regulated as GMOs. The country's regulatory framework is based on its membership of a number of international groups and committees (incl. FAO/-WHO/CODEX,[9] WTO/SPS,[10] FAO/IPPC[11]), as well as its effort to ratify the CPB.[12]

After a three-year country-internal debate concerning the applicability of the GMO regulation to products obtained through the application of NBTs, Argentina issued a regulation (resolution 173/2015) that aimed to establish case-by-case procedures to determine if a crop obtained by breeding techniques involving modern biotechnology does not fall under GMO regulations [9]; Figure 2 illustrates the underlying decision-making process, whose approach is based on the following components [8]:

1. All NBTs involve recombinant DNA techniques, which leads to the presumption of GMOs.
2. If the NBT in question does not involve a new combination of genetic material (e.g., does not use a transgene/uses a transgene, which is removed in the final product), a non-GM regulatory classification is applied.
3. If the NBT in question involves a new combination of genetic material (e.g., uses a transgene, which remains in the final

[8] In fact, the inability to retrospectively differentiate between a naturally occurring mutation and one that may have been induced through SDN1 genome-editing techniques presents a major discussion point vis-à-vis the basic "enforceability"-principle of good regulation; this will be discussed in more detail below (cf. section Considerations of adequate Regulation and Governance of Genome Editing), together with the other principles of "proportionality," "non-discrimination," and "predictability."

[9] Website: Food and Agriculture Organization of the United Nations (FAO) and World Health Organization (WHO) CODEX ALIMENTARIUS – International Food Standards (accessed: 10th March 2021).

[10] Website: World Trade Organization (WTO) Sanitary and Phytosanitary Measures (accessed: 10th March 2021).

[11] Website: FAO International Plant Protection Convention (IPPC) (accessed: 10th March 2021).

[12] According to Argentina's Country Profile at the Biosafety Clearing House for the CPB, Argentina signed the CPB in May 2000, but has not yet ratified it.

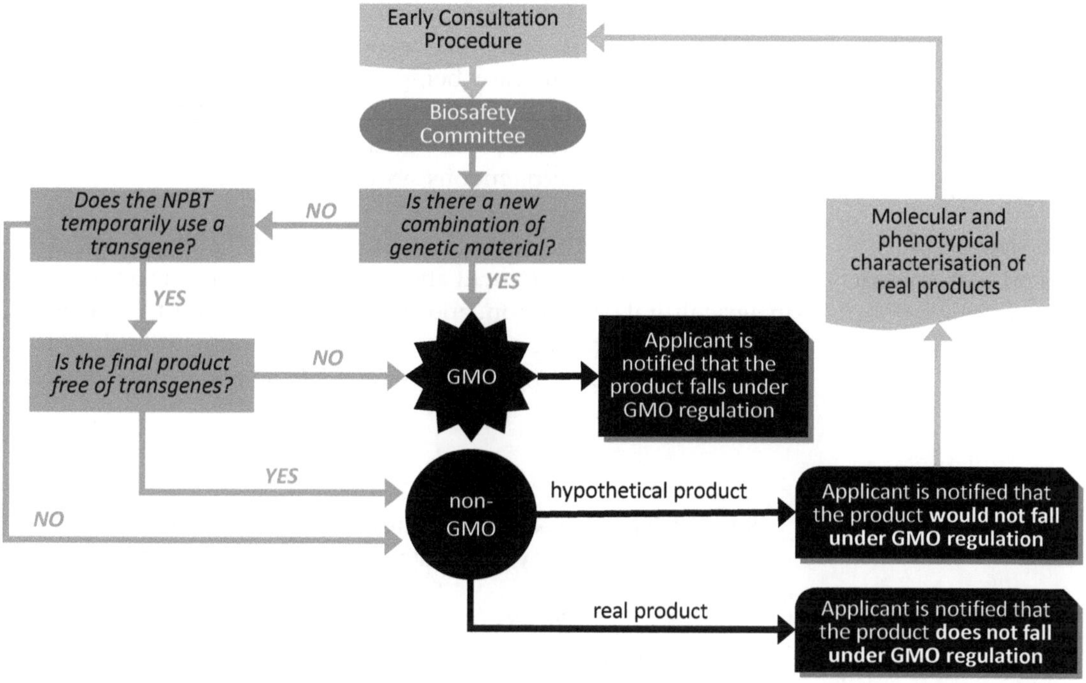

Fig. 2 Flow map of the analysis and decision-making process pertaining to NBTs in Argentina (after [5], updated according to a presentation by M. Lema (presented at an expert meeting of 'ReImagine Europa', Brussels, February 2021))

product), the regulatory classification stipulates that the final product falls under GM classification.

The line-by-line process described above can be (and has been) applied to both real products and hypothetical products; it asks basic information on the overall breeding process, genetic changes, traits, bred-out of helper transgenes, etc.

In a detailed description of the regulation, Whelan and Lema [9] explain that

> *[...]*
>
> *To such end, applicants shall submit each product (NBTs-derived crop) to establish whether the result of the breeding process is a new combination of genetic material or not. A genetic change shall be always regarded as a new combination of genetic material when a stable and joint insertion of one or more genes or DNA sequences that are a part of a defined genetic construct have been introduced permanently into the plant genome.*
>
> *Also, if appropriate, it must be established if there exists enough scientific evidence to support the absence of the transgenes that may have been used transiently during the crop breeding process.*
>
> *[...]*
>
> *In case the crop is not required to be regulated as a GMO but its features and/or novelty lead to a significant risk hypothesis, this must be also reported by the regulatory commission and such report is channeled to the appropriate regulator of varieties obtained by "conventional" breeding for consideration.*
>
> Whelan and Lema [9]

In 2019, Lema reported that several crops obtained by NBTs had been presented for determination of their regulatory status in

Argentina; the reviewing regulatory authorities found that for SDN1-type NBTs, "the final product would not be a GMO, regardless of the nuclease being originated from a recombinant DNA introduced in the plant cell (which is lost later along the breeding process) or by introduction of the protein previously expressed in an heterogeneous system." [10].

Joint Statement by Selected South American Countries

In September 2018, Argentina, Chile, Brazil, Paraguay, and Uruguay issued a joint statement about the role of genome edited crops in agricultural production, and their countries' plans to collaborate with third countries on the continuing development and trade of gene edited crops (English translation of the Spanish original, see Box 4)[13]:

Box 4: Joint statement by the Ministers of Agriculture of Argentina, Brazil, Chile, Paraguay, and Uruguay, September 2018 (see footnote 13)

GENETIC EDITING TECHNIQUES

SOUTHERN AGRICULTURAL COUNCIL (CAS) XXXV REGULAR MEETING

CAS/Council, Declaration II (XXXV 2018} Buenos Aires, Argentina, September 20, 2018

Original. Spanish

Declaration of the Ministers:

The Ministers of Agriculture of Argentina, Brazil, Chile, Paraguay, and Uruguay, members of the Southern Agricultural Council (CAS), meeting in Buenos Aires, Argentina.

1. *Gene-improved crops have the potential to play a critical role in addressing the challenges faced by agricultural production, helping to increase the supply of food and other agricultural products in a sustainable way.*
2. *Gene editing can generate crops analogous to those obtained through other conventional breeding methods.*
3. *The CAS countries present an important public and private investment in the development of crops improved by genetic editing. This is because it can accelerate the access of the agricultural producer and the consumer to new characteristics of productive and agro-industrial interest; at the same time, it represents an opportunity for the transfer of technologies developed by the national institutes of agricultural research and biotechnological SMEs.*

(continued)

[13] Genetic Editing Techniques, Southern Agricultural Council (CAS) XXXV Regular Meeting, 20th September 2018 (accessed: 1st March 2021).

Box 4 (continued)

4. *Arbitrary and unjustifiable distinctions between agricultural products obtained by gene editing and those obtained through other improvement methods should be avoided.*

Declare that:

1. *They will exchange information on product development and existing regulatory frameworks applicable to them, exploring science-based opportunities for regional and international regulatory harmonization.*
2. *They will endeavor to work jointly and with third countries to avoid scientifically based obstacles to the trade in agricultural products improved by single publishing.*

Brazil

The corresponding Brazilian regulation defines the so-called Innovative Precision Improvement Techniques (TIMP), from the English term "Precision Breeding Innovation (PBI)," which also encompasses the NBTs [9]. In 2018, the Brazilian Ministry of Science, Technology, Innovations and Communications published a Normative Resolution 16/2018, stipulating that "TIMP encompass a set of new methodologies and approaches that differ from the genetic engineering strategy by transgene, as it results in the absence of recombinant DNA/RNA in the final product," and that "[the new technologies] may originate a product not considered as a Genetically Modified Organism (GMO) [...] are considered examples of Innovative Precision Improvement Techniques (TIMP), but not limited to them." [9]. In order to determine whether a product obtained by TIMP should be considered a GMO, the Brazilian regulation creates a case-by-case assessment procedure similar to that of Argentina.

Chile

Similar to Argentina's approach, Chile's Ministry of Agriculture's Agricultural and Livestock Services regulates plants modified through genome editing techniques as conventional plants, unless they contain foreign DNA.[14] The underlying decisions and required procedures have been laid out in a guidance, published in 2017, that describes exemptions from the Chilean regulation for genetically modified crops (1523/2001)[15] stating that gene edited crops that do not contain "a new combination of genetic material" are not subjected to GMO regulations.

[14] Website: Human and Agriculture Gene Editing: Regulations and Index – Chile: Crops / Food (accessed: 10th March 2021).

[15] Website: Chilean Applicability of Resolution No. 1523 / 2001 in propagation material developed by new plant breeding techniques (accessed: 10th March 2021).

Paraguay

Genome edited crops and food stuff are regulated as conventional plants, unless they contain foreign DNA; a resolution passed in 2019 outlines requirements for approval of crops developed using gene editing and other NBTs and stipulates that the National Commission on Agricultural and Forestry Biosafety reviewed dossiers on the relevant new crops or food on a case-by-case basis, in order to determine, if they are exempted from GMO regulation.

Uruguay

Uruguay has no specific regulatory provisions for genome edited crops or food stuff; three different genome edited products are currently known to be under development: herbicide-resistant soy-beans, soybeans with reduced lectins, and mandarins and tomatoes with higher lycopene content.[16]

It is noteworthy that the Uruguayan regulation of GMOs differs from that of its neighbors in that it is triggered by the application of a process (utilizing recombinant DNA) [11].

Ecuador

Since a referendum in 2008, Ecuador's constitution prohibits the cultivation of GMOs in the country, but permits their import, as long as they are labeled, supported by the President and approved by the National Assembly [12].[17] In May 2019, the Regulation of the Organic Code of the Environment was published, establishing a regulatory approach to GMOs, and detailing that "organisms resulting from genetic improvement which are not harboring recombinant or foreign DNA in the resulting genome are excluded from risk assessment" (article 230a) [12].

Colombia

In February 2018, the Colombian Agricultural Institute (ICA) notified the World Trade Organization (WTO) of a "Draft Colombian Agricultural Institute (ICA) Resolution establishing the procedure applicable to a cultivar where innovative plant breeding techniques deriving from modern biotechnology have been used at some stage of the plant-breeding process, and the end product does not contain foreign genetic material."[18,19]

The resolution foresees a case-by-case consultation process, under which an applicant submits a dossier to the ICA, and the "ICA shall carry out an assessment of the information received within a period no longer than sixty (60) working days, determining whether the new crop contains any foreign genetic material

[16] Website: Human and Agriculture Gene Editing Regulations and Index – Uruguay: Crops/Food (accessed: 1st March 2021).

[17] Website: Human and Agriculture Gene Editing Regulations and Index – Ecuador: Crops/Food (accessed: 1st March 2021).

[18] Website: Human and Agriculture Gene Editing: Regulations and Index – Colombia: Crops / Food (accessed: 1st March 2021).

[19] Website: WTO, G/SPS/N/COL/282, 26 February 2018, Committee on Sanitary and Phytosanitary Measures, Notification (accessed: 1st March 2021).

Table 4
Overview of regulatory agencies, products, and regulations pertaining to GMOs and genome edited plants in Canada (ECCC (Environment and Climate Change Canada), DFO (Fisheries and Oceans Canada), AAFC (Agriculture and Agri-Food Canada), GAC (Global Affairs Canada), ISED (Innovation, Science and Economic Development Canada) (source: (Friedrichs, et al. [8]))

Lead	Product	Regulation
Canadian Food Inspection Agency	Livestock feed Seeds Fertilizer Veterinary biologics	*Feeds Act* *Seeds Act* *Fertilizers Act* *Health of Animals Act*
Health Canada (HC)	Pesticides Novel foods, drugs, and biologics, medical devices	*Pest Control Products Act* *Food and Drugs Act*
ECCC, HC, DFO	Animals and all other substances	*Canadian Environmental Protection Act*
AAFC, GAC, ISED	Non-regulatory considerations	*Market access, industrial policy, socioeconomic impacts, trade*

inserted in its genome due to the use of modern biotechnology techniques. For a genetic change to be considered as foreign genetic material, it shall be analyzed whether a stable and joint exogenous insertion has been produced in one (1) or more genes or DNA sequences forming part of a specific genetic construction." [9].

1.2.2 Central America

In 2019, Honduras, Guatemala, and El Salvador published a "resolution to establish a streamlined authorization procedure for crops developed using new breeding techniques (NBTs)."[20]

1.2.3 North America

Canada

Canada applies a risk-based regulatory approach to both GMOs and genome edited plants, that is triggered by "novelty" and based on a rather complex biotechnology-related regulatory oversight consisting of no fewer than eight acts and policies, administered by eight agencies, pertaining to the different aspects of biotechnology products on the Canadian market (see Table 4).

The complexity of the Canadian model is afforded by the country's effort to use existing acts and regulations rather than developing any new regulation, when new breeding methodologies had started to be increasingly used in the 1980s and 1990s [13].

Genetically Modified and Genome Edited Plants

The Canadian regulatory approach relies on the assumption that "since there are many paths to the same result, the consistent and risk-based regulatory approach is to regulate products, not

[20] Website: Human and Agriculture Gene Editing Regulations and Index – Central America: Crops/Food (accessed: 9th March 2021).

processes"; it requires a pre-market safety-assessment based on a categorization into one of two groups only [13]:

(a) *pre-market safety assessment* for agriculture biotechnology products, including products produced through genome editing, only if they were novel (i.e., express a new characteristic) and could therefore pose a *new risk*

(b) *no pre-market safety assessment* for genome edited products that did not express a novel trait (i.e., "novel" meant "novel to the Canadian environment, or the food or feed supply in Canada").

Throughout the assessment, the Canadian regulatory process is considered risk-based, in that the responsible governmental agencies have flexible information requirements that are (a) not prescriptive, (b) case-specific, and (c) outcome-based; proponents are explicitly encouraged to contact the regulatory authorities early in the product development process, in order to discuss both the potential regulatory requirements (pre-submission consultations), and the novelty determination. The environmental safety of novel plants is assessed with regard to (a) potential for weediness, (b) consequences of gene flow, (c) potential to become a plant pest, (d) impact on non-target organisms, and (e) and impact on biodiversity [13].

Genetically Modified and Genome Edited Animals

Genetically modified animals are regulated under the Canadian Environmental Protection Act for potential environmental or indirect human health concerns; no specific genome editing policy is currently in place for genome edited animals, but several inquiries had been received, and the relevant authority is currently preparing to examine the resulting organisms for genetic changes on a case-by-case basis [13].

The United States

The United States applies a regulation under existing laws to all biotechnology products, providing a network of agency jurisdictions. In 2015, however, a modernization of the regulatory system for biotechnology products was initiated, whose main goal was to "[e]nsure public confidence in the regulatory system and improve transparency, predictability, coordination, and efficiency of the regulatory system." The review resulted in two key documents:

1. 2017 Update to the Coordinated Framework [14]
2. 2016 National Strategy for Modernizing the Regulatory System for Biotechnology Products [15]

In 2017, the US Administration initiated a review "[t]o identify legislative, regulatory, and policy changes to promote agriculture, economic development, job growth, infrastructure improvements, technological innovation, energy security, and quality of life in rural America." The resulting report by the Interagency Task Force on Agriculture and Rural Prosperity noted that "[a]dvancements in genome editing and genomic selection have produced favorable crop and livestock traits, including resistance to drought, disease, and heat; enhancements to nutritional value; and increased resource efficiency. Those technologies, combined with public and private research and development investments, have enabled U.S. farmers to increase the supply and quality of crop and livestock commodities using fewer resources and at lower costs of production." [16]

The responsibility for the implementation of the regulation is spread over three separate agencies [13]:

- US Department of Agriculture (USDA) Animal and Plant Health Inspection Service (APHIS):
  - Protection of Plant Health
  - Safety of Veterinary Biologics
- US Environmental Protection Agency (EPA):
  - Regulation of Plant Incorporated Protectants (PIPs) and Bio-pesticides
  - Safe use of New Pesticides
  - Safe use of Chemicals
- US Food and Drug Administration (FDA):
  - Safety of Food, Food Additives, and Feed
  - Safety of Veterinary and Human Drugs and Human Biologics

USDA APHIS

USDA Animal and Plant Health Inspection Service (APHIS) regulates biotechnology products through the control of movement (i.e., permits for, or notifications of, import, interstate movement, and environmental release) of regulated articles (living organisms that had been genetically engineered and involving a plant pest as a donor, recipient, or vector). Under the "Am I regulated" (AIR) process, APHIS encourages developers to submit letters of inquiry if they are not sure that their product falls under the relevant regulation; submissions to the process are considered on a case-by-case basis and answered [13].[21]

In May 2020, the USDA published a new rule on the "Movement of Certain Genetically Engineered Organisms" [17], clarifying its oversight of plants produced through innovative new breeding techniques, such as genome editing. This USDA's so-called SECURE (Sustainable, Ecological, Consistent, Uniform, Responsible, Efficient) Rule codifies new exemptions for genetic modifications in plants if:

[21] Website: USDA APHIS "Am I regulated" (AIR) process (accessed: 10th March 2021).

- "The genetic modification is solely a deletion of any size.
- The genetic modification is a single base pair substitution.
- The genetic modification is solely introducing nucleic acid sequences from within the plant's natural gene pool or from editing nucleic acid sequences in a plant to correspond to a sequence known to occur in that plant's natural gene pool.
- The plant is an offspring of a GE plant and does not retain the genetic modification in the GE plant parent."

These exemptions are based on the following principles:

- "Plants created through conventional breeding have a history of safe use related to plant pest risk.
- The types of plants that qualify for these exemptions can also be created through conventional breeding.
- There is no evidence that use of recombinant deoxyribonucleic acid (DNA) or genome editing techniques necessarily and in and of itself introduces plant pest risk, irrespective of the technique employed.

When a plant meets one of the above listed exemptions, it is not expected to pose plant pest risks greater than the plant pest risks posed by plants modified by conventional breeding [...]."

The rule explains that it exempts "plants generated using plant breeding technologies that have non-templated insertions and deletions and that have a single base pair substitution, because they could otherwise be created by conventional breeding and pose no increased plant pest risk relative to their conventionally bred counterparts," and that "APHIS does not believe it is necessary to regulate off-target effects of genome editing in plants because (1) the off-target mutation rate from genome editing is low relative to the background mutation rate that occurs in conventional breeding, and (2) whatever changes do occur are likely to be segregated away from the target mutation during the breeding process."

The rule emphasizes, however, that "APHIS has not, in this rulemaking [...], taken the position that genome editing does not constitute genetic engineering. Taking such a position would be inconsistent with the generally accepted scientific characterization of genome editing technology [...]." The rule also outlines the revised detailed information requirements of APHIS.

US EPA

Since 2001, the US EPA exempts plant-incorporated protectants (PIPs) from sexually compatible plants that occurred naturally in the plant or that were moved through conventional plant breeding (40 CFR 174.25) from the United States Federal Insecticide, Fungicide, and Rodenticide (FIFRA) requirements (e.g., for product registration/licensing and field testing) [13, 18]. It furthermore exempts residues of PIPs from sexually compatible plants that occurred naturally in the plant or that were moved through conventional plant breeding (40 CFR 174.508) from the United States Federal Food, Drug, and Cosmetic Act (FFDCA) tolerance requirements (e.g., for pesticide residues in food or feed, provided the residues are not present in food at levels that are injurious or deleterious to human health) [13, 19].

During a meeting at the Organisation for Economic Cooperation and Development (OECD) in June 2018, a US EPA representative noted that the Agency was evaluating the extent to which the current exemptions covered genome edited PIPs and considering approaches to clarify the regulatory status of these materials [13].

US FDA

Plants

The regulatory status of a food (and feed) in the United States is "dependent upon the objective characteristics of that food, independent of the methods used to develop the food," based on an underlying policy outlined in 1992 [13][22]:

Section 409—Food Additives:

- New components of food will be regulated as additives if they are not generally recognized as safe (GRAS), subject to certain exceptions.
- Food additives require premarket review and approval before they can be lawfully marketed. The safety standard for use of a food additive is reasonable certainty of no harm under the conditions of intended use in food.
- In order for use of a substance to be GRAS:
  - There must be reasonable certainty of no harm under the conditions of intended use and general recognition of that fact.

Animals

The regulation of genetically engineered animals is subject to the following rules [13]:

- Federal Food, Drug, and Cosmetic Act (FD&C Act), new animal drug provisions.
- National Environmental Policy Act (NEPA).
- 2009 FDA Guidance for Industry (GFI) #187 (revised in 2017 to cover genome edited animals—DRAFT): genome edited animals are evaluated as new animal drugs for the safety and effectiveness of the application.[23]

In 2017, the FDA launched two public consultations on (a) its regulatory approach to genome edited plant-derived foods, and

[22] Website: Determining the Regulatory Status of a Food Ingredient (accessed: 29th April 2021).

[23] Website: Q&A on FDA Regulation of Intentional Genomic Alterations in Animals (accessed: 29th April 2021).

(b) application of genome editing to animals but has not yet followed up with an official reaction to these consultations.

1.2.4 Europe

Norway

Norway maintains one of the most restrictive regulations on the cultivation and import of GMO food and feed; while the use of GMO soy as salmon feed is allowed, the industry avoids it, due to fear of a public backlash [20]. In December 2018, the Norwegian Biotechnology Advisory Board published "A forward-looking regulatory framework for GMO," recommending a tiered approach for regulatory assessment under the Gene Technology Act of 1993 [21, 22].[24] The proposed tiers differentiate the following levels of changes to the genome and their corresponding assessment:

- Level 0 (exempted): Temporary and simultaneously non-heritable changes.
- Level 1: Changes that exist or can arise naturally and can be achieved using conventional breeding methods: Obligation to notify (confirmation of receipt required).
- Level 2: Other species-specific genetic changes: Expedite assessment and approval.
- Level 3: Genetic changes that crosses species barriers or involve synthetic (artificial) DNA-sequence: Standard assessment and approval (current system).

European Union

At the time of writing, the regulatory situation in the European Union (EU), vis-à-vis genome editing, is in a state of flux for reasons which will become apparent. This may be resolved in the current weeks and months. In any event, it will be inevitable to come back to this discussion in more detail at a later stage. This chapter considers current Union law in the context of current discussions in the EU related to genome editing and current and upcoming developments.

According to the European Commission, the current EU legal framework related to GMOs aims to[25]:

- "Protect human and animal health and the environment by introducing a safety assessment of the highest possible standards at EU level before any GMO is placed on the market.
- Put in place harmonized procedures for risk assessment and authorization of GMOs that are efficient, time-limited, and transparent.

[24] Website: Human and Agriculture Gene Editing Regulations and Index – Norway: Crops/Food (accessed: 9th March 2021).

[25] Website: European Union: GMO Legislation (accessed: 20th March 2021).

- Ensure clear labeling of GMOs placed on the market in order to enable consumers as well as professionals (e.g., farmers and food feed chain operators) to make an informed choice.
- Ensure the traceability of GMOs placed on the market."

The Commission also identifies the following building blocks of GMO legislation:

- *Directive 2001/18/EC* on the deliberate release of GMOs into the environment [23].
- *Regulation (EC) 1829/2003* on genetically modified food and feed [24].
- *Directive (EU) 2015/412* amending Directive 2001/18/EC with regard to the possibility for the Member States to restrict or prohibit the cultivation of GMOs in their territory [25].
- *Regulation (EC) 1830/2003* concerning the traceability and labeling of genetically modified organisms and the traceability of food and feed products produced from genetically modified organisms [26].
- *Directive 2009/41/EC* on contained use of genetically modified microorganisms. Regulation (EC) 1946/2003 on transboundary movements of GMOs [27].

It also notes that these main pieces of legislation are supplemented by a number of implementing rules or by recommendations and guidelines on more specific aspects which we will not consider at this stage (see footnote 24).

In 2018, the question as to whether genome editing is covered by EU GMO law came before the Court of Justice of the European Union (CJEU), which is the ultimate judicial authority for dispute resolution and, as in this example, to interpret EU legislation. This case was raised through a request for clarification by the French *Conseil d'État*, which had previously received a plea at national level [13]. In July 2018, the CJEU delivered its ruling, which is described below (Case C-528/16) [28].

In November 2018, the European Commission's Scientific Advice Mechanism (SAM) Group of Chief Scientific Advisors published *A scientific perspective on the regulatory status of products derived from gene editing, and the implications for the GMO Directive* [29], in which they recommend "revising the existing GMO Directive to reflect current knowledge and scientific evidence, in particular on gene editing and established techniques of genetic modification. This should be done with reference to other legislation relevant to food safety and environmental protection."

The Commission's reaction to the CJEU's judgment is evident from their website: "In July 2018, the Court of Justice of the European Union (CJEU) clarified that organisms from new mutagenesis techniques fall within the scope of the EU GMO legislation.

The Commission is now working with EU countries and stakeholders to implement the Court's ruling."

In November 2019, the European Council requested "the Commission to submit a study in light of the Court of Justice's judgment in Case C-528/16 regarding the status of novel genomic techniques under Union law, and a proposal, if appropriate in view of the outcomes of the study" [30]. It is worth emphasizing that the request to the Commission referred not only to *Directive 2001/18/EC*, on which there has been much commentary, but also to the other legal instruments noted in Commission's building blocks described above [23]. The Commission's study is expected to be released at the end of April 2021. When it is published, it is certain to stimulate a major debate on the path forward in the EU.

In the meantime, there have been numerous responses both formal and informal to the CJEU judgment. It would be an onerous task to summarize this body of information. It would probably be more confusing rather than clarifying and clarity is more helpful at that this stage.

However, three documents are worthy of note. First, a report by the All European Academies (ALLEA) reviews the latest scientific evidence with respect to applications of genome editing in agriculture [31]. The report was based on a two-day symposium that was organized jointly by ALLEA and the Royal Flemish Academy of Belgium (KVAB) in November 2019 that aimed to establish a dialog between scientists and other relevant stakeholders. The resulting report presents the state of the art of scientific evidence in the field and explores what genome editing can deliver for agriculture. In addition, it explores the ethical and societal considerations related to agricultural applications of genome editing, traceability, and intellectual property issues, and possible paths to harmonize EU regulation and legislation with recent scientific developments.

Second, the European Group on Ethics in Science and New Technologies (EGE) published an opinion on the Ethics of Genome Editing in March 2021 [32]. The report addresses genome editing across a range of sectors including human health, animal livestock breeding, and gene drives. It has a chapter dedicated to genome editing in plants. It identifies and discusses a range of ethical questions raised across all sectors in which genome editing is expected to have an impact.

Finally, an article entitled *Status Under EU Law of Organisms Developed Through Novel Genomic Techniques* [4] is a comprehensive analysis which contributes to the discussion on the legal status of organisms developed through novel genomic techniques. It achieves this by considering the history of EU GMO Directives as well as the technical context of some of the terms in the Directive. It also goes on to consider the judgment of the CJEU in Case C-528/16 (July 2018).

1.2.5 Middle East

Israel

In Israel, GMOs are regulated by the "Seed Act (Genetically Modified Plants and Organisms)—2005," which defines [33]:

Genetically Modified Organism—An organism, including a microorganism, virus, viroid, and any single-celled or multi-celled entity, that has undergone a modification by genetic engineering and is involved with plants in any way during its life cycle.

In March 2017, the Israeli National Committee for Transgenic Plants (NCTP) published a decision that the cultivation of genome editing plants would not be subject to the country's GMO regulation, if no foreign DNA sequences were incorporated into the plant genome [34].

1.2.6 Asia

India

In January 2020, the Indian government published *Draft Document on Genome Edited Organisms: Regulatory Framework and Guidelines for Risk Assessment*, in order to seek comment on the proposed policy pertaining to genome editing through a public consultation exercise.[26] Issued by the Department of Biotechnology, of the Ministry of Science and Technology, the document is based on the biosafety regulatory framework established under "Manufacture, use, import, export and storage of hazardous micro-organisms/genetically engineered organisms or cells, Rules 1989 (Rules 1989) under Environment (Protection) Act (EPA), 1986," and clarifies up front that "[t]he definition of gene technology under Rules 1989 covers genome editing, process and product." In this course, the draft explicitly applies to both genome edited plants, animals, and human cells in one sweep of the proposed biosafety rules.

The document proposes the "tiered approach for the risk assessment of genome edited products/organisms" described in Table 5.

The Draft Document received a number of critical comments both as part of its public consultation, and in the scientific and trade press; Schmidt *et al.* note that "there is no differentiation between the health risks of off-target mutations in somatic cells for human therapeutics versus plant cells where deleterious effects are eliminated by crossing and pose no ethical issues." [35]. The authors furthermore highlight the immense bureaucratic burden of the proposed measures, under which "any market authorization will require review by no less than three agencies: the Institutional Biosafety Committee, the Ministry of Agriculture & Farmers' Welfare, and the Food Safety and Standards Authority of India." [35].

[26] Government of India (January 2020): *Draft Document on Genome Edited Organisms: Regulatory Framework and Guidelines for Risk Assessment* (accessed: 7th March 2021).

Table 5
Proposed tiered approach for the risk assessment of genome edited products/organisms in India (see footnote 25)

Proposed group of genome edited products/ organisms	Proposed risk assessment process
Genome Editing Group I: Single or few base pair edits/deletions/insertions leading to least complexity (Phenotype/Genotype). Changes leading to knock-down/knock-out of protein/ RNA that result in a new trait which may be familiar with prior knowledge. Chances of off-target effects.	[The] following biosafety concerns should be addressed for G[enome] Ed[ited] Group I organisms: • Changes resulting in altered expression/ activity of native protein. • Presence of vector/components used in the editing process. • Confirmation of intended trait efficacy. • Confirmation of phenotypic equivalence. • Changes leading to protein with new/altered functions. In rare cases, single base pair mutation (s) might result in the introduction of novel trait (e.g., Herbicide tolerance) which might pose additional biosafety concerns. The G[enome] Ed[ited] cells/organisms falling under Group I would be assessed mainly to confirm targeted edit(s) as well as the absence of any biologically significant off-target genomic changes. Also, they would be subjected to phenotypic equivalence analysis on case-by-case basis.
Genome Editing Group II: Several base pair edits leading to certain degree of complexity in Phenotype/Genotype (leading to improvement of an existing attribute or creation of a new attribute). Changes leading to gain of function with a new protein or RNA. May or may not be familiar with prior knowledge. Chances of off-target effects.	[The] following additional biosafety concerns should be addressed considering increased potential of off-target effects: • Insertion of exogenous DNA sequence leading to: – Altered expression of gene. – Deletion/knock-out of gene expression. – Modification of amino acid sequence of a native protein. • Insertion of allelic sequences having prior knowledge. • Introduction of foreign gene with novel trait (s). The G[enome] Ed[ited] cells/organisms falling under Group II would be assessed to confirm targeted edit(s) as well as absence of any biologically significant off-target genomic changes. Also, they would be assessed for phenotypic equivalence and trait efficacy through appropriate contained and/or confined field trials.
Genome Editing Group III: Insertion of foreign gene/DNA sequence leading to high degree of complexity in Phenotype/Genotype (leading to creation of a new attribute, new metabolic	The data requirement and biosafety assessment level would increase with the increase in the complexity of the modification. Nevertheless, regulatory approval for the G[enome] Ed[ited]

Table 5 (continued)

Proposed group of genome edited products/ organisms	Proposed risk assessment process
pathways, etc.). Changes leading to gain of function with new protein or RNA. May not have prior knowledge. Chances of off-target effects.	cell/organism will depend on biosafety assessment on a case-by-case basis. Group III G[enome] Ed[ited] cells/organisms harboring large or foreign DNA in the recipient cell/organism genome may represent similar biosafety concerns as that of genetically engineered (GE) cells/organisms with typical foreign gene insertion(s). Therefore, all the biosafety data requirements which are prescribed in existing food and environmental safety guidelines specific for GE. Cells/organisms on case-by-case basis where foreign genes are inserted would be envisaged.

Russia

In April 2019, Russia announced a large funding program, under which 30 new varieties of genome edited crops and animals would be created (i.e., 10 by 2020 and another 20 by 2027). The program furthermore suggested that some genome edited products would be exempt from a law passed in 2016 that prohibits the cultivation of genetically modified organisms in Russia, except for research purposes [36].[27]

China

No Chinese policy pertaining to genome edited plants is known although the country is by far the biggest generator of publications and patents for genome editing applications in agriculture [35, 37].[28]

Japan

Concluding a five-year review process, the Japanese Ministry of the Environment (MOE) published a decision regarding the regulation of genome edited plants in February 2019, that "some genome-edited organisms should be considered as LMOs (Box 5) while others are not subject to the Cartagena Act." [38]

[27] Website: Human and Agriculture Gene Editing Regulations and Index – Russia: Crops/Food (accessed: 10th March 2021).

[28] Website: Human and Agriculture Gene Editing Regulations and Index – China: Crops/Food (accessed: 10th March 2021).

Box 5: Excerpt from a flyer issued by the Japanese Ministry of the Environment, February 2019 [39]

To Genome Editing Technologies Users

- *Any organism that inserted intracellularly processed nucleic acid (including RNA) is regarded as a living modified organism (LMO), even those obtained using genome editing technologies, and is subject to the regulations stipulated in the Cartagena Act, in principle.*
- *Such organisms are subject to the Cartagena Act unless complete removal of the inserted nucleic acid (including RNA), or its replicated product, is confirmed.*
- *When using organisms obtained through genome editing technologies, please submit information to the competent government agencies [. . .], even when the organism is not subject to the Cartagena Act.*

Users of genome editing technologies are required to notify the government and provide information on unregulated end products created through genome editing technologies, including the details of their production and any knowledge of their impact on biodiversity prior to use [38].[29] No submission is required, if no change has been made to a previously submitted review, or if the genome edited organisms are used in an environment, in which containment measures have been taken [38].

1.2.7 Africa

With the exception of the support of the "*International Statement on Agricultural Applications of Precision Biotechnology*"[30] to the World Trade Organization (WTO) Committee in November 2018 by the Secretariat of the Economic Community of West African States (see the relevant section at the end of this country analysis), little is known about policies pertaining to genome editing in African countries. Schmidt *et al.* reports that South Africa currently treats genome edited plants as GMOs, but that a discussion about a policy amendment was ongoing, while in Kenya, the National Biosafety Authority was in the process of drafting guidelines for the regulation of genome editing technologies [35].

The African Union (AU) announced that it was looking at the possibility of harmonizing biosafety regulations to ensure more effective handling of biotechnology application in its 55 member states, with a specific view to "enhance regulatory framework by

[29] Website: Human and Agriculture Gene Editing Regulations and Index – Japan: Crops/Food (accessed: 10th March 2021).

[30] WTO, G/SPS/GEN/1658/Rev3, 1st November 2018, Committee on Sanitary and Phytosanitary Measures, International Statement on Agricultural Applications of Precision Biotechnology (accessed: 1st March 2021).

engaging themselves in addressing the issues of emerging technologies." [40] (see footnote 29).

Nigeria

In November 2019, Nigeria published an Amendment to its 2015 National Biosafety Management Agency Act [41, 42]. As a result of the amendment, the legislation now provides that "[a] person shall not carry out gene drive, gene editing and synthetic biology except with the approval of the Agency, and adds 'gene editing' means a type of genetic engineering in which DNA is inserted, deleted, modified, or replaced in the genome of a living organism" [42]. National Guidelines on Gene Editing (2020) have also been published, which make it clear that "if the gene editing process or the gene edited product does not lead to or have a new combination of genetic material (e.g. does not use a transgene/uses a transgene which is removed in the final product), a non-GM regulatory classification is applied.... If on the other hand, the gene editing process requires the use of recombinant DNA sequences or the gene edited product has a new combination of genetic material (e.g. uses a transgene which remains in the final product)" the final product will fall under GM regulation [43].

Senegal

Senegal reported that it was "reviewing and revising its new biosafety law which may include language for an expedited approval process for certain genetically engineered (GE) products." [44][31]

South Africa

In South Africa, all non-human modifications to genomes are governed by the 1997 "GMO" Act [45]. An expert report on the regulatory implications of new breeding techniques, commissioned by the Government's Department of Science and Technology, and authored by the Academy of Science of South Africa, noted "that new techniques may be more accurate and precise and may thus need a lower/different level of regulatory scrutiny." The Department of Science and Technology may investigate the need for possible regulatory amendments [45] (see footnote 29).

1.2.8 Oceania

Australia

Australia's response to genome edited plants and their products builds on regulatory frameworks created for GMOs. All dealings with GMOs and the sale of GM food must be authorized, the form of authorization corresponding with potential risks. In summary, recent amendments clarify that organisms and products produced using SDN1 are not regulated as GMOs although they may be regulated by other agencies such as that for food or therapeutic products. Similarly, the food regulator has determined that food from plants with no new DNA present in their genome, even if the

[31] Website: Human and Agriculture Gene Editing Regulations and Index – Africa: Crops/Food (accessed: 10th March 2021).

genome has been changed, are not regulated under the current GM standard (FSANZ webpage "Foods Derived using new breeding techniques—review" (April 2020)). Additional modernization of these two regulatory frameworks is underway to ensure that regulation remains proportionate to potential risk.

More specifically, the question whether GMO and/or food regulations apply to genome edited plants and their products is determined by the use of "gene technology" although the meaning of gene technology differs between GMO regulation and food regulation.

For the regulatory framework around GMOs,[32] gene technology is defined very broadly to mean "any technique for the modification of genes, or other genetic material" and it is irrelevant whether the modification produces a novel combination of DNA or not. Genome editing is unsurprisingly included as gene technology given it modifies the plant's genetic material. Because regulations prohibit all dealings with organisms produced using gene technology (i.e., a "GMO") unless authorized by the independent regulator created by the framework, genome editing of plants requires authorization. That authorization may require only approval by an institution's internal biosafety committee or further steps including case-by-case assessment by the regulator. These regulations apply to all stages of research and development, importation into Australia, laboratory work, field trials, and commercial release and are intended to manage risks to human health and safety and to the environment in light of the proposed use of the GMO.

In September 2019, following extensive review, the regulations were updated to address genome editing. Importantly, although genome editing is gene technology, the final plant may not be a GMO. If a plant is not a GMO, use of it will not require further authorization. When the regulations were created, organisms produced using conventional breeding (such as selective or protoplast breeding) or mutagenesis, and the processes themselves, were expressly excluded from regulation because they are considered to pose little risk. Following the review, these exceptions were replaced to exclude organisms where the mutational event does not involve introduction of foreign nucleic acid and also organisms modified by repair of single- or double-stranded breaks induced by a site-directed nuclease provided that no nucleic acid repair template was used. However, other amendments mean plants produced using a nucleic acid repair template are now expressly included as GMOs. As the regulator has observed in regard to plants produced using SDN2 or SDN3, "The number of resulting nucleotide

[32] That framework comprises the *Gene Technology Act 2000 (GT Act)* and its accompanying regulations (*Gene Technology Regulations 2001*) as well as an inter-governmental between the Australian Federal, State and Territory governments. Website: Office of the Gene Technology Regulator (accessed: 15th March 2021).

changes, whether insertions or deletions, or whether the resulting nucleotide sequence may be found in sexually compatible species, is not a deciding factor" ("Overview of amendments to the Gene Technology Regulations 2001," published by the Government of Australia in September 2019).

Despite the amendments referred to above, a parallel review of Australia's broader policy around GMO regulation has found that some genome edited plants continue to be regulated despite them being of no greater risk than plants produced using random and unregulated mutagenesis or being identical to natural variants. This parallel review, begun in 2017, has confirmed that while the use of the process trigger will continue, further regulatory changes will be made to modernize and future proof the regulations. This is likely to include a risk tiering model but in any case is intended to allow streamlining of low risk dealings with GMOs or dealings that are under the remit of other agencies, such as vaccines, as well as the creation of new authorization categories [46]. Legislative definitions of "gene technology," "GMO," and "deal with" are also to be updated to enable a faster response to the changing science. These changes are intended to be implemented by 2023.

Like the GMO regulatory framework, Australia's food regulations have also been reviewed in response to genome edited plants. While Australia and New Zealand have separate regulations for GMOs, the two countries share a *Food Standards Code* under the arrangements of a binational food system. In 1999, a new Standard on "Food produced using gene technology," was included in the Code. This standard establishes a pre-market safety assessment and approval system for food produced using gene technology and mandatory labeling requirements for some of that food. Like the GMO regulatory framework, a process trigger is used such that the use of gene technology in the production of food causes the Standard to apply. In such cases, the food undergoes a formal risk assessment to obtain pre-market approval prior to sale.

However, the Standard predates the GMO framework and thus uses a different definition of gene technology. The definition in the Food Standards Code is limited to "recombinant DNA techniques," which the food regulator Food Standards Australia New Zealand (FSANZ) has explained as being the joining of DNA from two or more sources (including from the same species) and then insertion into an organism. The trigger is therefore the inclusion of "new" DNA, with new in this context meaning in terms of its nucleotide sequence, genome location, or orientation of insertion [47]. Conventional breeding techniques (e.g., cross-breeding and selection, mutation breeding of plants, cell culture techniques), which have a presumption of safety based on their history of safe use are not regulated under this standard.

Because of the ambiguity of the place of genome edited plants and their products within the scope of the Standard, FSANZ initiated a review in June 2017. The final report, delivered in December 2019, found that the definitions used in the Standard are outdated and unclear and no longer fit for purposes. Whether food from genome edited plants is included within the Standard or not, the report recommended consideration of whether pre-market safety assessment is justified based on the risks posed either by the modification of the genome or changes to a food's characteristics [48]. FSANZ is now considering how to modernize the definitions in the Code to make them clearer and ensure they better reflect existing and emerging genetic technologies.[33] Public consultation is still to be sought on this (expected in mid 2021) followed by another round of public consultation (expected in 2022) once draft changes are released.

At present then, food from plants with no new DNA present in the genome, even if the genome has been changed, is not considered to be food produced using gene technology and is therefore not regulated under the standard, although general principles of food safety apply (FSANZ webpage "Foods Derived using new breeding techniques—review" (April 2020)). Where a plant's genome does contain new DNA, such as where transgenesis, cisgenesis, or intragenesis have been used, food from these plants is regulated as "food produced using gene technology."

New Zealand

Using a stringent process-driven regulatory framework,[34] New Zealand currently regulates all genome edited plants as GMOs that are subject to an approval process by the Environmental Protection Authority (EPA).[35] New Zealand regulations are triggered by the use of a GMO, defined as "any organism in which any of the genes or other genetic material have been modified by in vitro techniques" unless expressly excluded. The definition of GMO was amended in 2016 to clarify that while plants created using chemical and radiation mutagenesis techniques in use before 29 July 1998 or using conventional plant breeding techniques (such as selective breeding, selection and propagation of somaclonal variants, and cell fusion) are not GMOs, plants resulting from other newer techniques including genome editing are GMOs. As discussed above in relation to Australia, New Zealand shares a binational food regulatory agency with Australia meaning that food from plants with no new DNA present in their genome, even if the genome has been changed, is not considered to be

[33] Website: Food Standards Australia New Zealand: *Proposal P1055 – Definitions for gene technology and new breeding techniques* (accessed: 15th March 2021).

[34] Hazardous Substances and New Organisms Act 1996 and Hazardous Substances and New Organisms (Organisms Not Genetically Modified) Regulations 1998.

[35] Website: New Zealand Environmental Protection Authority (NZ EPA) (accessed: 15th March 2021).

GM food but food from plants whose genome does contain new DNA is regulated and requires pre-approval.

1.2.9 International Organizations and Collaborations

Statement by a Coalition of 14 Countries and Regions in Support of Precision Biotechnology

At a WTO meeting in November 2019, a coalition of 13 countries (i.e., Argentina, Australia, Brazil, Canada, Colombia, the Dominican Republic, Guatemala, Honduras, Jordan, Paraguay, the United States, Uruguay, Vietnam) and Secretariat of the Economic Community of West African States[36] announced its "support [of] policies that enable agricultural innovation, including genome editing" [49]; in a previous joint statement by 10 of the 13 countries (see footnote 30), the coalition noted "the potential positive contributions of precision biotechnology to global agriculture," and emphasized "the importance of early action to identify avenues to minimize the trade impacts of differing regulatory approaches" (see Box 6) (see footnote 30).

Box 6: International Statement on Agricultural Applications of Precision Biotechnology, November 2018 (see footnote 30)

World Trade Organization, Committee on Sanitary and Phytosanitary Measures, 1st November 2018

International Statement on Agricultural Applications of Precision Biotechnology

Communication from Argentina, Australia, Brazil, Canada, the Dominican Republic, Guatemala, Honduras, Paraguay, the United States, and Uruguay

The following communication, received on 26 October 2018, is being circulated at the request of the delegations of Argentina, Australia, Brazil, Canada, the Dominican Republic, Guatemala, Honduras, Paraguay, the United States, and Uruguay.

1 OVERVIEW

1.1. Precision biotechnology techniques, as a whole, constitute an essential tool for agricultural innovation. Their use provides farmers with access to products that increase productivity while preserving environmental sustainability.

1.2. Global environmental challenges, pest and disease pressures, food insecurity, and changes in consumer preferences, among other factors, have made the use and fostering of tools such as precision biotechnology vital for increasing the production of safe food.

1.3. In this light, in April 2018, the countries participating in the "Seminar on Genome Editing for Regulators," organized by the Inter-American Institute for Cooperation on Agriculture

(continued)

[36] Website: Secretariat of the Economic Community of West African States (accessed: 15th March 2021).

Box 6 (continued)

(IICA), shared a draft statement on the applications of precision biotechnology.

1.4. The primary objective is to coordinate efforts to ensure that the regulatory approaches for these techniques, which include gene editing, are scientifically based and internationally harmonized.

1.5. The final text of the international statement is non-binding on supporting countries, but provides the necessary guidelines for preventing regulatory asymmetries and, in turn, potential trade disruption.

2 TEXT OF THE "INTERNATIONAL STATEMENT ON AGRICULTURAL APPLICATIONS OF PRECISION BIOTECHNOLOGY"

2.1. Agricultural innovation has played an essential role in increasing yields and productivity in support of growing, prosperous civilizations. Innovations in precision biotechnology, such as gene editing, have brought the promise of major improvements in terms of the ease and precision of introducing desirable traits into agricultural organisms, as compared to other breeding methods. Farmers continually need to broaden access to new tools to improve productivity, plant and animal health, and environmental sustainability, and need to help address global challenges such as climate change, pest and disease pressures, and the safety and security of worldwide food supplies, as well as meet consumer preferences and demands for healthier, higher quality foods at affordable prices. Government policies must continue to foster innovation, including in the public sector and by small- and medium-sized enterprises (SMEs), and mitigate unintended, unnecessary barriers to the entry of agricultural products.

2.2. In some cases, precision biotechnology, such as gene editing, may generate organisms with characteristics similar to those obtainable through conventional breeding. In other cases, the organisms generated may have characteristics similar to those introduced into organisms using recombinant-DNA technologies. In either case, the food, animal, and environmental safety of such products can be adequately addressed by existing regulatory frameworks for agricultural products and existing safety standards based on the characteristics of the product or organism.

2.3. Governments are engaging in policy discussions on regulatory frameworks and global regulatory compatibility to encourage cross-border research collaboration and minimize potential disruptions to trade. Differing domestic regulatory approaches for products derived from precision biotechnology

(continued)

Box 6 (continued)

may result not only in international asynchronicity in approvals, but also in asymmetry in regulatory approaches, and create potential trade issues that could impede innovation. Recognizing the potential positive contributions of precision biotechnology to global agriculture and emphasizing the importance of early action to identify avenues to minimize the trade impacts of differing regulatory approaches, the undersigned governments acknowledge that:

- *Precision biotechnology products have the potential to play a critical role in addressing the challenges facing agricultural production, including by contributing to increasing the supply of foods and other agricultural products, in a sustainable way.*
- *Collaborative research efforts and the ability to introduce useful products into the market, especially by SMEs and public sector researchers, are necessary to fully realize the potential of precision biotechnology.*
- *Given the differences internationally in approaches used to assess agricultural biotechnology, due consideration should be exercised by governments to avoid arbitrary and unjustifiable distinctions between end products derived from precision biotechnology and similar end products obtained through other production methods.*
- *To ensure appropriate science- and risk-based approaches consistent with the protection of human, animal, and plant health and the environment, due consideration should be given to available scientific and technical information when updating existing regulatory frameworks or applying these frameworks to products of precision biotechnology, and when using available flexibility within existing regulatory frameworks for agricultural products.*
- *Regulatory approaches necessary to help ensure safety (of humans, animals, plants, and the environment) in respect of products derived from precision biotechnology should be science- and risk-based, transparent, predictable, timely, and consistent with relevant international trade obligations.*
- *Cooperative work by governments to minimize unnecessary barriers to trade related to the regulatory oversight of products of precision biotechnology, including the exploring of opportunities for regulatory and policy alignment, should be pursued where possible.*
- *This collaborative work should promote constructive dialog with trading partners and agricultural stakeholders on potential trade issues related to precision biotechnology, so as to support open and fair trade and encourage research and innovation.*

(continued)

Box 6 (continued)

Public communication efforts can build trust in regulatory frameworks and improve the acceptability of future agricultural innovations that will help farmers address global challenges with a view to the production of abundant, safe and affordable food, feed, fibers, and energy in the twenty-first century.

1.3 Considerations of Adequate Regulation and Governance of Genome Editing

The universal principles of all good regulation are:

A. Proportionality

B. Non-discrimination

C. Predictability

D. Enforceability

Many jurisdictions discussed above have concluded their review processes of the most adequate regulation of genome editing processes and their products by deciding to accept those genome edited plants that do not contain foreign DNA as conventional plants, while those that do contain foreign DNA are subjected to more detailed risk assessment processes under existing GMO regulation.

It seems notable that this approach fulfils all four principles of good regulation listed above:

1.3.1 Proportionality

In general terms, the "proportionality" of a regulation refers to the flexibility of a set of regulatory requirements with regard to the specific status and situation of the subject (e.g., a company or a legal person) that needs to comply with it; especially in the field of financial and banking regulation, the concept of "proportionality" often considers a firm's size, systemic importance, and risk profile.

The case-by-case approach to regulating genome editing processes and their products, and the scientific-evidence-based decisions to consider genome edited plants that do not contain foreign DNA as "conventional plants" represent key element of proportionality in genome editing (de-)regulations.

1.3.2 Non-Discrimination

The EU's legislation of GMOs is widely regarded as discriminating against small- and medium-sized enterprises (SMEs) that cannot afford the lengthy and onerous process to obtain regulatory market approval in Europe; this effect is worsened by the incredible shrinking of the possible European market since 19 EU Member States have decided to issue restrictions of the geographical scope of GMO applications. As a result, big agricultural companies have moved their advanced breeding programs out of Europe, which accounts for only 8% of CRISPR patents in agriculture, while 60% originate from China and 26% from the United States.

[50]. Nevertheless, the EU is an important market, comprising over 500 million consumers, as well as a big seed market, which is worth 7 billion Euros per year, and thus contributes to a strongly positive trade balance in the agricultural sector, even if Europe is extremely dependent on imports of GM feed [13].

Even in countries where the cultivation and trade of GMOs is widely accepted and practiced; however, authorization processes of GMOs pose high barriers to SMEs, and thus inadvertently favor large, multinational companies. The increasingly widespread use of genome editing technologies visibly levels this playing field, by allowing SMEs and micro-enterprises and start-ups the possibility to commercialize their processes and products.

Argentina: A Case in Point for the Beneficial Socioeconomic Impact of Genome Editing

Argentina is one of the "six founder" countries, which, in 1996, began with the commercial cultivation of genetically modified (GM) crops [10]; according to the Argentinian Ministry of Agroindustry, the adoption of GM crops resulted in a number of economic and production impacts in Argentina and contributed to a more sustainable agriculture in the country. Today, Argentina is the world's leading exporter of soya oil and meal, and second highest exporter of corn grains and third highest exporter of soy grains [10].

The number of applications for NBT products presented to the Argentinian authorities steadily increased from three in 2016 to eight in 2019 [51]. By the beginning of 2020, over 20 applications had been received; some of these had been classified by the authorities as GMOs (and would thus need to undergo "the GMO deregulation process, which would take several years for a subsequent approval," but the majority had "obtained a determination of being non-GMO (i.e. conventional) organisms" [51].

The origin of applicants for "non-GMO NBT products" differ notably from that of GMO-product applicants (see Fig. 3); while the latter cohort is populated mainly by multinational companies (MNCs) (i.e., by 90%), the former is dominated by local companies and public research laboratories (i.e., 59%) and foreign small- and medium-sized companies (SMEs) (i.e., 32%). Whelan et al. [51] report that in the first two decades, all GMO-deregulation applications submitted to the Argentinian authorities had come from MNCs, and only in the past 5 years had applications been received from some local companies and public research laboratories, and from a few foreign small- and medium-sized companies (SMEs) [51].

Socioeconomic Potential and Impacts in Chile

Chile is the fifth largest producer of seeds in the world. For plants falling under the GMO regulation, seed developers and researchers can use genetic modification technologies for research and export only (see footnote 12). The increasing development of genome

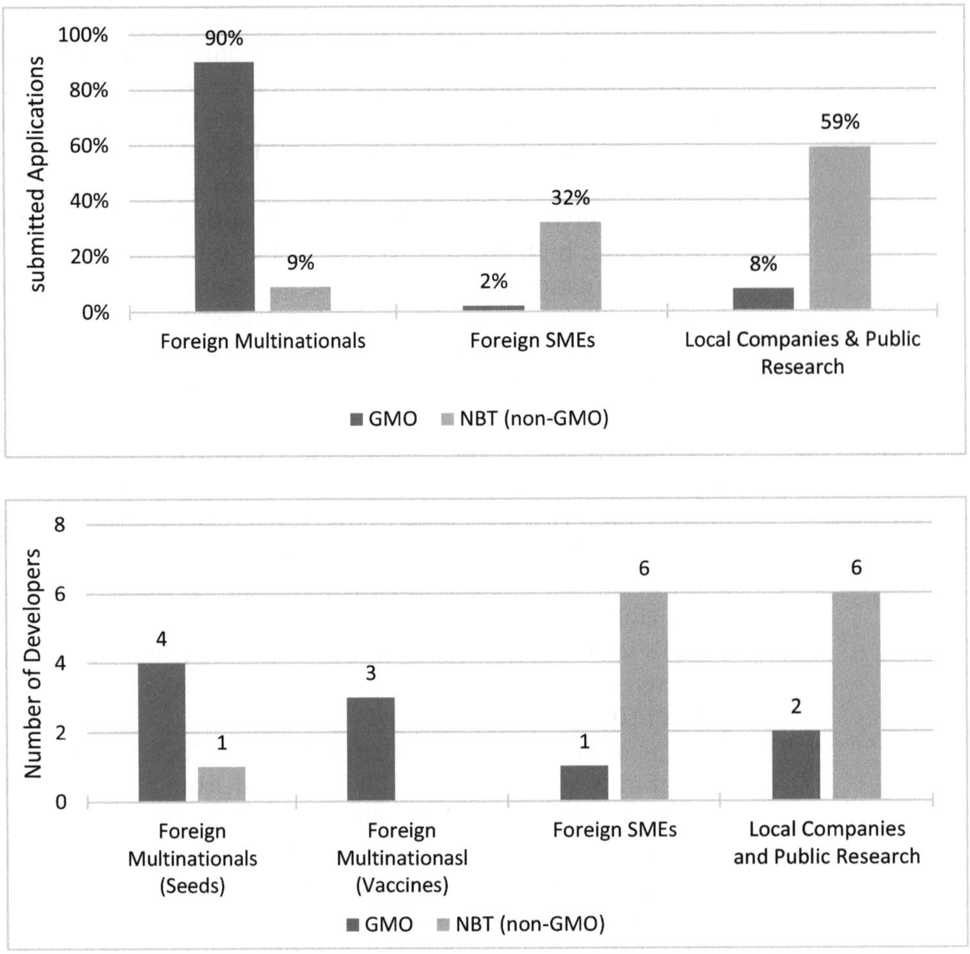

Fig. 3 Statistics of the origins of applications for (**a**) GMOs, and (**b**) non-GMO NBT products (after [51])

edited crops that do not contain foreign DNA (and that are thus exempt for restrictions to GMOs) harbor enormous potential to cultivate plants with improved properties in Chile.

Socioeconomic Potential and Impacts in Paraguay

In 2018, Paraguay was reported to be the sixth largest producer of transgenic crops in the world (i.e., based mainly on soy, corn, and cotton), but no genome edited crops have yet been approved for commercial production.[37] The deregulation of genome edited plants promises to both make the plant breeding process faster and accessible to more local and smaller-sized companies.

[37] Website: Human and Agriculture Gene Editing Regulations and Index – Paraguay: Crops/Food (accessed: 1st March 2021).

Socioeconomic Factors in New Zealand

New Zealand may be called "the exception that confirms the rule" in this listing of socioeconomic consideration during the development of policies pertaining to genome editing. Exports are critical to the island nation's economy: New Zealand is the world's largest dairy exporter and accounts for a third of the world's dairy trade [52]. The country's sheep and beef sector exports 90% of its produce, and its horticultural sector exports 95% of its kiwifruits, while wine, apple, and pear are also exported in smaller, but nevertheless significant volumes [53]. In this, however, New Zealand regards genome editing not as a chance to increase its market footprint, but as a threat to the (perceived) niche and quality of its exports. Bryan and Roberts note that "[t]he globally traded cash crops (corn, soybean, canola and cotton) that make up the majority of current GM plantings are not widely grown and do not provide a compelling value proposition for NZ. In contrast, NZ aims to supply high value innovative products that are not cultivated on a global large scale, e.g., kiwifruit and radiata pine." [53]. The Environment Minister of New Zealand admits that "[t]he definition of what is and is not a GMO is difficult because humans have been changing the genetic content of organisms for over a century. [...] The rationale for our cautious approach is that New Zealand is an exporter of billions of dollars of food products and we need to be mindful of market perceptions as well as the science."[38]

1.3.3 Predictability

In most cases of gradual, evolutionary rule-making, jurisdictions tend to adapt their existing regulations (to socioeconomic, political, and technological changes); this has been no different in the case of the rapid introduction of genome editing technologies (see Box 3): even before the first reporting of the CRISPR/Cas technology that is now understood as the epitome of easy, readily-available, widespread genome editing, policy-makers had started to review the applicability and adequacy of existing GMO regulations for what soon became known as "New (Plant) Breeding Techniques" [15]. But while the initial drive for these reviews was a potential differentiation from existing GMO regulations and the (transgenic) processes and products they had typically been applied to, the existing GMO regulations remained the basis for delineation of N(P)BTs and their products. The majority of jurisdictions that conducted such a review concluded that the delineation between GM(O)s and N(P)BTs (and their products) was the presence of foreign DNA in the resulting plant, and subsequently (de-) regulated genome editing processes and/or their products within or based upon the existing GMO regulations, based on the scientific evidence gathered during the review process. This evolution of regulations constitutes a predictability of the policy-making process

[38] Website: GMO regulations clarified (accessed: 1st March 2021).

that has altogether beneficial socioeconomic effects: it provides re-assurance to both the public and to the industrial sector (vis-à-vis plans for future investments or (R&D) activities), and it saves resources both during the development and the implementation of new policies.

1.3.4 Enforceability

The element of enforceability of a regulation represents its ultimate litmus test; a regulation that relies on measures of notification or self-declaration to trigger an onerous and expensive regulatory process, while lacking the means of identifying culprits that may have dodged such measures to gain fast and cheap access to local markets, exerts an undue burden on locally regulated industry. Scientific evidence indicates that genome edited plants that do not contain foreign DNA are not distinguishable from conventional plants, rendering the enforceability of genome editing regulations, such as those currently in place in Europe and New Zealand, close to impossible. The ongoing discussions surrounding the regulation of genome editing processes and their products in the European Union need to review the EU's vulnerability to questionable imports that look like and are classified as "conventional plants" in most other countries, as these would add injury to the insult that European plant breeders already have to endure.

2 Conclusions

The regulation of genome editing as a new (plant) breeding technique has given rise to engaged discussions around the globe that are in many ways, a continuation of the debates concerning GMO crops in the 1990s and early 2000s. Of the 25 countries and regions discussed in this chapter, 16 have published policies or signed statements that exempt gene edited plants from GMO-regulations, as long as no foreign DNA or transgene remains in the final product, while five countries/regions have decided against exemptions of gene edited plants from the prevailing GMO regulation.

It needs to be recognized that the existing GMO regulation and the acceptance of GMO crops in a jurisdiction is probably the most influential local factor that determines the outcome of the debate concerning the regulation of genome editing. While many stakeholders (i.e., experts in the field of agriculture, climate change, biodiversity, biotechnology, food security) may regard the advent of genome editing as a chance to overcome onerous GMO regulations, no country or region that has pre-existing heavily restrictive regulations or even bans of GMOs has so far been able to swing the boat around and look at genome editing techniques as something separate from the existing, 20-year-old rules.

Those countries that have published outspoken support of the deregulation of genome editing techniques, stress the importance of precision biotechnology in addressing "global environmental challenges, pest and disease pressures, food insecurity, and changes in consumer preferences" (see the *International Statement on Agricultural Applications of Precision Biotechnology* (November 2018) above). These countries note the responsibilities of adequate policies to "foster innovation, including in the public sector and by small and medium-sized enterprises (SMEs), and mitigate unintended, unnecessary barriers to the entry of agricultural products"—a concept that genome editing has already successfully demonstrated in the form of beneficial socioeconomic impacts in Argentina.

While it is the right of countries to develop their own regulations independent from that of their neighbors (and under adequate considerations of the resulting trade barriers), it is highly unlikely that genome edited products, that are recognized by many countries to be akin to the products of conventional breeding techniques, will observe country borders; while GMO products are relatively easy to recognize in the laboratory and can even be traced back to the specific laboratories of their creation, this will not be possible for SDN1 (and some SDN2) products of genome editing. The limitation of the cultivation of genome editing in a minority of countries is therefore likely to cause significant damage to the agricultural sectors of those countries in multiple ways.

The primary objective outlined by the *International Statement on Agricultural Applications of Precision Biotechnology* "to coordinate efforts to ensure that the regulatory approaches for these techniques, which include gene editing, are scientifically based and internationally harmonized" thus remains the ultimate goal and a formidable challenge for years to come.

References

1. Cohen J (2019) To feed its 1.4 billion, China bets big on genome editing of crops. Science
2. Secretariat of the Convention on Biological Diversity, W. T (2000) Cartagena protocol on biosafety to the convention on biological diversity. Secretariat of the Convention on Biological Diversity, Montreal. Obtenido de http://bch.cbd.int/protocol/text/
3. Ahmad S, Wei X, Sheng Z, Hu P, Tang S (2018) CRISPR/Cas9 for development of disease resistance in plants: recent progress, limitations and future prospects. Brief Funct Genomics 19(1):26–39
4. van der Meer P, Angenon G, Bergmans H, Buhk H-J, Callebaut S, Chamon M et al (2021) The status under EU law of organisms developed through novel genomic techniques. SSRN Electr J. Obtenido de https://papers.ssrn.com/abstract=3730116
5. Whelan A, Lema M (2015) Regulatory framework for gene editing and other new breeding techniques (NBTs) in Argentina. GM Crops Food *6*(4):253–265. https://doi.org/10.1080/21645698.2015.1114698
6. Friedrichs S, Takasu Y, Kearns P, Dagallier B, Oshima R, Schofield J, Moreddu C (2019c)

Policy considerations regarding genome editing. Trends Biotechnol 37(10):1029–1032

7. Ricroch A (2019) Global developments of genome editing in agriculture. Transgenic Res *28*(Suppl. 2):45–52. Obtenido de https://pubmed.ncbi.nlm.nih.gov/31321682/
8. Friedrichs S, Takasu Y, Kearns P, Dagallier B, Oshima R, Schofield J, Moreddu C (2019a) An overview of regulatory approaches to genome editing in agriculture. Biotechnol Res Innov 3(2):208–220
9. Whelan A, Lema M (2019) Regulation of genome editing in plant biotechnology: Argentina. Regul Genome Editing Plant Biotechnol:19–62. https://doi.org/10.1007/978-3-030-17119-3_2
10. Lema, M. (8 de 2019). Regulatory aspects of gene editing in Argentina. Transgenic Res, *28*(2), 147-150. Obtenido de https://link.springer.com/article/10.1007/s11248-019-00145-2
11. Alejandra M, Gardini F (2013) Establishment of the Uruguayan biosafety framework and a regulatory perspective of environmental risk assessment of transgenic crops engineered with complex traits
12. Gatica-Arias A (2020) The regulatory current status of plant breeding technologies in some Latin American and the Caribbean countries. Plant Cell Tissue Org Culture 141(2): 229–242. Obtenido de https://link.springer.com/article/10.1007/s11240-020-01799-1
13. Friedrichs S, Takasu Y, Kearns P, Dagallier B, Oshima R, Schofield J, Moreddu C (2019b) Meeting report of the OECD conference on "Genome editing: applications in agriculture—implications for health, environment and regulation". Transgenic Res 28(3–4):419–463
14. OSTP. (2017) Modernizing the regulatory system for biotechnology products: final version of the 2017 update to the coordinated framework for the regulation of biotechnology. Obtenido de https://obamawhitehouse.archives.gov/sites/default/files/microsites/ostp/2017_coordinated_framework_update.pdf
15. OECD. (2016) Report of the OECD workshop on environmental risk assessment of products derived from new plant breeding techniques.
16. USDA. (2017) Report to the president of the United States from the task force on agriculture and rural prosperity. Obtenido de https://www.usda.gov/sites/default/files/documents/rural-prosperity-report.pdf
17. USDA. (2020) DEPARTMENT OF AGRICULTURE Animal and Plant Health Inspection Service 7 CFR Parts 330, 340, and 372. Obtenido de https://www.govinfo.gov/content/pkg/FR-2020-05-18/pdf/2020-10638.pdf
18. EPA. (2015) 40 CFR § 174.25—Plant-incorporated protectant from sexually compatible plant.—Content Details—CFR-2015-title40-vol24-sec174-25. Obtenido de https://www.govinfo.gov/app/details/CFR-2015-title40-vol24/CFR-2015-title40-vol24-sec174-25
19. EPA. (2010). 40 CFR 174.508—Pesticidal substance from sexually compatible plant; exemption from the requirement of a tolerance.—Content details—CFR-2010-title40-vol23-sec174-508. Obtenido de https://www.govinfo.gov/app/details/CFR-2010-title40-vol23/CFR-2010-title40-vol23-sec174-508
20. Heggdal, Ø. (2016). How Norway became an anti-GMO powerhouse—food and farm discussion Lab. Obtenido de http://fafdl.org/blog/2016/10/14/how-norway-became-an-anti-gmo-powerhouse/
21. Government of Norway. (1993). Gene Technology Act. Obtenido de https://www.regjeringen.no/en/dokumenter/gene-technology-act/id173031/
22. The Norwegian Biotechnology Advisory Board. (2018) A forward-looking regulatory framework for GMO. Obtenido de https://www.bioteknologiradet.no/filarkiv/2018/12/2018-12-03-Sammendrag-genteknologiloven-Bioteknologir%C3%A5det-ENGELSK-for-web.pdf
23. European Union. (2001) Directive 2001/18/EC of the European Parliament and of the Council of 12 March 2001 on the deliberate release into the environment of genetically modified organisms and repealing Council Directive 90/220/EEC—Commission Declaration (EUR-Lex-32001L0018.
24. European Union. (2003b) Regulation (EC) No 1829/2003 of the European Parliament and of the Council of 22 September 2003 on genetically modified food and feed (Text with EEA relevance) (EUR-Lex - 32003R1829 - EN - EUR-Lex). Obtenido de https://eur-lex.europa.eu/legal-content/EN/ALL/?uri=CELEX:32003R1829
25. European Union. (2015) Directive (EU) 2015/412 of the European Parliament and of the Council of 11 March 2015 amending Directive 2001/18/EC as regards the possibility for the Member States to restrict or prohibit the cultivation of genetically modified organisms (GMOs) in their

26. European Union. (2003a) Genetically modified organisms—traceability and labelling (EUR-Lex - l21170 - EN - EUR-Lex). Obtenido de https://eur-lex.europa.eu/legal-content/EN/TXT/?uri=LEGISSUM:l21170
27. European Union. (2009) Directive 2009/41/EC of the European Parliament and of the Council of 6 May 2009 on the contained use of genetically modified micro-organisms (Recast) (Text with EEA relevance) (EUR-Lex - 32009L0041 - EN - EUR-Lex). Obtenido de https://eur-lex.europa.eu/legal-content/EN/TXT/?uri=CELEX:32009L0041
28. Court of Justice of the European Union (CJEU). (2018) Judgement of the Court of Justice of the European Union (Case C-528/16). Obtenido de http://curia.europa.eu/juris/document/document.jsf;jsessionid=F54A1CE4758F9FACEA9BB8F4A3C68439?text=&docid=204387&pageIndex=0&doclang=EN&mode=lst&dir=&occ=first&part=1&cid=293900
29. European Commission Group of Chief Scientific Advisors. (2018) A scientific perspective on the regulatory status of products derived from gene editing and the implications for the GMO Directive—Publications Office of the EU. Obtenido de https://op.europa.eu/en/publication-detail/-/publication/a9100d3c-4930-11e9-a8ed-01aa75ed71a1/language-en/format-PDF/source-94584603
30. European Council. (2019) Council Decision (EU) 2019/1904. Recuperado el https://eur-lex.europa.eu/legal-content/EN/TXT/HTML/?uri=CELEX:32019D1904&from=EN
31. ALLEA. (2020) Genome editing for crop improvement. Obtenido de www.kvab.be
32. EGE. (2021) EGE opinion on the ethics of genome editing—Publications Office of the EU. Obtenido de https://op.europa.eu/en/web/eu-law-and-publications/publication-detail/-/publication/6d9879f7-8c55-11eb-b85c-01aa75ed71a1
33. Israel. (2005) Israel—Seed Regulations (Genetically Modified Plants and Organisms)—2005 (unofficial translation). Obtenido de https://www.moag.gov.il/ppis/Laws/Regulation/Documents/gmoregulations2005_2009.pdf
34. Floral Daily. (2017). Israel will not require additional regulations for gene-edited plants | Genetic Literacy Project. Obtenido de https://geneticliteracyproject.org/2017/05/18/israel-will-not-require-additional-regulations-gene-edited-plants/
35. Schmidt S, Belisle M, Frommer W (2020) The evolving landscape around genome editing in agriculture. Obtenido de https://www.embopress.org/doi/full/10.15252/embr.202050680
36. Dobrovidova O (2019) Russia joins in global gene-editing bonanza. Nature 569(7756): 319–320
37. Cohen J, Desai N (2019) With its CRISPR revolution, China becomes a world leader in genome editing. Science
38. Tsuda M, Watanabe K, Ohsawa R (2019) Regulatory status of genome-edited organisms under the Japanese Cartagena Act. Front Bioeng Biotechnol 7:387. Obtenido de /pmc/articles/PMC6908812/
39. Japanese Ministry of the Environment. (2019). Flyer 'To genome editing technologies users'.
40. Alliance for Science. (2019) African Union mulls harmonized biosafety system framework—Alliance for Science. Obtenido de https://allianceforscience.cornell.edu/blog/2019/07/african-union-mulls-harmonized-biosafety-system-framework/
41. Nigerian National Biosafety Management Agency. (2015) Nigeria: National Biosafety Management Agency Act, *2015.*
42. Nigerian National Biosafety Management Agency. (2019) Nigeria: National Biosafety Management Agency (Amendment) Act., 2019.
43. Federal Republic of Nigeria. (2020) National guidelines for the regulation of gene editing.
44. Osinski J, Sylla F (2020) Agricultural biotechnology annual. Senegal, Country. Obtenido de https://apps.fas.usda.gov/newgainapi/api/Report/DownloadReportByFileName?fileName=Agricultural%20Biotechnology%20Annual_Dakar_Senegal_10-20-2019
45. USDA Foreign Agricultural Service. (2019) Republic of South Africa—Agricultural biotechnology annual. Obtenido de https://apps.fas.usda.gov/newgainapi/api/Report/DownloadReportByFileName?fileName=Agricultural%20Biotechnology%20Annual_Pretoria_South%20Africa%20-%20Republic%20of_2-5-2019
46. Commonweath of Australia. (2020) Modernising and future-proofing the National Gene Technology Scheme: Proposed regulatory framework to support implementation of the Third Review of the Scheme. Consultation Regulation Impact statement.
47. FSANZ. (2018) Consultation Paper: food derived using new breeding techniques.
48. Food Standards Australia New Zealand. (2019) Final report—review of food derived

using new breeding techniques. Obtenido de https://www.foodstandards.gov.au/consumer/gmfood/Documents/NBT%20Final%20report.pdf

49. USDA. (2018) WTO Members support policy approaches to enable innovation in agriculture
50. Martin-Laffon J, Kuntz M, Ricroch A (2019) Worldwide CRISPR patent landscape shows strong geographical biases. Nat Biotechnol *37*(6):613–620. Obtenido de https://www.nature.com/articles/s41587-019-0138-7
51. Whelan A, Gutti P, Lema M (2020) Gene editing regulation and innovation economics. Front Bioeng Biotechnol *8*:15. https://doi.org/10.3389/fbioe.2020.00303/full
52. Chobtang J, McLaren S, Ledgard S, Donaghy D (2017) Consequential life cycle assessment of pasture-based milk production: a case study in the Waikato Region, New Zealand. J Indust Ecol 21(5):1139–1152. https://doi.org/10.1111/jiec.12484
53. Fritsche, S., Poovaiah, C., MacRae, E., & Thorlby, G. (9 de 2018). A New Zealand perspective on the application and regulation of gene editing. Front Plant Sci 9, 1323. doi: https://doi.org/10.3389/fpls.2018.01323/full

Index

A

B

C

D

E

Paul John Verma, Huseyin Sumer and Jun Liu (eds.), *Applications of Genome Modulation and Editing*, Methods in Molecular Biology, vol. 2495, https://doi.org/10.1007/978-1-0716-2301-5,

R

S

T

W

Z

MIX
Papier aus verantwortungsvollen Quellen
Paper from responsible sources
FSC® C105338
FSC
www.fsc.org

If you have any concerns about our products,
you can contact us on
ProductSafety@springernature.com

In case Publisher is established outside the EU,
the EU authorized representative is:
Springer Nature Customer Service Center GmbH
Europaplatz 3, 69115 Heidelberg, Germany

Printed by Libri Plureos GmbH
in Hamburg, Germany